Turkey Diseases and Disorders Volume 2

Hafez M. Hafez • Awad A. Shehata
Editors

Turkey Diseases and Disorders Volume 2

Infectious and Nutritional Diseases, Diagnostics and Control Strategies

Springer

Editors
Hafez M. Hafez
Institute of Poultry Diseases
Free University of Berlin
Berlin, Germany

Awad A. Shehata
TUM School of Natural Sciences
Technical University of Munich
Garching, Germany

ISBN 978-3-031-63324-9 ISBN 978-3-031-63322-5 (eBook)
https://doi.org/10.1007/978-3-031-63322-5

This Springer imprint is published by the registered company Springer Nature Switzerland AG
The registered company address is: Gewerbestrasse 11, 6330 Cham, Switzerland

Preface

The global turkey business strives for increased production, higher quality, and competitive pricing. Turkey production has increased in recent decades due to the progress made in artificial brooding, genetics, nutrition, and management. However, the rising demand for turkey meat requires a consistent, practical, and goal-oriented healthcare system to minimize and control the emergence and spread of infections in turkey farms. Nevertheless, the production and health of turkeys are also being impacted by various factors and problems. Among these factors are intense global competition between producing countries and permanent changes in social, political, and consumer perceptions regarding food safety, animal welfare, and environmental protection. Several human foodborne infections are linked to poultry and poultry products, causing a serious challenge because it is difficult to control. Moreover, contamination of turkey meat and products with antibiotic-resistant bacteria is a constant public health hazard. The loss of consumer confidence and trust in turkey meat product safety and quality will also be a major concern.

The current and future turkey health concepts should cover the control of diseases in birds and the relationship between birds' health, welfare, and environmental protection. Additionally, the emergence and re-emergence of infectious turkey diseases will remain an important and never-ending challenge. Only a few authorized pharmaceuticals and veterinary products are available to treat turkeys. Developing efficient vaccines and natural antimicrobials against bacterial infections will reduce antibiotic use and resistant bacteria's development. Genetic selective breeding to improve production traits and health is a long-standing goal of the turkey industry. Furthermore, rearing technology, management, and feeding will help maintain the birds healthy and comfortable. Finally, all other partners involved in the production chain, including farmers, veterinarians, and stockholders, need to collaborate to meet consumer expectations for high-quality and safe products.

The book *Turkey Diseases and Disorders* aims to address the main challenges facing turkey production and is organized into two volumes: Volume 1 covers the main bacterial and fungal diseases of turkeys in 22 chapters. Volume 2 covers viral and parasitic diseases and nutritional disorders in 20 chapters. The book is designed to be a handbook for undergraduate students and a valuable source for researchers, practical poultry specialists, and nutritionists. Additionally, this book may be instrumental as a guide for production and health problems in turkeys. At the end of each chapter, we provide the reader with selected literature that covers the topic.

Comprehensive citation of the references is minimized, and the presentation of literature data is based on the interpretation or correlation of research findings.

This book can serve as a textbook, a research reference, and a valuable guide to the knowledge of turkey management and diseases. We hope that readers will find this book useful and interesting to read.

Berlin, Germany — Hafez M. Hafez
Garching, Germany — Awad A. Shehata

Contents

Editors and Contributors

About the Editors

Awad A. Shehata gained his bachelor's degree in veterinary medicine from the Faculty of Veterinary Medicine, Alexandria University, Egypt, and his license as veterinarian from Ludwig-Maximilians-University, Munich, Germany. In 2005, he obtained a master's degree in avian diseases at the Faculty of Veterinary Medicine, Sadat City University, Egypt. In 2011, he completed his Ph.D. (Dr. med. vet.) in virology at the Faculty of Medicine, Leipzig University, Germany. In 2015, he obtained his Dr. habilitatus (Dr. med. vet. habil.) in bacteriology and poultry diseases at Leipzig University, Germany. In 2022, he was honored with the title of professor of avian diseases from Sadat City University, Egypt. Besides his academic work at several universities, Dr. Shehata has six years of industrial experience in vaccine production and the development of alternatives to antimicrobials. He is a certified quality manager, auditor, project manager, European Business Competence License level (EBCL), Good Manufacturing Practice (GMP), FELASA-B and -C, and phytotherapist. Dr. Shehata's research interests lie mainly in developing alternatives or complementary to antimicrobials and diagnostic strategies and studying avian pathogens' epidemiology and molecular epidemiology. Alternatives include developing and evaluating recombinant vaccines, live attenuated bacterial vaccines, inactivated vaccines, probiotics, and prebiotics. Dr. Shehata teaches avian diseases and microbiology courses in English and German at several universities worldwide. He is currently a project leader at the Structural Membrane Biochemistry, Technical

University of Munich, Garching, Germany. His research focuses on the analysis and mechanism of action of natural products, metabolic pathways, and fluxes in various organisms using 13C-labeling experiments and profiling.

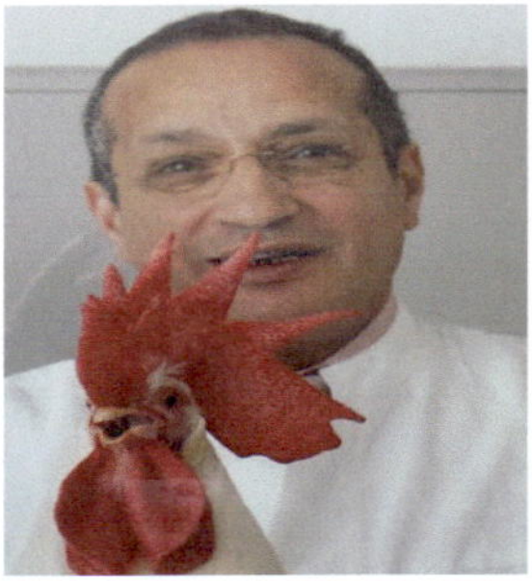

Hafez M. Hafez was Head of the Institute of Poultry Diseases of the Free University in Berlin from 1997 until 2016. He is currently a guest senior professor at the same institute. Hafez gained his Master of Veterinary Science (MVSc) at the Department of Poultry Diseases from Cairo University, Egypt, in 1975. In 1981, he completed his Doctorate Degree (Dr. med. vet.) at the Department of Poultry Diseases, Giessen University, Germany. In 1994, he finished the Habilitation (Dr. med. vet. habil.) thesis at the Department of Poultry Diseases, Munich University, Germany. Dr. Hafez has been recognized as a veterinary poultry specialist since 1982, veterinary microbiology specialist since 1989, and veterinary animal hygiene specialist since 1996. He became the diplomat of the European College of Veterinary Public Health (Dipl. ECVPH) in 2005 and the European College of Poultry Veterinary Science (Dipl. ECPVS) in 2009. He is currently the honorary life president of the World Veterinary Poultry Association (WVPA). He was the past-president of the ECPVS, the chairman of the poultry scientific committee of the German Veterinary Chamber, the chairman of the German branch of the WVPA, the chairman of Working Group 10 (Turkey), and European Branch of World Poultry Science Association (WPSA). Besides, he has been an honorary professor at the University of Hohenheim, Germany, since 1996 and an honorary professor at Alexandria University, Egypt, since 2009. Since 2015, he has been an advisor to the Arab Federation for Food Industries (AFFI). Dr. Hafez' research interest focused on poultry disease diagnosis and control in general and food-borne diseases, management, animal welfare, and hygiene.

Contributors

Shereen Basiouni Institute of Molecular Physiology, Johannes-Gutenberg University, Mainz, Germany

Inkar Castellanos-Huerta Department of Poultry Science, University of Arkansas, Fayetteville, AR, USA

Bussarakam Chuppava Foundation Institute for Animal Nutrition, University of Veterinary Medicine Hannover, Hannover, Germany

John Dunn United States Department of Agriculture, Southeast Poultry Research Laboratory, US National Poultry Research Center, Agricultural Research Service, Athens, GA, USA

Wolfgang Eisenreich Bavarian NMR Center, Structural Membrane Biochemistry, Department of Chemistry, TUM School of Natural Sciences, Technical University of Munich, Garching, Germany

Hosny El-Adawy Institute of Bacterial Infections and Zoonoses, Friedrich Loeffler Institut, Jena, Germany

Saeed El-Ashram College of Life Science and Engineering, Foshan University, Foshan, Guangdong, China

Faculty of Science, Kafrelsheikh University, Kafr El-Sheikh, Egypt

Amr Abd El-Wahab Foundation Institute for Animal Nutrition, University of Veterinary Medicine Hannover, Hannover, Germany

Hafez M. Hafez Institute of Poultry Diseases, Faculty of Veterinary Medicine, Free University of Berlin, Berlin, Germany

TUM School of Natural Sciences, Bavarian NMR Center (BNMRZ), Structural Membrane Biochemistry, Technical University of Munich, Garching, Germany

Rüdiger Hauck School of Veterinary Medicine, University of California, Davis, USA

Daniel Hernandez-Patlan Laboratory 5: LEDEFAR, Multidisciplinary Research Unit, National Autonomous University of Mexico-Superior Studies Faculty at Cuautitlan (UNAM-FESC), Cuautitlan Izcalli, Mexico State, Mexico

Nanotechnology Engineering Division, Polytechnic University of the Valley of Mexico, Tultitlan, Mexico State, Mexico

Xochitl Hernandez-Velasco Departamento de Medicina y Zootecnia de Aves, Facultad de Medicina Veterinaria y Zootecnia, UNAM, Mexico City, Mexico

Juan D. Latorre Department of Poultry Science, University of Arkansas, Fayetteville, AR, USA

Dörte Lüschow Institute of Poultry Diseases, Faculty of Veterinary Medicine, Free University of Berlin, Berlin, Germany

Jesús A. Maguey-González Department of Poultry Science, University of Arkansas, Fayetteville, AR, USA

Víctor M. Petrone-García Departamento de Ciencias Pecuarias, FESC, UNAM, Cuautitlán, Estado de Mexico, Mexico

Awad A. Shehata TUM School of Natural Sciences, Bavarian NMR Center (BNMRZ), Structural Membrane Biochemistry, Technical University of Munich, Garching, Germany

Institute of Poultry Diseases, Faculty of Veterinary Medicine, Free University of Berlin, Berlin, Germany

Bruno Solis-Cruz Laboratory 5: LEDEFAR, Multidisciplinary Research Unit, National Autonomous University of Mexico-Superior Studies Faculty at Cuautitlan (UNAM-FESC), Cuautitlan Izcalli, Mexico State, Mexico

Nanotechnology Engineering Division, Polytechnic University of the Valley of Mexico, Tultitlan, Mexico State, Mexico

Guillermo Tellez-Isaias Department of Poultry Science, University of Arkansas, Fayetteville, AR, USA

Department of Poultry Science, University of Arkansas Agricultural Experiment Station, Fayetteville, USA

Part I

Viral Diseases

Avian Influenza

1

Awad A. Shehata and Hafez M. Hafez

Abstract

The first report on the isolation of the influenza virus from turkeys with respiratory signs was in the 1960s in the United States. Since then, several outbreaks caused by low-pathogenic (LP) and highly pathogenic avian influenza (HPAI) subtypes, that is, H1, H3, H5, H6, H7, and H9, have been observed in many countries worldwide. Differences in susceptibilities between turkeys and chickens have been identified. It was found that H7N2 was more infectious for turkeys than chickens. Sinusitis is a common sign in turkeys infected with LPAI. In addition, turkeys play an important role in the evolution of avian influenza viruses for several reasons: (i) Turkeys have both avian- and human-type receptors, making them highly probable mixing vessels for avian influenza viruses. Both avian-type and human-type receptors are expressed in the nasal cavity, lung, kidney, esophagus, and intestine. (ii) Turkey breeders especially can also be infected with swine influenza viruses such as H1N1, H1N2, and H3N2, causing a severe drop in egg production and severe economic losses as well as increasing the probability of reassortments and virus evolution in turkey hosts. (iii) Interspecies transmission of swine influenza viruses to turkeys is common. Interspecies transmission between ducks to turkeys has been reported but less frequently. Mixing different poultry species and outdoor rearing could favor the adaptation of LPAI viruses and pose a serious health risk.

A. A. Shehata (✉)
TUM School of Natural Sciences, Bavarian NMR Center (BNMRZ), Structural Membrane Biochemistry, Technical University of Munich, Garching, Germany
e-mail: awad.shehata@tum.de

H. M. Hafez
Institute of Poultry Diseases, Faculty of Veterinary Medicine, Free University of Berlin, Berlin, Germany
e-mail: hafez.mohamed@fu-berlin.de

H. M. Hafez, A. A. Shehata (eds.), *Turkey Diseases and Disorders Volume 2*,
https://doi.org/10.1007/978-3-031-63322-5_1

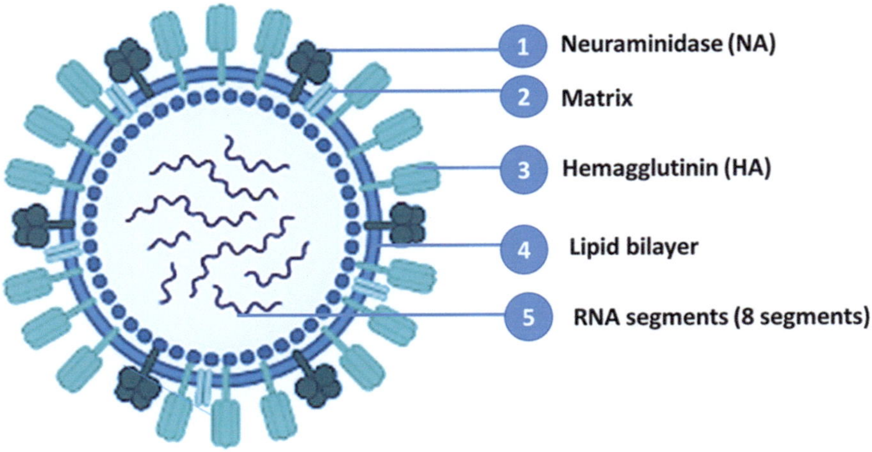

Fig. 1.1 Avian influenza virus (Generated by Biorender)

Keywords

Turkeys · Influenza · HPAI · LPAI · Virus evolution · Vaccination

1.1 Etiology

AI viruses are members of the *Orthomyxoviridae* family. Four types of influenza viruses, A, B, C, and D, are known within this family, based on nucleoproteins and matrix proteins (Kuhn et al. 2020). The structure of avian influenza is illustrated in Fig. 1.1. The virus genome is a single-stranded segmented RNA; the genome of influenza A and B viruses comprises eight gene segments, while the influenza C and D genome comprises seven segments (Abdelwahab and Mettenleiter 2023).

The eight segments of influenza A viruses encode at least ten viral proteins: PB2, PB1, PA, HA, NP, NA, M1, M2, and NEP are included in the virion, while the non-structural protein NS1 is expressed only in host cells after infection. Type A viruses can infect birds, humans, and mammals such as horses, pigs, seals, and whales. The virus is RNA, enveloped, and sensitive to ether, chloroform, and different chemical disinfectants. Type B and C affect only humans. Type D infects a broad range of mammalian species.

1.2 Influenza Pandemics

AIV causes periodic epidemics in humans, horses, pigs, seals, whales, and several birds (Swayne et al. 2020).

(i) The Russian flu pandemic emerged between 1889 and 1890 in Russia. It was likely caused by H3N8 and H2N2 strains (Ryan 2008).

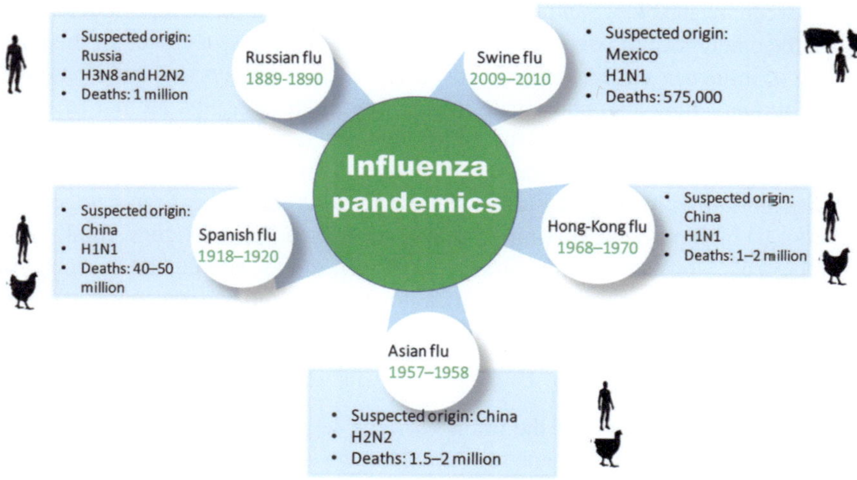

Fig. 1.2 Pandemic and epidemic events of influenza (Generated from Parvin et al. 2022)

(ii) Spanish flu was the first documented influenza pandemic in humans. About 500 million people were infected, and about 50 million deaths were reported over 2 years. The cause of Spanish flu is H1N1 of avian origin, which emerged in four waves during 1918–1920 (Bassareo et al. 2020).

(iii) Asian flu and Hong Kong flu pandemics emerged in 1968 and 1970 and were caused by H2N2 and H1N1 viruses. The two viruses had reassortment events comprising human and avian-origin gene segments (Martini et al. 2019).

(iv) Swine flu was caused by a novel H1N1pdm09 virus in 2009 (Riley et al. 2011; Dawood et al. 2012). This virus emerged from triple reassortment between avian, swine, and human influenza viruses (Tewawong et al. 2015). Various pandemic and epidemic events of influenza are illustrated in Fig. 1.2.

1.3 Evolution of Influenza Viruses

There are three mechanisms of influenza viruses' evolution:

(i) **Antigenic drift:** The virus genome experiences a single mutation that alters the amino acid sequence. The reason for these mutations is a lack of the ability to proofread RNA-dependent RNA polymerase (RdRP), which causes a rate of 10^{-3} and 10^{-4} integration of false nucleotides (Drake 1993; Shao et al. 2017). By altering or hiding the immunogenic epitopes of a circulating virus, it may cause immunological escape. The primary cause of vaccination failure in humans and poultry is antigenic drift, which necessitates regular updates to influenza virus vaccines (Grund et al. 2011; Wen et al. 2021).

(ii) **Antigenic shift (Reassortment):** When a host cell is co-infected with numerous viruses, the genome segments may be shifted to generate progeny viruses with novel genome combinations (Marshall et al. 2013). Antigenic shift between avian, swine, and human influenza viruses has been reported, leading to the emergence of new subtypes. Although pigs serve as a vessel for the mixing of influenza viruses, other host species, such as turkeys and quails, as well as humans, may also participate in this way; therefore, pigs are not the only species that participate in the production of reassortant influenza viruses (Hennig et al. 2022). More recently, potential "mixing vessels" based on the distribution of avian and human sialic acid receptors have been categorized into: (i) high probable mixing vessels hosts, such as humans, pigs, minks, ferrets, seals, dogs, cats, birds, turkeys, chickens, quails, and ducks; (ii) medium probable mixing vessel hosts such as nonhuman primates, raccoons, camels, pikas, horses, and zoo animals, including tigers and lions; and (iii) low probable mixing vessels hosts such as foxes, bats, and whales (Abdelwahab and Mettenleiter 2023).

(iii) **Recombination:** Parts of the influenza gene segments, or host cellular RNA, are integrated into other gene segments, known as homologous recombination and/or nonhomologous recombination, respectively. HPAI emerged from LPAI due to recombination in the HA cleavage site (Gultyaev et al. 2021).

1.4 Subtypes and Pathotypes of Avian Influenza

To date, there are 18 H subtypes and 11 N subtypes of Influenza A viruses. Avian influenza viruses (AIVs) contain H1 to H16 and N1 to N9 subtypes, while H17N10 and H18N11 are detected or isolated only in bats (Gamblin and Skehel 2010; Suarez 2016). According to the pathogenicity, influenza viruses are classified into LAPI and HPAI (Table 1.1).

Representatives of all the different subtypes of Influenza A viruses have been isolated from several species of birds, mainly from aquatic species such as ducks, geese, and gulls. The viruses are known to have a high mutation rate. The mutation

Table 1.1 Features of high pathogenic (HPAI) and low pathogenic (LPAI)

Criteria	HPAI	LPAI
The cleavage site of HA0	Multiple basic amino acids, such as lysine and arginine	Monobasic amino acid
HA0 cleavage	Ubiquitous cell proteins are found in most cells of the body	Only trypsin-like proteases found in the respiratory and digestive systems
Virus replication	Pantropic (all cells and all organs)	Epithelial cells of respiratory and digestive tracts
Virus replication in cell culture	Without trypsin	With trypsin
Subtypes	H5 and H7[a]	Other subtypes
Lethality in 4–6 weeks chicks (IV-infection)	6–8/8	

[a] H5 and H7 are usually HPAI with few exceptions

of the virus and recombinations between strains lead to ongoing dynamic changes in the virus surface structures (Swayne et al. 2020). Several features are used to describe new influenza viruses, including antigenic type (A, B, C, D), animal host, geographical location (city, state, or country), laboratory or reference ID, the year of isolation, and HA and NA types. An example is A/chicken/Egypt/SCU20/2014 H9N2.

1.5 Avian Influenza in Wild and Domestic Birds

AIV subtypes H1 to H11 and H13 have been detected and/or isolated from domestic birds. The most frequently isolated subtypes in domestic birds are subtypes H5Nx, H6N2, H7N3, H7N9, and H9N2. However, subtypes H12 and H14-H16 have not yet been detected in domesticated birds. Mixed infections with several influenza subtypes were reported. Mixed viral infections were observed in both broiler and layer chickens. The detected triple H5N1, H9N2, and H5N8 influenza co-infection raises the concern of potential AI epidemic strain emergence (Shehata et al. 2019). AIVs that are highly pathogenic in ducks are also highly pathogenic in chickens, but the reverse is not true. Most HPAIV H5/H7 strains in ducks are non-virulent, unlike in chickens and turkeys. Mallard ducks are known to be the primary carriers of avian influenza viruses, although several H5N1 and H5N8 viruses are also highly virulent in mallards. Muscovy ducks are more sensitive than Pekin ducks (reviewed in Abdelwahab and Mettenleiter 2023). The diversity of host-specific influenza viruses is shown in Tables 1.2 and 1.3.

Table 1.2 Diversity of host-specific influenza viruses. The H17N10 and H18N11 subtypes were identified in bats

Subtype	Human	Horse	Pig	Wild birds	Domestic birds	Spill over to other animals*
Diversity of HA						
H1	■		■	■	■	
H2	■			■	■	
H3	■	■	■	■	■	
H4				■	■	
H5	■		■	■	■	■
H6				■	■	
H7	■	■	■	■	■	
H8				■	■	
H9	■			■	■	
H10	■			■	■	
H11				■	■	
H12				■		
H13				■	■	
H14				■		
H15				■		
H16				■		
Diversity of NA						
N1	■		■	■	■	■
N2	■			■	■	
N3			■	■	■	
N4				■	■	
N5				■	■	
N6			■	■	■	
N7	■	■		■	■	
N8		■		■	■	
N9				■	■	

* Mice, ferrets, monkeys, tigers, cats, dogs, marten, donkeys, and civet cat.

Table 1.3 Pathogenicity of HPAI in chicken/turkeys versus ducks

	Chickens/turkey	Ducks
HPAI subtypes H5 and H7	Most strains are pathogenic and can lead to 100% morbidity and mortality in poultry	Most strains are avirulent to Mallard ducks Muscovy ducks are more sensitive compared to Pekin ducks
HPAI isolated from chickens	Pathogenic in chickens and turkeys	Not all pathogenic in ducks
HPAI isolated from ducks	Pathogenic in chickens and turkeys	Pathogenic in ducks

1.5.1 Avian Influenza in Turkeys

Turkeys are considered a bridging host for adapting wild-bird AIVs to infect poultry (Pillai et al. 2010). They are naturally susceptible to H1N1, triple reassortant H3N2 viruses. Due to the expression of both avian-type receptors in turkeys (Kimble et al. 2010), they are considered highly probable mixing vessels. Both avian- and human-type receptors are expressed in the nasal cavity, lung, kidney, esophagus, and intestine; however, only an avian-type receptor is expressed in the trachea (Pillai et al. 2010; Costa et al. 2012).

The first report on the isolation of influenza virus from turkeys with respiratory signs was in 1963 (Lang et al. 1968). In 1999, low pathogen Influenza A subtype H7N1 was isolated from turkey flocks in Italy, accompanied by a high mortality rate and a turn to virulence (HPAI) by the end of 2000 (Capua and Alexander 2004).

From December 2001 to January 2002, AI outbreaks were observed in three turkey flocks reared in the central-west region of Germany. In all cases, sudden onset of depression, decreased feed and water intake, and respiratory signs accompanied by high mortality were observed. Postmortem lesions revealed pericarditis, petechial hemorrhages in pericardial fat, fibrinous airsacculitis, lung congestion, and pneumonia (Hafez 2003).

Since 2006, HPAIV H5N1 of clade 2.2.1 infected a wide range of poultry, including turkeys, and caused severe economic losses. The virus is endemic in poultry in several countries and has diversified into two genetic clades: clade 2.2.1.1 and clade 2.2.1.2. The 2.2.1.1 clade represents immune-escape variants in vaccinated commercial poultry from 2007 to 2011.

The 2.2.1.2 clade was detected in humans, non-vaccinated backyard poultry, and, recently, commercial poultry in Egypt (Salaheldin et al. 2022). The low-pathogenic AIV A/turkey/Ontario/6213/1966 (H5N1) was isolated and proved that it is a progenitor of HPAI A/turkey/Ontario/7732/1966 (H5N9) (Ping et al. 2012).

Currently, the major turkey-producing countries have a problem at one time or another with AI (LPAI and HPAI). Although transmission from birds to humans is rare, there is a risk that these viruses may adapt and become able to infect and gain the ability to spread from person to person. Therefore, early detection and identification of human infection is of great public health importance. Focusing testing of people who develop symptoms should be tested to reduce the risk of further spread to public health. Such investigations increase our knowledge of the zoonotic risk of influenza A viruses and provide vital evidence to help strengthen One Health responses, particularly given the unusual infection pressure in avian populations and the extensive global spread of H5N1.

1.6 Transmission and Source of Infection

The disease is transmitted directly via direct contact with infected birds and/or indirectly via contaminated equipment. Infected birds shed the virus in fecal and/or oculo-nasal discharges. Recovered flocks will intermittently shed the virus and should be considered as infected for an extended period if not all the life. The virus can survive in the contaminated environment for long periods at moderate temperatures and longer in frozen materials. The infection can be easily spread by people via contaminated shoes and clothing, crates of egg flats, vehicles, rodents, and insects that may mechanically carry the virus from infected to susceptible poultry. There is little or no evidence of vertical transmission (egg-borne infection) in poultry. However, eggshell surfaces can be contaminated with the virus. Wild and domesticated waterfowl are the primary natural reservoirs of influenza viruses. They may be infected with multiple subtypes without clinical signs, excrete the virus for long periods, and do not develop detectable antibodies.

1.7 Course of Infection

The severity of clinical signs, course of the disease, and mortality after infection with AI are incredibly variable and vary from a very mild or even inapparent form to a highly acute form, depending on the virulence of the virus, the species, age, the immune status of the host, concurrent diseases, and management.

1.8 Clinical Signs

Clinical signs after infection with HPAI may include high mortality, ruffled feathers, depression, diarrhea, sudden drop in egg production in breeder flocks, cyanosis of the snood, oedema, swelling of the head, blood-tinged discharge from nostrils, respiratory distress, incoordination, and pinpoint hemorrhages mostly seen on the feet and shanks. LPAI in turkeys is characterized mainly by respiratory manifestations.

1.9 Postmortem Lesions

The main postmortem lesions of avian influenza in turkeys are shown in Fig. 1.3. In turkeys, lesions consist of sinusitis, tracheitis pericarditis, petechial hemorrhages in pericardial fat, fibrinous airsacculitis, lung congestion, pneumonia, enlargement of

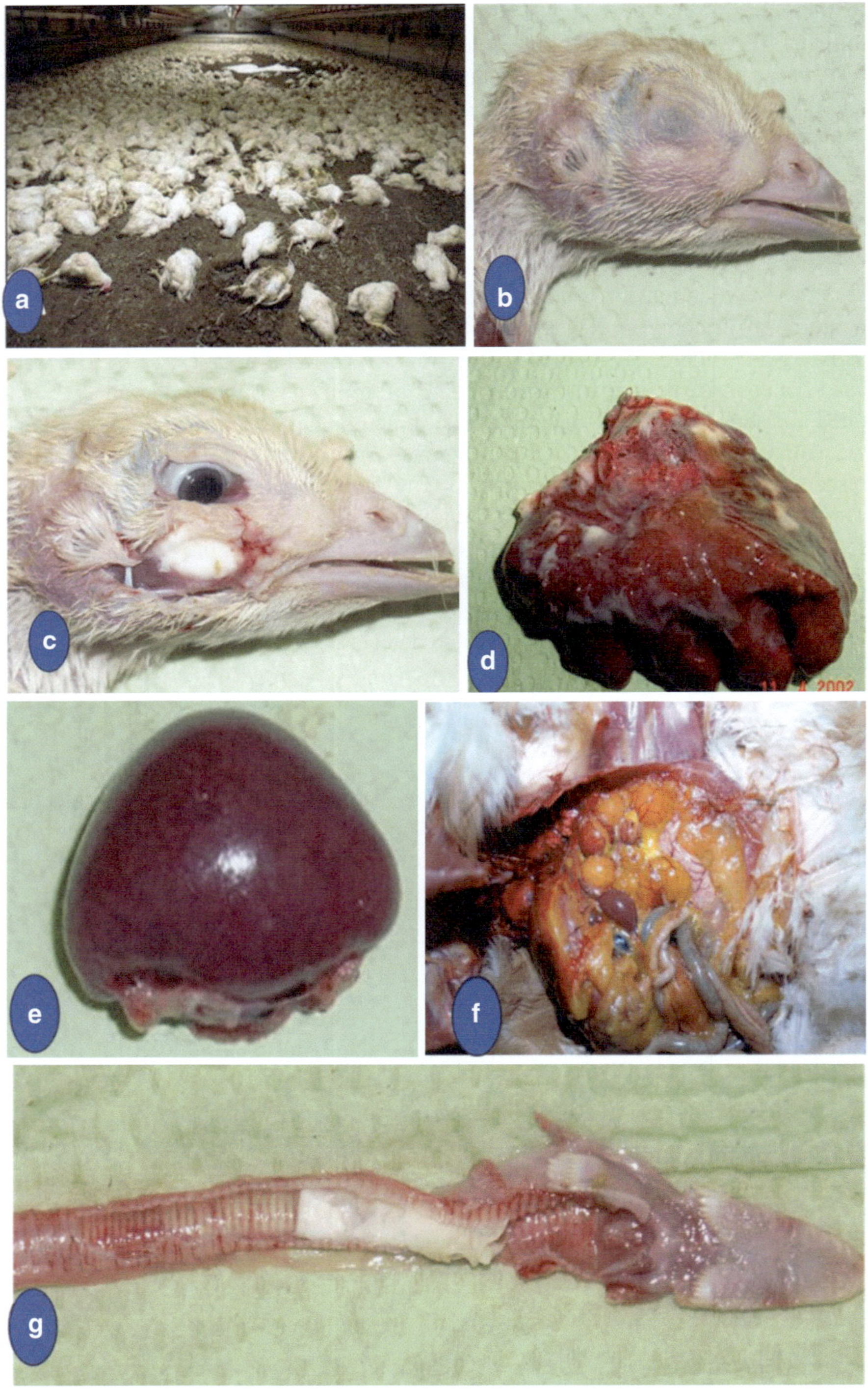

Fig. 1.3 Postmortem lesions of avian influenza in turkeys. (**a**) sudden death, (**b**, **c**) sinusitis, (**d**) pneumonia, (**b**) **s**plenomegaly, (**f**) peritonitis and hemorrhages on the ovarian follicles, (**g**) tracheitis, (**h**) healthy pancreas, (**i**) hemorrhagic pancreatitis (©Hafez, H.M. Fu-Berlin)

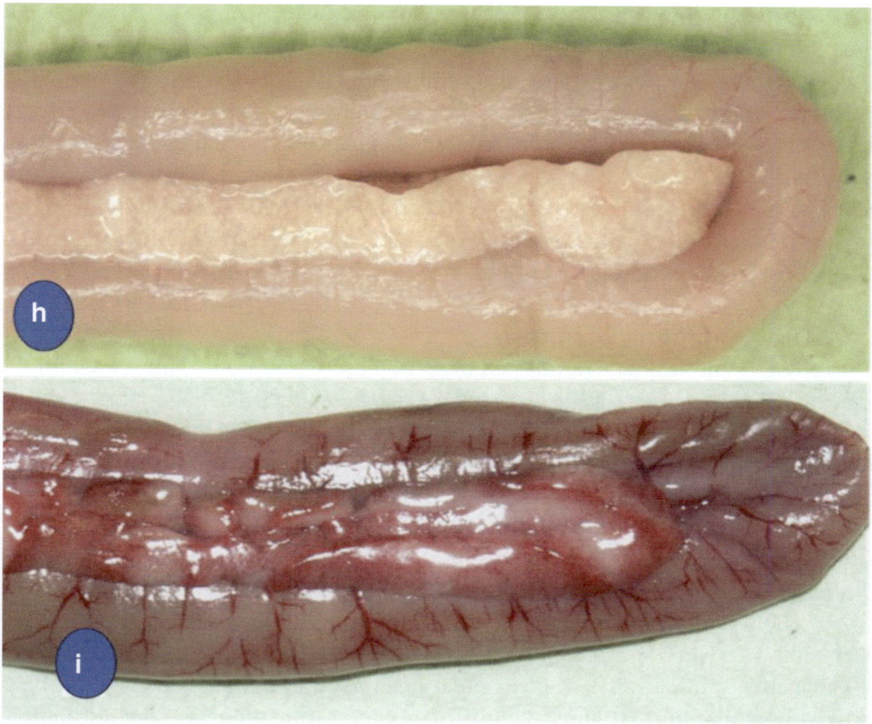

Fig. 1.3 (continued)

the spleen, and inflammation of the pancreas. Blood vessels are usually engorged. Hemorrhage may occur in the trachea, proventriculus, and intestines. HPAI can also cause necrotizing myocarditis. The lining of the gizzard may be easily removed. Young broilers may show signs of severe dehydration with other less pronounced or absent lesions.

1.10 Diagnosis

1.10.1 Sampling

Sampling tracheal swabs from poultry flocks to detect avian influenza viruses is time-consuming and involves a high workload for the staff and a lot of stress for the sampled animals. Swab samples in/on drinkers can easily be taken. When taking water, every animal with a respiratory infection leaves small amounts of its respiratory secretions (mucus) in/on the drinkers. Even before the first clinical signs appear in a flock, the influenza virus is already in small amounts of mucus particles in the drinkers present. By sampling swabs in/on many drinkers in the barn, the smallest amounts of respiratory murus from infected birds in the flock can be detected by PCR technique.

In contrast to the fecal samples, the samples from the drinkers contain only a small amount of DNA and RNA from microorganisms (Sieverding and Hafez 2023).

The PCR results from this field study showed that swabs from the drinkers in the barn yield very reliable and very important information about the presence of avian influenza viruses in a poultry population. However, taking swab samples from many drinkers in the barn is important. Influenza monitoring with swab samples from drinkers is rapid, sensitive, specific, reliable, and inexpensive and can be taken easily by individuals, regardless of the age and the number of birds in the flock. Detecting the avian influenza virus in infected flocks with swab sampling in/on drinkers is an animal-friendly process with an improvement in the statistical significance of the infection status of a flock.

1.10.1.1 Laboratory Diagnosis

Clinical signs and lesions are not pathognomonic. Therefore, isolation, identification, and characterization of the virus involved are essential. The laboratory diagnosis of AI is done in three steps, including (i) identification of influenza type, (ii) identification of subtype, and (iii) pathotyping using animal experiments or sequence analysis of the hemagglutinin cleavage side (Fig. 1.4).

AI virus can be isolated in embryonated chicken eggs by the allantoic sac route or using several cell lines. Depending on the pathotype, the embryos may or may not die within a 5-day observation period, and usually, there are no characteristic lesions to be seen in either the embryo or the allantoic membrane. Hatching eggs inoculated with HPAIV-containing material usually die within 48 h.

The hemagglutination test can detect a hemagglutinating agent in the harvested allantoic fluid; it must be differentiated from other hemagglutinating viruses, such as the Newcastle disease virus and egg drop syndrome. Chicken RBCs are commonly used; however, H1 and H3 influenza viruses isolated from turkeys agglutinate turkey, horse, and guinea pig RBCS may not agglutinate chicken RBCs.

The agar gel immunodiffusion (AGID) test using influenza A antigen can confirm that the isolated virus is influenza A, depending on the matrix and nucleocapsid antigens. PCR can also be used for the detection of the circulating subtype. HI and NI tests can be used for subtyping influenza viruses using specific sera for H5 and H7 subtypes.

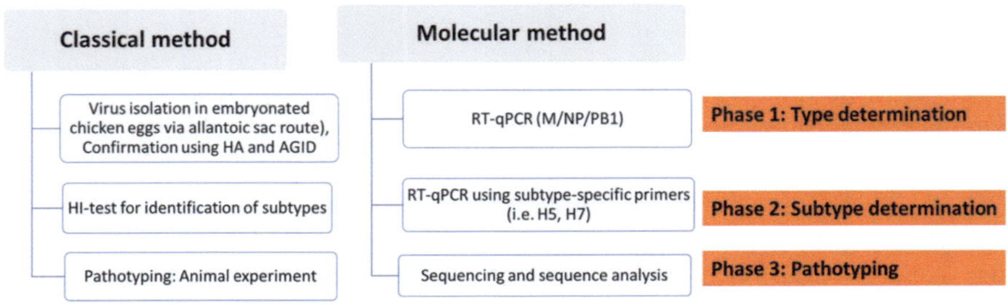

Fig. 1.4 Diagnosis of avian influenza. The laboratory diagnosis of avian influenza is done in three steps. (i) isolation of avian influenza A, (ii) identification of subtype, and (iii) pathotyping using animal experiment or IV: sequence analysis of the hemagglutinin cleavage side

Pathotyping can be determined using the in vivo pathogenicity test and/or sequencing and phylogenetic analysis. In vivo, the pathogenicity test shall be conducted according to the guideline published by the OIE manual by inoculating 4- to 8-week-old specific pathogen-free chicks intravenously. HPAI causes the mortality of 75% of birds within 10 days. Isolates with an IVPI >1.2 are considered HPAI. The cleavage site of HPAI is multiple basic amino acids in the case of HPAI. However, LPAI viruses have monobasic amino acids.

Serological detection. AGID and ELISA can be used to identify antibodies in flocks exposed to infection. AGID detects antibodies toward M1 and nucleocapsid proteins for all type A viruses (group-specific). The AGID test requires large quantities of reagents and takes 24–48 h for results to be obtained, while most commercial ELISAs detect only antibodies toward nucleoproteins. Furthermore, the AGID test may not be suitable as a universal assay for some other species of birds; serum samples from waterfowl do not contain good precipitating antibodies. The HI test is more sensitive and rapid than the AGID test, but it is complicated due to the existence of 16 HA subtypes of AIV.

1.11 Prevention and Control

Since AIVs in nature are maintained in wild aquatic birds, eradicating the infections seems complicated and even impossible. However, every effort should be taken to prevent direct and/or indirect contact between domestic poultry and wild waterfowl as well as vaccination of poultry flocks.

In conjunction with strict quarantine, several vaccination programs have been used to control the disease in commercial turkey flocks. In infected flocks with HPAI, strict quarantine and rapid depopulation of infected flocks remain the only effective methods of stopping AI. The success of vaccination programs depends on the course of the infection, governmental regulations, veterinarians, and the poultry industry. However, using inactivated vaccines against highly pathogenic influenza viruses in some countries has revealed promising results.

1.11.1 Pros and Contras Against the Need for Vaccination

Vaccines against LPAI viruses were successfully used in turkey farms in the United States, demonstrating their potential effectiveness against HPAI viruses (Halvorson 2009). After that, vaccination was implemented in several countries, including Italy (LPAI), Mexico (HPAI), and China (HPAI). In influenza-endemic countries, such as China, Egypt, Indonesia, and Vietnam, vaccination against HPAI has been used after the ineffective implemented stamping out policy. Vaccination of turkeys is commonly used to control HPAI H5N1 infections in many countries (Halvorson 2002; Swayne et al. 2014).

The decision to vaccinate against AI remains challenging (Sims et al. 2016), Fig. 1.5. The outbreaks' severity and economic consequences have led to debates on

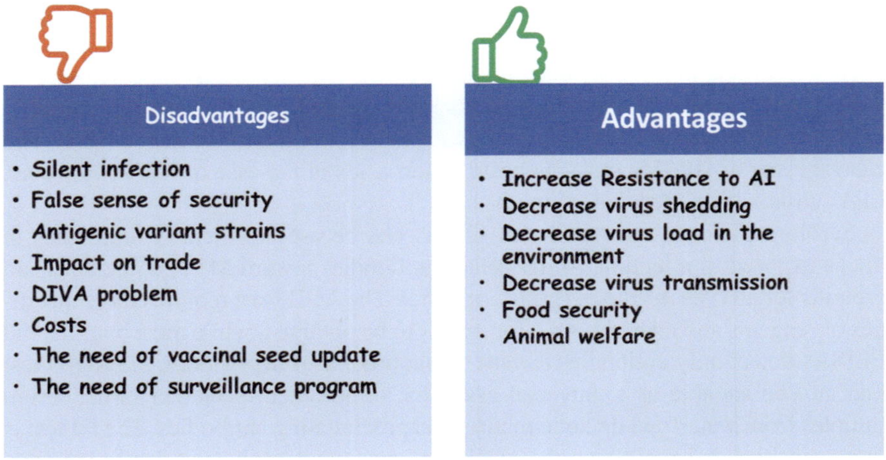

Fig. 1.5 Pros and cons against HPAI influenza vaccines

whether vaccination should be allowed alongside eradicating infected flocks. Poultry farmers advocate for vaccination, citing several reasons: (i) The stamping out method is insufficient in controlling the spread of the deadly AI virus. (ii) The affected regions are too vast to prevent further virus transmission. (iii) Vaccination is vital in safeguarding valuable flocks, especially breeding stock.

However, some scientists argue against vaccination for the following reasons: (i) Vaccination does not completely prevent the infection of flocks and can lead to the development of clinically silent virus carriers. (ii) It can be challenging to distinguish between infected and noninfected birds in vaccinated flocks. (iii) Vaccinating all flocks in affected regions can result in significantly high expenses. (iv) Vaccination may result in additional trade restrictions. (**v**) Virus evolution under vaccination pressure leads to the emergence of antigenically variant strains. In Egypt, since 2006, HPAIV H5N1 of clade 2.2.1 infected a wide range of poultry, including turkeys, and caused severe economic losses. The virus is endemic in poultry in several countries and has diversified into two genetic clades: clade 2.2.1.1 and clade 2.2.1.2. The 2.2.1.1 clade represents immune-escape variants in vaccinated commercial poultry from 2007 to 2011. The 2.2.1.2 clade was detected in humans, non-vaccinated backyards, and commercial poultry (Salaheldin et al. 2017). In 2022, The 2.2.4.4b clade emerged. Generally, it is recommended to vaccinate against influenza with vaccines that provide the Differentiating Infected from Vaccinated Animals (DIVA) principle to detect active infection in vaccinated animals. Capua et al. developed and validated a serological test, "DIVA "(Capua et al. 2003).

Over the years, advancements in vaccine technology and extensive research have proven the safety and effectiveness of vaccines against HPAI. Innovative vaccination strategies, like vector-based vaccines and recombinant technologies, have overcome previous limitations and strengthened confidence in immunization and control measures. The collaborative efforts between international organizations, governments, and research institutions have also played a vital role in establishing standardized vaccine development, distribution, and administration protocols. These efforts have harmonized global strategies and improved preparedness and response mechanisms for potential HPAI outbreaks. Consumer attitudes have evolved, driven by increased awareness of the public health implications of avian influenza and growing demand for sustainable and ethically produced poultry products. This shift in consumer sentiment has encouraged industry stakeholders to embrace vaccination as a proactive measure to safeguard avian welfare and human health.

In 2023, vaccination against HPAI was approved in several EU countries for several poultry species, such as France (domestic ducks), the Netherlands (layers), Hungary (geese), Italy (turkeys), the Czech Republic (geese), and Belgium.

1.11.1.1 Types of Vaccines

(i) **Inactivated vaccines** prepared in embryonated chicken eggs, used as oil emulsions, and administered intramuscularly or subcutaneously. These vaccines produce humoral immune antibodies.
(ii) **Vectored recombinant** vaccines by inserting the HA gene into the Fowlpox virus vaccine strain **or** herpesvirus turkey (HVT) vectors (Kapczynski et al. 2016).

The host develops an immune response against both influenza and the used vector. Vector vaccines are characterized by the induction of both cell-mediated and humoral immune responses. Also, this type of vaccine can differentiate between vaccinated and infected birds (DIVA), with no antibodies against NP. The disadvantage of these vaccines is that birds exposed to the fowlpox virus will not develop antibodies toward AI (Brugère-Picoux et al. 2015). Several types of H5 vaccines are used, including local and nonlocal field strains and inactivated and recombinant vaccines from historic and recent H5Nx viruses.

Vaccination programs are highly variable; co-infections with viral and bacterial infections are widespread and can negatively influence the vaccine's efficacy. Also, maternal immunity in day-old poults interferes with vaccination at an early age (Abdelwahab and Hafez 2011). A substantial antigenic drift has been reported as an immune evasion mechanism due to mutations in immunogenic epitopes of the hemagglutinin gene 2.2.1.1 (Abdelwahab et al. 2016). Nevertheless, the infection in vaccinated poultry continues (Abdelwahab et al. 2016). However, little is known about HPAI H5N1 infections in vaccinated turkeys (Salaheldin et al. 2017).

1.11.1.2 Vaccination Against Low-Pathogenic Avian Influenza

Generally, the inactivated H9N2 vaccine does not effectively control the transmission of the LPAI virus in poultry (Cui et al. 2021). Inactivated H9N2 vaccines mainly induce humoral immunity, which makes it difficult to interrupt virus infection and shedding in the chicken upper respiratory tract. H9N2 AIV strains are still circulated between vaccinated chickens (Zhong et al. 2014), highlighting the urgent need to develop more effective vaccines that provide cellular and mucosal immunity against H9.

However, there are some trials for developing other vaccine types. Shehata et al. (2020) evaluated the efficacy of H9 plasmid-based DNA targeting the HA gene of H9N2 A/CK/Egypt/SCU8/2014 in turkey poults. The effectiveness of DNA (pVAX-H9 and pCR- H9) vaccine, naked or saponin-adjuvanted, was evaluated in turkey poults at third week of age intramuscularly and challenged 3 weeks later with the same life isolate at a dose level of 10^6 EID_{50}/bird. None of the birds vaccinated with naked or saponin-adjuvanted pVAX-H9 or pCR-H9 showed any clinical signs. However, the pVAX-H9 and pCR-H9 alone did not prevent cloacal and oropharyngeal virus shedding.

On the other hand, saponin-adjuvanted pVAX1-H9 and pCR-H9 prevented cloacal and oropharyngeal virus shedding at the third and fifth days post-challenge, respectively. All vaccinated birds showed high antibody titers in HI (7–8 $\log_2$) in the third week post-vaccination. In conclusion, DNA vaccination with pVAX1-H9 and pCR-H9 could protect turkeys from the H9N2 virus, but vaccination regimes should be improved.

1.11.1.3 Considerations and Essential Components

Several considerations should be considered for preventing and controlling avian influenza (Halvorson 2002, 2009; Swayne et al. 2014; Sims et al. 2016).

1. HPAI should be eradicated from poultry, and vaccination should be used only to deliver eradication. However, stamping out programs are complex, expensive, and labor-intensive. In several countries, many farms are not registered. Passive surveillance systems and farmer compensation schemes should be implemented, particularly if virus elimination is still the immediate goal. Compensation for culled birds is essential to encourage the owner to report the disease as early as possible. It must be paid at or near the actual market value of the birds.
2. Biosecurity is the first line to prevent the introduction and spread of infection. In simple terms, keep pathogens away from poultry and poultry away from pathogens. However, practically, biosecurity can reduce but does not eliminate risks.
3. Vaccination is recommended for countries where there is a risk that stamping out may result in the removal of a major source of food for rural communities and damage the commercial viability of the local poultry industry. If vaccination is to be an option in the control of AI, then it must be used in parallel with enforced biosecurity measures. The objectives and strategies for vaccination should be clear, consistent, and regularly reviewed to align with prevention and control plans.
4. High-quality vaccines that are registered by national authorities should be used if vaccination is implemented. Vaccines should exhibit a good antigenic match to the circulating field strain(s). However, it's important to note that not all virus strains are equally effective in stimulating immunity, which may be attributed to the glycosylation patterns of the HA protein. Therefore, the best approach for evaluating vaccine efficacy is to challenge experiments.

5. The efficacy of the influenza vaccines should be evaluated in the target species. Generally, higher antibodies correlate with efficacy. The potency of vaccines should also be considered. The vaccine should contain 1–5 μg HA/dose or 512–1024 HA unit/dose. Usually, HI titers of >1:32 can prevent mortality, while >1:128 prevent oropharyngeal shedding. The efficacy should be tested in vivo against newly emergent mutants. Turkeys should be vaccinated at least three to four times.
6. The merits of proper vaccination against AI are increasing the resistance of birds to influenza infection and reducing the virus shedding by reducing virus replication in the respiratory and gastrointestinal tracts. However, vaccines cannot prevent infection, virus replication, virus shedding, spread from farm to farm, and problems in diagnosing infected flocks (DIVA strategy).
7. When selecting vaccines, DIVA should be considered. Four strategies can achieve DIVA: (i) sentinel birds, where 20 susceptible birds are reared with a vaccinated flock and examined for seroconversion and/or presence of AI antigen; (ii) heterologous neuraminidase strategy, in which neutralization inhibition (NI) is available for all nine NA, but the availability of diagnostics is an issue; and (iii) nonstructural protein I (NS1) DIVA strategy. Theoretically, killed vaccines do not contain NS1, while naturally infected birds develop Ab toward NS1. However, inactivated vaccines are contaminated with NS1 during preparation, so dilution of serum before testing may decrease the nonspecific reactions. (iv) recombinant/subunit vaccine: AGID and ELISA can be used (subunit vaccines lack AI nucleoprotein).
8. Recombinant influenza vaccines using fowlpox or herpesvirus turkeys (HVT) induce both cellular and humoral immune response and provide DIVA strategy (HA-only based vaccines supporting serological DIVA strategies). However, the fowlpox vector vaccine is ineffective in birds exposed previously to fowlpox or birds with immunity to the vector.
9. Vaccination alone without culling affected birds to reduce virus load in the environment will probably not be successful.
10. Continuous updates and development of new diagnostics are crucial in the face of genetic mutations of the virus in endemic countries.
11. Genetic monitoring of the field virus under vaccine pressure is essential. The efficacy of the vaccines should be tested in vivo using the newly emergent mutants. Principally, the closer the amino acid similarity of the vaccine strain, the greater the virus replication and shedding reduction.
12. Particular attention should be given to the household and backyard poultry production systems; the improper disposal of dead birds and wastes is another important factor in disease control failure.
13. Any vaccination policy should include an exit strategy so that countries do not rely on costly long-term vaccination campaigns. It is important to continue using vaccination until infection is controlled and prevented. Vaccination may be required for an extended period in areas where the virus is widespread and elimination is unlikely. Vaccination can be stopped once the disease is under control and the risk of recurrence is low. Rapid detection and management of new outbreaks are crucial to ensure successful control.

References

Abdelwahab EM, Hafez HM (2011) An overview of the epidemic of highly pathogenic H5N1 avian influenza virus in Egypt: epidemiology and control challenges. Epidemiol Infect 139(5):647–657. https://doi.org/10.1017/S0950268810003122

Abdelwahab EM, Mettenleiter TC (2023) Zoonotic animal influenza virus and potential mixing vessel hosts. Viruses 15(4):980. https://doi.org/10.3390/v15040980

Abdelwahab EM, Veits J, Tauscher K, Ziller M, Grund C, Hassan MK et al (2016) Progressive glycosylation of the haemagglutinin of avian influenza H5N1 modulates virus replication, virulence and chicken-to-chicken transmission without significant impact on antigenic drift. J Gen Virol 97(12):3193–3204. https://doi.org/10.1099/jgv.0.000648

Bassareo PP, Melis MR, Marras S, Calcaterra G (2020) Learning from the past in the COVID-19 era: rediscovery of quarantine, previous pandemics, origin of hospitals and national healthcare systems, and ethics in medicine. Postgrad Med J 96(1140):633–638. https://doi.org/10.1136/postgradmedj-2020-138370

Brugère-Picoux J, Vaillancourt J-P, Bouzouaïa M (2015) Manual of poultry diseases, Version anglaise. AFAS, Paris

Capua I, Alexander DJ (2004) Avian influenza: Recent developments. Avian Pathol 33(4):393–404. https://doi.org/10.1080/03079450410001724085

Capua I, Terregino C, Cattoli G, Mutinelli F, Rodriguez JF (2003) Development of a DIVA (Differentiating Infected from Vaccinated Animals) strategy using a vaccine containing a heterologous neuraminidase for the control of avian influenza. Avian Pathol 32(1):47–55. https://doi.org/10.1080/0307945021000070714

Costa T, Chaves AJ, Valle R, Darji A, van Riel D, Kuiken T et al (2012) Distribution patterns of influenza virus receptors and viral attachment patterns in the respiratory and intestinal tracts of seven avian species. Vet Res 43(1):28. https://doi.org/10.1186/1297-9716-43-28

Cui H, de Jong MC, Beerens N, van Oers MM, Teng Q, Li L et al (2021) Vaccination with inactivated virus against low pathogenic avian influenza subtype H9N2 does not prevent virus transmission in chickens. J Virus Erad 7(3):100055. https://doi.org/10.1016/j.jve.2021.100055

Dawood FS, Iuliano AD, Reed C, Meltzer MI, Shay DK, Cheng P-Y et al (2012) Estimated global mortality associated with the first 12 months of 2009 pandemic influenza A H1N1 virus circulation: a modelling study. Lancet Infect Dis 12(9):687–695. https://doi.org/10.1016/S1473-3099(12)70121-4

Drake JW (1993) Rates of spontaneous mutation among RNA viruses. Proc Natl Acad Sci USA 90(9):4171–4175. https://doi.org/10.1073/pnas.90.9.4171

Gamblin SJ, Skehel JJ (2010) Influenza hemagglutinin and neuraminidase membrane glycoproteins. J Biol Chem 285(37):28403–28409. https://doi.org/10.1074/jbc.R110.129809

Grund C, Abdelwhab E-SM, Arafa A-S, Ziller M, Hassan MK, Aly MM, Hafez HM, Harder TC, Beer M (2011) Highly pathogenic avian influenza virus H5N1 from Egypt escapes vaccine-induced immunity but confers clinical protection against a heterologous clade 2.2.1 Egyptian isolate. Vaccine 29(33):5567–5573. https://doi.org/10.1016/j.vaccine.2011.01.006

Gultyaev AP, Spronken MI, Funk M, Fouchier RAM, Richard M (2021) Insertions of codons encoding basic amino acids in H7 hemagglutinins of influenza A viruses occur by recombination with RNA at hotspots near snoRNA binding sites. RNA 27(2):123–132. https://doi.org/10.1261/rna.077495.120

Hafez HM (2003) Geflügelpest: Alte Krankheit mit ständiger Gefahr für Geflügel. Tierarztl Umsch 58:343–351

Halvorson DA (2002) The control of H5 or H7 mildly pathogenic avian influenza: a role for inactivated vaccine. Avian Pathol 31(1):5–12. https://doi.org/10.1080/03079450120106570

Halvorson DA (2009) Prevention and management of avian influenza outbreaks: experiences from The United States of America. Rev Sci Tech 28(1):359–369

Hennig C, Graaf A, Petric PP, Graf L, Schwemmle M, Beer M, Harder T (2022) Are pigs overestimated as a source of zoonotic influenza viruses? Porcine Health Manag 8(1):30. https://doi.org/10.1186/s40813-022-00274-x

Kapczynski DR, Dorsey K, Chrzastek K, Moraes M, Jackwood M, Hilt D et al (2016) Vaccine protection of turkeys against H5N1 highly pathogenic avian influenza virus with a recombinant Turkey herpesvirus expressing the hemagglutinin gene of avian influenza. Avian Dis 60(2):413–417. https://doi.org/10.1637/11267-090115-Reg

Kimble B, Nieto GR, Perez DR (2010) Characterization of influenza virus sialic acid receptors in minor poultry species. Virol J 7:365. https://doi.org/10.1186/1743-422X-7-365

Kuhn JH, Adkins S, Alioto D, Alkhovsky SV, Amarasinghe GK, Anthony SJ et al (2020) 2020 Taxonomic update for phylum Negarnaviricota (*Riboviria*: *Orthornavirae*), including the large orders Bunyavirales and Mononegavirales. Arch Virol 165(12):3023–3072. https://doi.org/10.1007/s00705-020-04731-2

Lang G, Narayan O, Rouse BT, Ferguson AE, Connell MC (1968) A new influenza A virus infection in turkeys II. A highly pathogenic variant, a/Turkey/Ontario 772/66. Can Vet J 9(7):151–160

Marshall N, Priyamvada L, Ende Z, Steel J, Lowen AC (2013) Influenza virus reassortment occurs with high frequency in the absence of segment mismatch. PLoS Pathog 9(6):e1003421. https://doi.org/10.1371/journal.ppat.1003421

Martini M, Gazzaniga V, Bragazzi NL, Barberis I (2019) The Spanish influenza pandemic: a lesson from history 100 years after 1918. J Prev Med Hyg 60(1):E64–E67. https://doi.org/10.15167/2421-4248/jpmh2019.60.1.1205

Parvin R, Hossain I, Hasan A, Afrin SZ, Shehata AA (2022) Influenza and coronavirus zoonoses: an overview on pandemic events, viral genome, replication and emergency preparedness. Ger J Microbiol 2(3):1–11. https://doi.org/10.51585/gjm.2022.3.0016

Pillai SPS, Pantin-Jackwood M, Yassine HM, Saif YM, Lee CW (2010) The high susceptibility of turkeys to influenza viruses of different origins implies their importance as potential intermediate hosts. Avian Dis 54(1 Suppl):522–526. https://doi.org/10.1637/8770-033109-Review.1

Ping J, Selman M, Tyler S, Forbes N, Keleta L, Brown EG (2012) Low-pathogenic avian influenza virus a/Turkey/Ontario/6213/1966 (H5N1) is the progenitor of highly pathogenic a/Turkey/Ontario/7732/1966 (H5N9). J Gen Virol 93(8):1649–1657. https://doi.org/10.1099/vir.0.042895-0

Riley S, Kwok KO, Wu KM, Ning DY, Cowling BJ, Wu JT et al (2011) Epidemiological characteristics of 2009 (H1N1) pandemic influenza based on paired sera from a longitudinal community cohort study. PLoS Med 8(6):e1000442. https://doi.org/10.1371/journal.pmed.1000442

Ryan JR (ed) (2008) Pandemic influenza: emergency planning and community preparedness, 1st edn. CRC Press. ISBN: 9781420060874

Salaheldin AH, Veits J, Abd El-Hamid HS, Harder TC, Devrishov D, Mettenleiter TC et al (2017) Isolation and genetic characterization of a novel 2.2.1.2a H5N1 virus from a vaccinated meat-turkeys flock in Egypt. Virol J 14(1):48. https://doi.org/10.1186/s12985-017-0697-5

Salaheldin AH, Elbestawy AR, Abdelkader AM, Sultan HA, Ibrahim AA, Abd El-Hamid HS et al (2022) Isolation of genetically diverse H5N8 avian influenza viruses in poultry in Egypt, 2019–2021. Viruses 14(7):1431. https://doi.org/10.3390/v14071431

Shao W, Li X, Goraya MU, Wang S, Chen J-L (2017) Evolution of influenza A virus by mutation and re-assortment. Int J Mol Sci 18(8):1650. https://doi.org/10.3390/ijms18081650

Shehata AA, Sedeik ME, Elbestawy AR, Zain El-Abideen MA, Ibrahim HH, Kilany WH, Ali A (2019) Co-infections, genetic, and antigenic relatedness of avian influenza H5N8 and H5N1 viruses in domestic and wild birds in Egypt. Poult Sci 98(6):2371–2379. https://doi.org/10.3382/ps/pez011

Shehata AA, Basiouni S, Ali A, Fawzy M, Hafez HM, Ulbert S et al (2020) Immunization of turkeys with a DNA vaccine expressing the haemagglutinin gene of low pathogenic avian influenza virus subtype H9N2. J Virol Methods 284:113938. https://doi.org/10.1016/j.jviromet.2020.113938

Sieverding E, Hafez HM (2023) An animal-friendly collection of samples from the upper respiratory tract via drinker-swabs for the detection of avian influenza (H5N1) in poultry flocks: Recogida de muestras del tracto respiratorio superior mediante hisopos bebederos para la detección de la gripe aviar (H5N1) en manadas de aves de corral de forma inocua para los animales. SFJEAS 3(2):80–100. https://doi.org/10.53499/sfjeasv3n2-004

Sims L, Tripodi A, von Doschuetz S, Garnder E, Aguanno R (2016) Rational use of vaccination for prevention and control of H5 highly pathogenic avian influenza. FOCUS ON, No. 10, May 2016. Rome

Suarez DL (2016) Influenza A virus. In: Swayne DE (ed) Animal influenza. John Wiley, Hoboken, NJ, pp. 1–30

Swayne DE, Spackman E, Pantin-Jackwood M (2014) Success factors for avian influenza vaccine use in poultry and potential impact at the wild bird-agricultural interface. EcoHealth 11:94–108

Swayne DE, Suarez DL, Sims LD (2020) Diseases of poultry. In: Swayne DE, Boulianne M, Logue CM, McDougald LR, Nair V, Suarez DL (eds) Poultry diseases. Iowa State Press, Ames, IA, pp 210–256

Tewawong N, Prachayangprecha S, Vichiwattana P, Korkong S, Klinfueng S, Vongpunsawad S et al (2015) Assessing antigenic drift of seasonal influenza A(H3N2) and A(H1N1)pdm09 viruses. PLoS One 10(10):e0139958. https://doi.org/10.1371/journal.pone.0139958

Wen S, Wu Z, Zhong S, Li M, Shu Y (2021) Factors influencing the immunogenicity of influenza vaccines. Hum Vaccin Immunother 17(8):2706–2718. https://doi.org/10.1080/21645515.2021.1875761

Zhong L, Wang X, Li Q, Liu D, Chen H, Zhao M et al (2014) Molecular mechanism of the airborne transmissibility of H9N2 avian influenza A viruses in chickens. J Virol 88(17):9568–9578. https://doi.org/10.1128/JVI.00943-14

Avian Paramyxoviruses

2

Hafez M. Hafez and Awad A. Shehata

Abstract

Avian paramyxoviruses (APMVs) have been detected in a wide range of bird species worldwide. In chickens and turkeys, APMVs are associated with respiratory manifestations, high mortality and morbidity, and/or a decrease in egg production, accompanied by severe economic losses. These viruses are members of the genus *Avulavirus* and the family *Paramyxoviridae*. Twenty-one serotypes of APMVs (PMV-1 to PMV-12) have been recognized. Newcastle disease virus (NDV), which belongs to APMV-1, is the most important pathogen for poultry. APMVs other than PMV-1 were isolated from wild birds and occasionally from poultry. APMV-2, 3, 6, and 7 infections are associated with respiratory manifestations and a drop in egg production in turkeys. APMVs can be isolated in the allantoic cavity of 8- to 10-day-old embryonated chicken eggs and identified using hemagglutination inhibition tests (HI) with specific antiserum. Currently, no vaccines are available against PMV-2, PMV-6, and PMV-7. Indeed, inactivated oil-emulsion vaccines against PMV-3 have been developed and used in turkey breeder flocks.

Keywords

Paramyxoviruses · Avulavirus · Newcastle disease · PMV · Classification · Diagnosis · Vaccination

H. M. Hafez
Institute of Poultry Diseases, Faculty of Veterinary Medicine, Free University of Berlin, Berlin, Germany
e-mail: hafez.mohamed@fu-berlin.de

A. A. Shehata (✉)
TUM School of Natural Sciences, Bavarian NMR Center (BNMRZ), Structural Membrane Biochemistry, Technical University of Munich, Garching, Germany
e-mail: awad.shehata@tum.de

H. M. Hafez, A. A. Shehata (eds.), *Turkey Diseases and Disorders Volume 2*,
https://doi.org/10.1007/978-3-031-63322-5_2

2.1 Classification of *Paramyxoviruses*

Currently, the *Paramyxoviridae* contains four subfamilies, namely, *Orthoparamyxovirinae*, *Metaparaymyxovirinae*, *Rubulavirinae*, and *Avulavirinae*, and 17 genera. Paramyxoviruses are classified within the genera *Metaavulavirus*, *Orthoavulavirus*, and *Paraavulavirus*, according to the International Committee on Taxonomy of Viruses (ICTV) Report 2019 (ICTV 2019).

The *Pneumovirinae* subfamily has two genera: Metapneumovirus and Pneumovirus. Paramyxoviruses are pleomorphic, enveloped, and non-segmented ssRNA with a genome size of 10–17 Kb (Lanb and Parks 2007). *Paramyxoviridae* contains viruses that cause diseases in mammals, birds, reptiles, and fish. However, paramyxoviruses are host-specific, that is, measles, mumps, Nipah, and Hendra, and parainfluenza viruses are pathogenic for humans (Rima et al. 2019). The avian Paramyxovirus (APMV) can be classified serologically using a hemagglutination inhibition assay (HI). Serological tests, including serum neutralization (NT) and agar gel diffusion tests (AGP), can also be used for virus serotyping. Recently, the viral genome sequence can also be used as a classification criterion (Cattoli et al. 2011). To date, there are 21 serotypes described worldwide of APMV, designated PMV-1 to PMV-21 (ZKBS 2021). According to the International Committee on Taxonomy of Viruses (ICTV) Report 2019 (ICTV 2019), the subfamily Avulavirus comprises three genera: (i) *Metaavulavirus* (APMV-2, 5, 6, 7, 8, 10, 11, 14, 15, and 20), (ii) *Orthoavulavirus* (APMV-1, 9, 12, 13, 16, 17, 18, 19, and 21), and *Paraavulavirus* (PMV-3 and 4).

Paramyxoviruses have limited cross-reactivity, and little information is available since APMV isolates of the same serotype might differ. However, it cannot be excluded that the host range of APMVs also extends to humans. APMV-1 can cause conjunctivitis or influenza-like symptoms in humans. APMV-2 has also been isolated from Javanese monkeys (*Macaca fascicularis*) (Nelson et al. 1952; Nishikawa et al. 1977).

2.2 Avian Paramyxoviruses

APMV-1 to PMV-9 viruses were first isolated in the 1970s (Bankowski et al. 1960) and then identified worldwide. Since 2003, APMV-10 to PMV-21 have been identified in wild and domestic birds (Dinter et al. 1964; Shortridge et al. 1978; Alexander and Senne 2008). However, the new serotypes, particularly APMV-10–21, are poorly characterized. APMVs can infect a broad host range, such as chicken birds (*Galliformes*), geese (*Anseriformes*), passerines (*Passeriformes*), or parrots (*Psittaciformes*) (Cloud and Rosenberger 1980; Capua et al. 2004). Most recently identified APMVs have been isolated from Penguins (*Sphenisciformes*) (Miller et al. 2010; Neira et al. 2017). Experimentally, mice and hamsters can also be infected with various APMVs, showing no or only mild clinical symptoms (Samuel et al. 2011).

APMV-1 (Newcastle disease virus) is the most pathogenic serotype for avian species, causing significant economic losses. APMV-2, APMV-3, APMV-6, and APMV-7 cause mild to moderate respiratory manifestations, drop egg production, and reduced hatchability in turkeys. APMV-8, APMV-9, APMV-10, and APMV-11 cause no clinical symptoms in wild and or domesticated bird species. The different APMVs and pathogenicity are shown in Table 2.1.

Table 2.1 Avian paramyxoviruses, year of isolation, and pathogenicity

Serotype	Species	Pathogenicity
PMV-1	Chicken	• NDV
PMV-2	Chicken, turkeys, passerines, parrots	• Mild to moderate respiratory manifestations • Reduction of egg production and hatchability
PMV-3	Turkey and parakeet	
PMV-4	Wild duck, chicken, geese, and mallard duck	
PMV-5	Budgerigar	• Encephalitis in parrots (*Neophema* sp., and *Psephites* sp.)
PMV-6	Duck, geese, turkeys, and mallard duck	• Mild to moderate respiratory manifestations • Reduction in egg production • Reduction in hatchability
PMV-7	Hunter-killed dove, turkey, and ostrich	
PMV-8	Feral Canadian goose and pintail	• No clinical diseases in wild or domesticated bird species
PMV-9	Domestic ducks and feral ducks	
PMV-10	Rockhopper penguin	
PMV-11	Common snipe	
PMV-12	Eurasian Wigeon	• Not associated with any diseases in wild birds, • exhibit low pathogenicity in 1-day-old chicks inoculated intracerebrally
PMV-13	Wild birds	
PMV-14	Wild migratory geese, duck	• No clinical diseases in wild or domesticated bird species
PMV-15	Migratory birds	
PMV-16	Wild birds	
PMV-17	Wild birds	
PMV-18	Wild birds	
PMV-19	Wild birds	
PMV-20	Wild birds	
PMV-21	Wild birds	

2.3 Newcastle Disease: PMV-1

It will be discussed in the next chapter in detail.

2.4 Yucaipa Disease: PMV-2

APMV-2 was first reported in chickens in Yucaipa, USA, in 1956 (Bankowski et al. 1960). The natural hosts are turkeys, racing pigeons, and passerines. However, chickens, psittacines, and rails might be infected accidentally. The passerine and psittacine species frequently serve as reservoirs for APMV-2 (Zhang et al. 2007).

In turkeys, APMV-2 infection is associated with respiratory manifestations and decreased egg production (Lipkind et al. 1979; Bankowski et al. 1981). It can also negatively affect hatchability. The clinical disease is more serious, particularly in turkeys, during secondary infections with other viruses (Lang et al. 1975). APMV-2 cannot propagate in neural tissues and has an affinity for the epithelial linings of the intestinal and respiratory systems (Subbiah et al. 2010).

PMV-2 can be isolated in 9-day-old SPF chicken eggs and several cell lines such as baby hamster kidney (BHK21), Madin–Darby canine kidney (MDCK), and African green monkey kidney (Vero cells). The virus causes cell rounding of single cells without syncytia formation. The mean death time (MDT) and intracerebral pathogenicity index (ICPI) values of different strains of APMV-2 were mostly observed to be >168 h and 0, respectively, suggesting the strains' avirulent nature in chickens, similar to lentogenic NDV strains. Serotyping of APMV-2 can be done serologically or using sequence analysis (Subbiah et al. 2010). Birds vaccinated against NDV show a rise in hemagglutination inhibition titers to both viruses if subsequently infected with PMV-3. Avian avulavirus-2 does not agglutinate the RBCS of chickens and turkeys.

2.5 Wisconsin Disease: PMV-3

APMV-3 was initially detected in turkeys in Ontario and then in Wisconsin in 1967 (Tumova et al. 1979). The virus has a diverse host range. It has been identified in domestic, wild birds, turkeys, and ostrich. Two strains of PMV-3 have been described, namely, turkey, psittacine, and passerine strains. Currently, numerous serologically related APMV-3 strains have been detected in turkeys in several European countries. The Netherlands and Wisconsin APMV-3 strains revealed that the Netherland APMV-3 strain is slightly pathogenic (MDT = 112 h), while the Wisconsin APMV-3 strain has MDT of more than 168 h (Kumar et al. 2010). The cleavage site of F-protein in the Netherlands and Wisconsin APMV-3 strains has three or two basic residues, respectively, similar but not identical to the pattern of virulent and avirulent NDV strains (Kumar et al. 2010).

The main clinical signs in turkeys are mild to moderate respiratory manifestations such as nasal discharge, coughing, and swelling of the infraorbital sinus

(Redmann et al. 1991; Samal 2011). It is also associated with a drop in egg production. In psittacine and passeriform species, APMV-3 causes nervous manifestations such as encephalitis. Intranuclear and intracytoplasmic inclusion bodies have been identified in glial cells. Also, myocarditis and pancreatitis with intranuclear inclusion bodies have been reported.

2.6 Avian Paramyxovirus Serotype-6 (APMV-6)

APMV-6 was first isolated from a healthy duck in Hong Kong in 1977 (Shortridge et al. 1980). The virus was isolated/detected from geese and wild birds (Stanislawek et al. 2002; Warke et al. 2008). In chickens, APMV-6 is avirulent. However, it causes mild to moderate respiratory manifestations and reduces egg production in turkeys (Alexander 1997). In a recent study, PMV-6 was isolated from wild waterfowl in America. Not all PMV-6 viruses can hemagglutinate chicken red blood cells (Hisanaga et al. 2021).

2.7 Avian Paramyxovirus Serotype-7 (APMV-7)

APMV-7 was first isolated in 1975 from hunter-killed doves in the United States (Alexander et al. 1981). APMV-7 was detected in chickens, turkeys, and pigeons. Occasionally, APMV-7 has also been linked to other avian species (Woolcock et al. 1996). In 1997, a natural epidemic of AMPV-7 was reported in Ohio in turkeys with respiratory symptoms, hepatomegaly, and splenomegaly (Saif et al. 1997). In turkeys, APMV-7 is associated with moderate respiratory symptoms and a drop in egg production; however, it is avirulent in chickens.

References

Alexander DJ (1997) Newcastle disease and other avian *Paramyxoviridae* infections. In: Calnek BW (ed) Diseases of poultry. Iowa State University Press, Ames, IA

Alexander D, Senne D (2008) Newcastle disease and other avian paramyxovirus and pneumovirus infection. In: Saif YM, Fadly AM, Glisson JR (eds) Diseases of poultry. Iowa State University Press, Ames, IA, pp 75–115

Alexander DJ, Hinshaw VS, Collins MS (1981) Characterization of viruses from doves representing a new serotype of avian paramyxoviruses. Arch Virol 68(3–4):265–269. https://doi.org/10.1007/BF01314580

Bankowski RA, Corstvet RE, Clark GT (1960) Isolation of an unidentified agent from the respiratory tract of chickens. Science 132(3422):292–293. https://doi.org/10.1126/science.132.3422.292

Bankowski RA, Almquist J, Dombrucki J (1981) Effect of paramyxovirus Yucaipa on fertility, hatchability, and poult yield of turkeys. Avian Dis 25(2):517. https://doi.org/10.2307/1589944

Capua I, De Nardi R, Beato MS, Terregino C, Scremin M, Guberti V (2004) Isolation of an avian paramyxovirus type 9 from migratory waterfowl in Italy. Vet Rec 155(5):156

Cattoli G, Susta L, Terregino C, Brown C (2011) Newcastle disease: a review of field recognition and current methods of laboratory detection. J Vet Diagn Invest 23(4):637–656. https://doi.org/10.1177/1040638711407887

Cloud SS, Rosenberger JK (1980) Characterization of nine avian paramyxoviruses. Avian Dis 24(1):139. https://doi.org/10.2307/1589773

Dinter Z, Hermodsson S, Hermodsson L (1964) Studies on myxovirus Yucaipa: its classification as a member of the paramyxovirus group. Virology 22(3):297–304. https://doi.org/10.1016/0042-6822(64)90020-0

Hisanaga T, Soos C, Lewis N, Lung O, Suderman M, Berhane Y (2021) Genetic and antigenic characterization of avian Avulavirus type 6 (AAvV-6) circulating in Canadian wild birds (2005–2017). Viruses 13(4):543. https://doi.org/10.3390/v13040543

ICTV (2019) Family: *Paramyxoviridae*. https://ictv.global/report/chapter/paramyxoviridae/paramyxoviridae

Kumar S, Militino Dias F, Nayak B, Collins PL, Samal SK (2010) Experimental avian paramyxovirus serotype-3 infection in chickens and turkeys. Vet Res 41(5):72. https://doi.org/10.1051/vetres/2010042. Epub 2010 Jul 23

Lanb RA, Parks GD (2007) *Paramyxoviridae*: the viruses and their replication. In: Fields BN, Knipe DN, Howley PM (eds) Fields virology. Lippincott, Williams, and Wilkins

Lang G, Gagnon A, Howell J (1975) The occurrence of *paramyxovirus* Yucaipa in Canadian poultry. Can Vet J 16(8):233–237

Lipkind MA, Weisman Y, Shihmanter E, Shoham D, Aronovici A (1979) The isolation of yucaipa-like paramyxoviruses from epizootics of a respiratory disease in Turkey poultry farms in Israel. Vet Rec 105:577–578

Miller PJ, Afonso CL, Spackman E, Scott MA, Pedersen JC, Senne DA et al (2010) Evidence for a new avian paramyxovirus serotype 10 detected in rockhopper penguins from The Falkland Islands. J Virol 84(21):11496–11504. https://doi.org/10.1128/JVI.00822-10

Neira V, Tapia R, Verdugo C, Barriga G, Mor S, Ng TFF et al (2017) Novel avulaviruses in penguins, Antarctica. Emerg Infect Dis 23(7):1212–1214. https://doi.org/10.3201/eid2307.170054

Nelson CB, Pomeroy BS, Schrall K, Park WE, Lindemann RJ (1952) An outbreak of conjunctivitis due to Newcastle disease virus (NDV) occurring in workers. Am J Public Health 42:672

Nishikawa F, Sugiyama T, Suzuki K (1977) A new paramyxovirus isolated from cynomolgus monkeys. Jpn J Med Sci Biol 30(4):191–204. https://doi.org/10.7883/yoken1952.30.191

Redmann T, Zeydanli MM, Herbst W, Kaleta EF (1991) Isolation of a paramyxovirus-3 from turkeys with respiratory tract disease in Germany. Dtsch Tierarztl Wochenschr 98(4):138–141

Rima B, Balkema-Buschmann A, Dundon WG, Duprex P, Easton A, Fouchier R et al (2019) ICTV virus taxonomy profile: *Paramyxoviridae*. J Gen Virol 100(12):1593–1594. https://doi.org/10.1099/jgv.0.001328

Saif YM, Mohan R, Ward L, Senne DA, Panigrahy B, Dearth RN (1997) Natural and experimental infection of turkeys with avian paramyxovirus-7. Avian Dis 41(2):326–329

Samal SK (2011) Newcastle disease and related avian paramyxoviruses. In: Samal SK (ed) The biology of paramyxoviruses. Caister Academic Press, Norfolk, UK

Samuel AS, Subbiah M, Shive H, Collins PL, Samal SK (2011) Experimental infection of hamsters with avian paramyxovirus serotypes 1 to 9. Vet Res 42(1):38. https://doi.org/10.1186/1297-9716-42-38

Shortridge KF, Alexander DJ, Hu LY, Kam SL (1978) Isolation of Newcastle disease virus from *Phasianidae* birds in Hong Kong. J Comp Pathol 88(4):633–636. https://doi.org/10.1016/0021-9975(78)90017-8

Shortridge KF, Alexander DJ, Collins MS (1980) Isolation and properties of viruses from poultry in Hong Kong which represent a new (sixth) distinct group of avian paramyxoviruses. J Gen Virol 49(2):255–262. https://doi.org/10.1099/0022-1317-49-2-255

Stanislawek WL, Wilks CR, Meers J, Horner GW, Alexander DJ, Manvell RJ et al (2002) Avian paramyxoviruses and influenza viruses isolated from mallard ducks (*Anas platyrhynchos*) in New Zealand. Arch Virol 147(7):1287–1302. https://doi.org/10.1007/s00705-002-0818-2

Subbiah M, Xiao S, Khattar SK, Dias FM, Collins PL, Samal SK (2010) Pathogenesis of two strains of avian paramyxovirus serotype 2, Yucaipa and Bangor, in chickens and turkeys. Avian Dis 54(3):1050–1057. https://doi.org/10.1637/9380-041910-Reg.1

Tumova B, Robinson JH, Easterday BC (1979) A hitherto unreported paramyxovirus of turkeys. Res Vet Sci 27(2):135–140

Warke A, Appleby L, Mundt E (2008) Prevalence of antibodies to different avian paramyxoviruses in commercial poultry in the United States. Avian Dis 52(4):694–697. https://doi.org/10.1637/8390-070308-RESNOTE.1

Woolcock PR, Moore JD, McFarland MD, Panigrahy B (1996) Isolation of paramyxovirus serotype 7 from ostriches (*Struthio camelus*). Avian Dis 40(4):945–949

Zhang G-Z, Zhao J-X, Wang M (2007) Serological survey on prevalence of antibodies to avian paramyxovirus serotype 2 in China. Avian Dis 51(1):137–139. https://doi.org/10.1637/0005-2086(2007)051[0137:SSOPOA]2.0.CO;2

ZKBS (2021) Stellungnahme der ZKBS zur Risikobewertung der Aviären Paramyxoviren 2—21 (APMV-2 bis -21) als Spender- oder Empfängerorganismen gemäß § 5 Absatz 1 GenTSV -Az. 6790-05-02-0072

Newcastle Disease in Turkeys

3

Awad A. Shehata and Hafez M. Hafez

Abstract

Newcastle disease (ND), caused by avian paramyxovirus serotype 1 (APMV-1), is a viral disease of many poultry species characterized by respiratory manifestations followed by/or accompanied by nervous manifestations. ND is ranked as the major virus disease of poultry in many countries worldwide, and it is challenging to assess the prevalence of ND worldwide. It is a worldwide enzootic disease. Based on its pathogenicity, NDV is classified into five pathotypes: apathogenic, lentogenic, mesogenic, viscerotropic velogenic, and neurotropic velogenic. The velogenic strains are the most virulent and can cause 100% mortality in susceptible birds. Based on the complete genomic analysis, they are grouped into classes (Class I and II); the latter class is subdivided into at least 21 distinct genotypes. The presumptive diagnosis of NDV is based on the clinical signs and postmortem lesions and confirmed by isolation and identification. Isolation can be done by inoculating the suspected samples into the allantoic cavity of 9- to 11-day-old embryonated specific pathogen-free chicken eggs. The virus can be identified using the hemagglutination test, PCR, and the hemagglutination inhibition test using PMV-1-specific serum. The pathogenicity can be identified using the intracerebral pathogenicity index (ICPI), mean death time, intravenous pathogenicity index (IVPI), and sequence analysis of the cleavage site of F-protein. ND is a notifiable disease in several countries. The prevention

A. A. Shehata (✉)
TUM School of Natural Sciences, Bavarian NMR Center (BNMRZ), Structural Membrane Biochemistry, Technical University of Munich, Garching, Germany
e-mail: awad.shehata@tum.de

H. M. Hafez
Institute of Poultry Diseases, Faculty of Veterinary Medicine, Free University of Berlin, Berlin, Germany
e-mail: hafez.mohamed@fu-berlin.de

H. M. Hafez, A. A. Shehata (eds.), *Turkey Diseases and Disorders Volume 2*,
https://doi.org/10.1007/978-3-031-63322-5_3

of ND is based on good biosecurity and vaccination. Several vaccines are available, including live, inactivated, and recombinant vaccines.

Keywords

APMV-1 · NDV · Pathotypes · Clinical signs · Postmortem lesions · Diagnosis · Prevention and control

3.1 Etiology

Newcastle disease (ND) is caused by NDV, family *Paramyxoviridae*, genus *Orthoavulavirus* (Rima et al. 2019; Swayne 2020). Like other members of this group, the NDV possesses two surface proteins vital to the identification and pathogenesis of the virus. Hemagglutinin (HA) and neuraminidase (HN) are essential for the attachment and release of the virus from the infected host cells. In addition, the HN allows serologic identification of the virus. The fusion (F) protein plays a critical role in the pathogenesis of the disease. The virus is enveloped, sensitive to lipid solvents and heat, and unstable at high and low pH.

3.2 Pathotypes of NDV

Currently, 21 serologically distinguishable groups of *avian paramyxoviruses* (APMV-1 to APMV-21), based on hemagglutination inhibition (HI) and neuraminidase inhibition (NI) assays, have been described**.** The established serotypes are pretty distinct, but some cross-reactions have been shown by HI and NI tests, particularly between the serotypes APMV-1 and APMV-3 (Lipkind and Shihmanter 1986).

Strains of the ND virus have been grouped into five pathotypes based on the clinical signs in the infected chickens (Beard and Hanson 1984), Table 3.1.

(i) **Viscerotropic velogenic**: highly pathogenic form in which hemorrhagic intestinal lesions are frequently seen

Table 3.1 Pathotypes of NDV

Type	Examples	Mortality	Hemorrhage	Respiratory signs	Nervous signs
Viscerotropic velogenic	Herts 33, NY, Parrot	++++	++++	–	–
Neurotropic velogenic	Texas GB	++++	–	++	+++
Mesogenic	Roakin, Komarov, Mukteswar, and H	+	–	+	+
Lentogenic	Hitchner B1, LaSota, Clone 30, Ulster 2C, V4	–	–	(+)	–
Avirulent	MC110	–	–	–	–

(ii) **Neurotropic velogenic:** accompanied by high mortality and usually following respiratory and nervous signs
(iii) **Mesogenic:** a form that presents with respiratory signs and occasional nervous manifestations but low mortality
(iv) **Lentogenic** or respiratory: a form that presents with a mild and/or subclinical respiratory infection
(v) **Asymptomatic enteric**: a form that usually consists of subclinical enteric infection

Although these classifications are useful for differentiation purposes, some overlapping occurs, and some viruses are difficult to place (Alexander 2000, Alexander and Gough 2003). According to Council Directive 92/66/EEC "Newcastle disease" means infection of poultry caused by any avian strain of paramyxovirus 1, with an intracerebral pathogenicity index (ICPI) in day-old chicks greater than 0.7 (EEC 1992) or mean death time (MDT) >90 h or intravenous pathogenicity index (IVPI) = 0, Table 3.2.

The activated fusion protein comprises two cleaved F1 and F2 subunits linked by a disulfide (-S-S-) bridge, cleavage of the F gene initiates cell fusion. The cleavage site is located between positions 112 and 117 of the fusion protein precursor (Fig. 3.1). Virulent NDVs have multibasic amino acids at the cleavage site (^{113}RQK/RR*F^{117}), and most virulent NDVs have an additional basic amino acid at position 112. All virulent NDVs have phenylalanine (F) at position 117 of the F1 N-terminus. However, low virulent NDVs have the sequence ^{113}K/RQG/ER*L^{117}.

The pathogenicity of NDV is based on the molecular basis of this cleavage site. Velogenic and mesogenic strains have multiple cleavage sites and at least three basic amino acids: arginine (R), lysine (K), and phenylalanine (F). Multiple basic amino acids of the cleavage site can be cleaved by furin-like enzymes, ubiquitous in the body, whereas a cleavage site containing ≤2 amino acids can be cleaved by trypsin-like enzymes located in the digestive and respiratory mucosa. This cleavage behavior explains why virulent strains cause systemic disease and lead to high mortality rates.

Table 3.2 Pathogenicity indices of Newcastle disease viruses using intracerebral pathogenicity index (ICPI) in day-old chicks, mean death time (MDT), or intravenous pathogenicity index (IVPI)

Test	Avirulent	Mesogenic	Velogenic
ICPI	<0.7	0.7–1.5	>1.5
MDT	>90 h	60–90 h	<60 h
IVPI	0	0.5–2	2–3

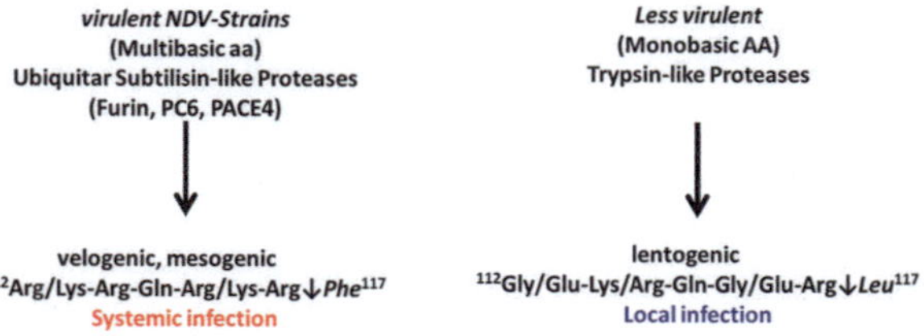

Fig. 3.1 The cleavage site of the F gene. Newcastle disease virus is infectious only when the glycoprotein F0 cleaves into F1 and F2

3.3 Genotypes of NDV

Sequence analysis of the F gene revealed that NDVs are classified into six distinct lineages (1–6); however, based on the complete genomic sequence analysis, they are grouped into two classes (Class I and Class II). **Class I** contains a single genotype and three sub-genotypes, while **class II** strains are grouped into at least 21 distinct genotypes (I–XXI) (Dimitrov et al. 2019) and several sub-genotypes. These high genetic variations could impact antigenicity, subsequently negatively impacting the traditional vaccines' effectiveness. Although NDV is endemic in several countries, such as most African countries, the Middle East, Asia, Central America, and northern South America, only sporadic outbreaks occur in Western Europe and the United States, highlighting the need to implement strict biosecurity and good vaccination programs.

3.4 Transmission

The primary introduction of ND into the bird population of a country can occur via migratory birds and the importation of captive and domestic poultry or poultry products. The infection can be transmitted primarily through direct contact between healthy and infected birds. The disease can also be secondarily spread by mechanical means, by vaccination and debeaking crews, as well as manure haulers, rendering truck drivers, and feed delivery personnel. Pet birds pose a significant risk of introducing exotic AND into poultry flocks. A large variety of wild birds and captive caged birds has been considered to have been infected naturally with NDV (Lancaster and Alexander 1975).

3.5 Clinical Signs

The course of the disease varies according to the virulence of the involved strain, species, age, and immune status of the birds as well as the general health and environmental conditions. Turkeys infected with vvNDV (velogenic) strains are mainly accompanied by a mortality rate between 30% and 50%. Clinical signs can include ruffled feathers, depression, diarrhea, and respiratory signs in the form of nasal discharge, coughing, rales, and dyspnea. These symptoms are accompanied mostly by nervous manifestations such as incoordination, tremors, twisting of the head and neck, abnormal movement (circling, rearing, somersaulting), paresis, and paralysis (Fig. 3.2). However, unlike chickens, infected turkeys can show oral exudate mixed with blood or bloody droppings. The clinical signs in commercial turkeys were less severe and of later onset than those observed in experimentally infected SPF turkeys (Aldous et al. 2010).

Fig. 3.2 Clinical signs of NDV in turkeys showing nervous manifestations (B) (©Hafez, FU-Berlin)

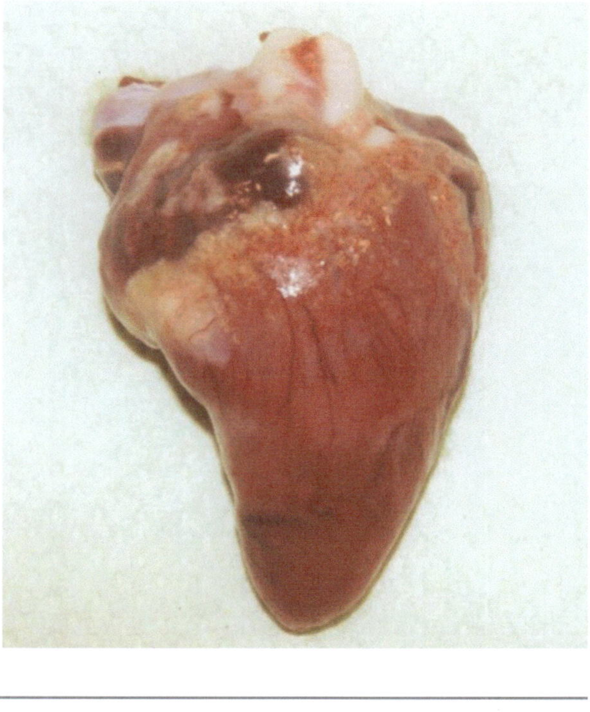

Fig. 3.3 Postmortem lesions of NDV in turkeys showing hemorrhages on the coronary fat (©Hafez, FU-Berlin)

3.6 Gross Lesions

Gross lesions of ND are characterized by petechial blood on the fat coronary of the heart (Fig. 3.3) and in the proventriculus, proventricular glands (Fig. 3.4) and intestine (Fig. 3.5), as well as under the horny lining of the gizzard. In many cases,

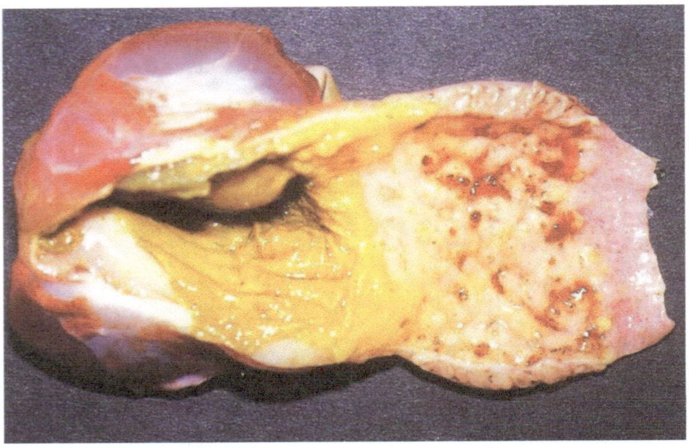

Fig. 3.4 Postmortem lesions of NDV in turkeys showing hemorrhages on the proventriculus glands (©Hafez, FU-Berlin)

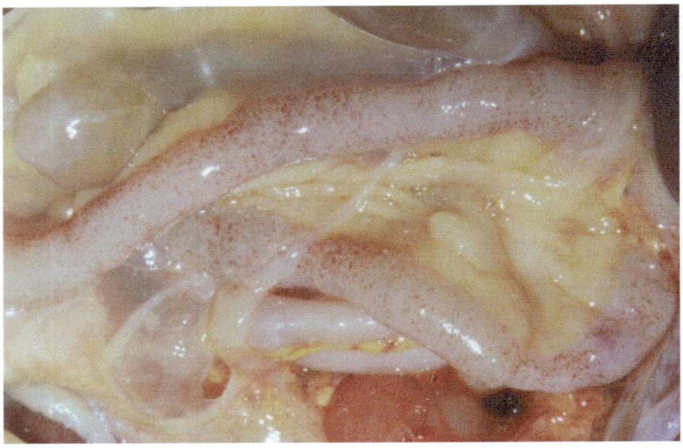

Fig. 3.5 Postmortem lesions of NDV in turkeys showing hemorrhages on the intestine (©Hafez, FU-Berlin)

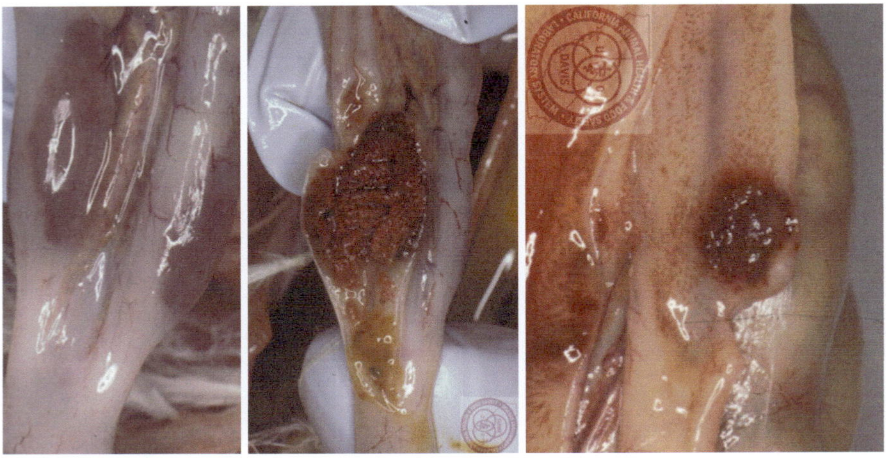

Fig. 3.6 Postmortem lesions of NDV in turkeys showing hemorrhages of ileocecal tonsils (©Hafez, FU-Berlin)

hemorrhagic areas or ulcers could also be observed in various parts of the intestines and cecal tonsils (Fig. 3.6), particularly in the case of viscerotropic velogenic NDV infection. In addition, congested tracheitis, caseous materials in the tracheal lumen (Fig. 3.7), and inflammation of the air sacs can be seen.

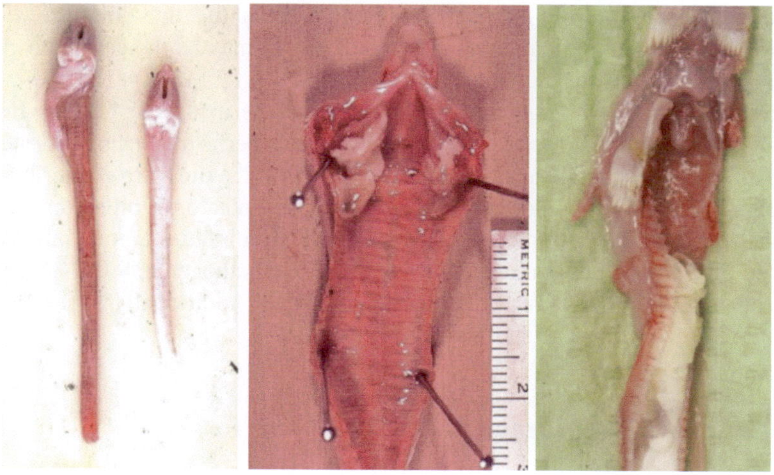

Fig. 3.7 Postmortem lesions of NDV in turkeys showing congested tracheitis and the presence of caseous materials in the tracheal lumen (©Hafez, FU-Berlin)

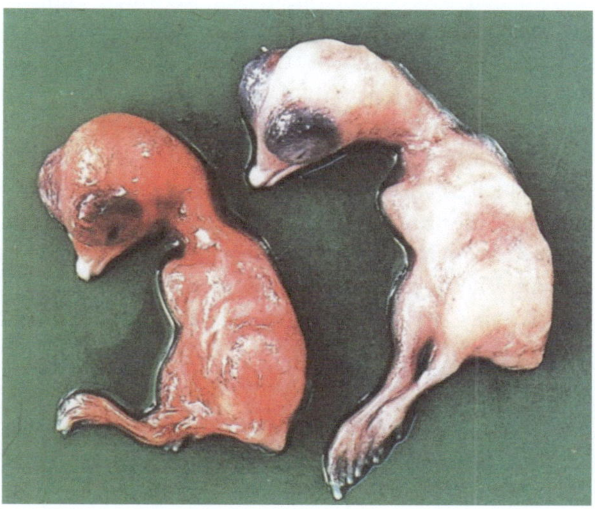

Fig. 3.8 Lesions of NDV in embryos showing hemorrhages (©Hafez, FU-Berlin)

3.7 Diagnosis

The conventional isolation technique recommended by the (OIE 2012) is based on inoculation of the suspected samples into the allantoic cavity of 9- to 11-day-old specific pathogen-free embryonated eggs from (SPF) chickens (Fig. 3.8).

The virus can be identified using the hemagglutination test, PCR, and the hemagglutination inhibition test using PMV-1 specific serum. However, the HI test can show cross-reactivity between PMV-1 and PMV-3 isolated from turkeys and psittacines. Therefore, monoclonal antibodies might be used to avoid cross-reactivity. Real-time PCR can be used to confirm virus identification (Creelan et al. 2002). Tracheal and oropharyngeal swabs are the most appropriate sample types for this type of analysis, as they do not contain organic material that may interfere with the reconstitution and amplification of the virus RNA, as may be the case with cloacal swabs, feces, or intestinal contents.

Serological diagnosis and monitoring can be done using the HI test. In non-vaccinated birds, the HI titer is $<1/8$ when the test is performed using four hemagglutinating units, while in vaccinated or infected birds, the titers are greater than 1/8. The distribution of antibodies in the flock is also an indicator of infection. Usually, vaccinated birds show a homologous distribution of antibodies among birds, while infected birds show a heterologous distribution. The pathogenicity of NDV can be assessed using MDT, ICPI, and ICPI tests (OIE 2012). Additionally, sequence analysis of the F-gene cleavage site is also widely used.

3.8 Control Measures

According to the EU-directive (Council Directive 92/66/EEC), an infected flock means any poultry in which the presence of NDV can be officially confirmed following an examination by an approved laboratory investigation or in the case of second and subsequent outbreaks, in which clinical symptoms or postmortem lesions consistent with ND are present.

Once the ND is officially confirmed on a holding, the member states shall ensure that all poultry on the holding, without delay, shall be killed on the spot. The poultry that has died and/or been killed and all eggs should be destroyed. These operations shall be carried out to minimize the risk of spreading disease; any substance or waste, such as animal feed and litter or manures liable to be contaminated, shall be destroyed or treated appropriately. This treatment is carried out following the instructions of the official veterinarian and shall ensure the destruction of any NDV present. The first line of disease control is to prevent the introduction and further spread via strict biosecurity, establishing and maintaining immunity, and vaccination.

In addition, disease control in food-producing animals is crucial for improving animal production, food safety, and human well-being. Monitor infectious diseases and establish a good vaccination program to control infectious diseases.

3.9 Vaccination

3.9.1 Live Vaccines

Several lentogenic and mesogenic strains are used for vaccination against NDV (Table 3.2). Live NDV vaccines originated from lentogenic strains such as LaSota, B1, and VG/GA vaccines belonging to genotype 2.II and are genetically and antigenically related with >98% nucleotide identity. The effectiveness of live ND vaccines is directly correlated with the dose of the administered vaccine, and under experimental conditions, the mean embryo infectious vaccine dose (EID_{50}) of 10^4–10^5 reliably achieves 100% protection against mortality. Live ND vaccines can

Table 3.3 List of live strains used for vaccination against Newcastle disease virus

Pathotype	Strain	Description
Lentogenic	F	Usually used in young chickens
	B1	Slightly more virulent than F, used as a vaccine in chickens of all ages and at 1 day old
	LaSota	Often causes post-vaccination respiratory signs, used as a booster vaccine in flocks vaccinated with F or B1
Avirulent	V4	Used in chickens of all ages
	V4-HR	Heat-resistant V4, thermostable, used in chickens of all ages
	I-2	Thermostable, used in chickens of all ages
	VG/GA	The Villegas Glisson/University of Georgia (VG/GA) strain of NDV has been proposed to replicate both in the respiratory and intestinal tract, with a preference for the intestine
Mesogenic	Mukteswar	An invasive strain used as a booster vaccine can cause adverse reactions (respiratory distress, loss of weight or drop in egg production, and even death) if used in partially immune chickens. Usually administered by injection
	Komarov	Less pathogenic than Mukteswar, used as a booster vaccine. Usually administered by injection

be administered by eye drop, intranasal, spray, or orally in drinking water. Hitchner B1 can be used for vaccinating one-day-old chicks by the coarse-spray method. In some countries, mesogenic vaccines, such as Komarov, Roakin, Mukteswar, and Hifa, are used. The mesogenic strains are more pathogenic and immunogenic than lentogenic strains. Indeed, strict quarantine, rapid diagnostics, biosecurity, stamping out, and other containment measures seem to keep ND under control. A description of live strains of NDV is shown in Table 3.3.

For the European Union (EU), only vaccines containing lentogenic strains are licensed that fulfill the criteria of an ICPI of <0.4 when tested, with not less than 10^7 EID_{50} /bird or <0.5 when tested with not less than 10^8 EID_{50}/per bird (93/152/ EEC). Live NDV vaccines have the advantage of mass application by drinking water and/ or sprays. This is less labor-intensive and inexpensive than inactivated vaccines, which must be injected into individual birds (Dortmans et al. 2014). Some lentogenic vaccines have been cloned to select viruses that produce lower post-vaccination reactions than the original virus (Hafez 2000).

Members of states shall ensure that (i) vaccination against ND with vaccines authorized by the competent authority for a prophylactic purpose or to supplement the control measures carried out when the disease appears and (ii) the only vaccines allowed are those that have received marketing authorization from the competent authority of the member state in which the vaccine is used.

3.9.2 Inactivated Vaccines

The live virus is propagated in SPF chicken embryos, and harvested allantoic fluid is inactivated by formalin, beta-propiolactone (BPL), or binary ethylenimine (BEI). Modern vaccine preparations are oil emulsions that proved more effective (Dortmans et al. 2014). The inactivated vaccines are applied intramuscularly or subcutaneously mainly to achieve long-lasting high antibody titers, which can also be passed from breeder to offspring.

Also, according to directive 93/152/EEC of the Commission of the European Council, inactivated vaccines are produced according to the guideline from ND virus strains, whose original seed virus at testing has an ICPI of <0.7 when administered no less than 10^8 EID_{50} to each bird in the ICPI test.

Commercially available recombinant ND vaccines are based mainly on fowlpox (rFPV) or herpesvirus of turkeys (rHVT) as a vector. Both types of recombinant vaccines have no side effects and can be successfully administered to chicks and turkeys with maternal antibodies (Weiss et al. 2018). A disadvantage of the rFPV vaccine is interference with maternally derived antibodies (MDA) (Hu et al. 2020). While this is not affected by rHVT vaccines, the vector virus is cell-associated and must be kept in liquid nitrogen and administered within an hour of being thawed. However, only one rHVT vaccine can be applied; otherwise, there would be interference between the different HVT vaccines (Dimitrov et al. 2017; Rauw et al. 2020). Several factors can cause vaccinal breaks, such as (i) birds already infected before vaccination, (ii) improper administration, (iii) early infection with an immunosuppressive virus, and (iv) mycotoxins in the feed.

3.10 Conclusion

ND infections in poultry are primarily associated with severe economic losses; early recognition and monitoring programs are essential in managing the infections. When diseases emerge or reemerge, evaluating their economic impact is essential to determine whether new control measures and, especially, new vaccines are needed. Generally, therapy or vaccination alone is of little value unless improvements in all aspects of management and biosecurity accompany them. Biosecurity is the cheapest, most effective means of disease control, and no disease prevention program will work without it. Since the success of any control program depends on the hygiene practices of the personnel, it is essential to incorporate education programs about microorganisms and their modes of transmission, as well as awareness of the reasons behind such control programs for all people involved throughout the poultry production chain. In the long term, the development of poultry lines that are genetically resistant to some pathogens should be progressed, and further attention must be paid to developing efficient vaccines against bacterial infections to reduce the use of antibiotics.

References

Aldous EW, Seekings JM, McNally A, Nili H, Fuller CM, Irvine RM et al (2010) Infection dynamics of highly pathogenic avian influenza and virulent avian paramyxovirus type 1 viruses in chickens, turkeys and ducks. Avian Pathol 39(4):265–273. https://doi.org/10.1080/03079457.2010.492825

Alexander DJ (2000) Newcastle disease and other avian paramyxoviruses. Rev Sci Tech 19(2):443–462. https://doi.org/10.20506/rst.19.2.1231

Alexander DJ, Gough RE (2003) Newcastle disease and other avian paramyxovirus infections. In: Saif YM, Barnes HJ, Glisson JR, Fadly AM, McDougal LR and Swayne DE, Eds., Disease of Poultry, 11th Edition, Iowa State University Press, Ames, 1A:63–87

Beard CW, Hanson RP (1984) Newcastle disease. In: Hofstad MS, Barnes HJ, Calnek BW, Reid WM, Yoder HW (eds) Diseases of poultry, 8th edn. Iowa State University Press, Ames, IA:452–470

Creelan JL, Graham DA, McCullough SJ (2002) Detection and differentiation of pathogenicity of avian paramyxovirus serotype 1 from field cases using one-step reverse transcriptase-polymerase chain reaction. Avian Pathol 31(5):493–499. https://doi.org/10.1080/0307945021000005860

Dimitrov KM, Afonso CL, Yu Q, Miller PJ (2017) Newcastle disease vaccines-a solved problem or a continuous challenge? Vet Microbiol 206:126–136. https://doi.org/10.1016/j.vetmic.2016.12.019

Dimitrov KM, Abolnik C, Afonso CL, Albina E, Bahl J, Berg M et al (2019) Updated unified phylogenetic classification system and revised nomenclature for Newcastle disease virus. Infect Genet Evol 74:103917. https://doi.org/10.1016/j.meegid.2019.103917

Dortmans JCFM, Venema-Kemper S, Peeters BPH, Koch G (2014) Field vaccinated chickens with low antibody titers show equally insufficient protection against matching and non-matching genotypes of virulent Newcastle disease virus. Vet Microbiol 172(1–2):100–107. https://doi.org/10.1016/j.vetmic.2014.05.004

EEC (1992) Council Directive 92/66/EEC of 14 July 1992 introducing community measures for the control of Newcastle disease.

Hafez HM (2000), Respiratory diseases of Turkey. World poultry—Elsevier special'00—Turkey Health 2000, pp 13–19

Hu Z, Ni J, Cao Y, Liu X (2020) Newcastle disease virus as a vaccine vector for 20 years: a focus on maternally derived antibody interference. Vaccine 8(2):222. https://doi.org/10.3390/vaccines8020222

Lancaster, Alexander DJ (1975) Newcastle disease: virus and spread. Monograph No. 11

Lipkind M, Shihmanter E (1986) Antigenic relationships between avian paramyxoviruses. I. Quantitative characteristics based on hemagglutination and neuraminidase inhibition tests. Arch Virol 89(1–4):89–111. https://doi.org/10.1007/BF01309882

OIE (2012) Biological standards commission. In: World Organisation for Animal Health (ed) 2012. Manual of diagnostic tests and vaccines for terrestrial animals: mammals, birds and bees, vol 1, part 2, chapter 2.3.14. World Organisation for Animal Health, Paris, France, pp 555–574

Rauw F, Ngabirano E, Gardin Y, Palya V, Lambrecht B (2020) Effectiveness of a simultaneous rHVT-F(ND) and rHVT-H5(AI) vaccination of day-old chickens and the influence of NDV- and AIV-specific MDA on immune response and conferred protection. Vaccines (Basel) 8(3):536. https://doi.org/10.3390/vaccines8030536

Rima B, Balkema-Buschmann A, Dundon WG, Duprex P, Easton A, Fouchier R et al (2019) ICTV Virus Taxonomy profile: *Paramyxoviridae*. J Gen Virol 100(12):1593–1594. https://doi.org/10.1099/jgv.0.001328

Swayne DE (ed) (2020) Diseases of poultry, 14th edn. Wiley-Blackwell, Hoboken, NJ

Weiss H, Schonewille E, Bolte AL, Kruse W (2018) Newcastle disease vaccination of fattening turkeys: a vectored vaccine as an alternative to "traditional" forms of vaccination? Praktische Tierarzt 99:906–912

Avian Metapneumovirus

4

Hafez M. Hafez and Awad A. Shehata

Abstract

Avian metapneumovirus (AMPV), a member of the genus *Metapneumovirus* of the family *Paramyxoviridae*, is an acute, highly contagious respiratory disease, causing respiratory manifestations and reproductive problems in turkeys, causing economic losses, particularly in case of secondary bacterial infections. In chickens, it is associated with swollen head syndrome (SHS), respiratory disease complexes, and a drop in egg production. Pheasant, guinea fowl, and ducks are also susceptible. The diagnosis of AMPV is based on clinical signs, pathological lesions, virus isolation, and /or molecular investigation. Six different antigenic subtypes, namely, A–D and two new subtypes, are based on nucleotides and deduced amino acids of the G-glycoprotein. Vaccination and implementation of biosecurity are recommended as a control strategy for AMPV, particularly in endemic regions.

Keywords

Pneumovirinae · Swollen head syndrome · Antigenic subtypes · Vaccination

H. M. Hafez
Institute of Poultry Diseases, Faculty of Veterinary Medicine, Free University of Berlin, Berlin, Germany
e-mail: hafez.mohamed@fu-berlin.de

A. A. Shehata (✉)
TUM School of Natural Sciences, Bavarian NMR Center (BNMRZ), Structural Membrane Biochemistry, Technical University of Munich, Garching, Germany
e-mail: awad.shehata@tum.de

H. M. Hafez, A. A. Shehata (eds.), *Turkey Diseases and Disorders Volume 2*,
https://doi.org/10.1007/978-3-031-63322-5_4

4.1 Etiology

The AMPV belongs to the genus Metapneumovirus, subfamily *Pneumovirinae of the Paramyxoviridae family*. AMPV genome is a non-segmented, single-stranded negative-sense RNA of approximately 14 kilobases. The genome encodes eight genes, namely, 3′-nucleoprotein (N), phosphoprotein (P), matrix (M), fusion (F), matrix 2 (M2), small hydrophobic (SH), attachment (G), and large polymerase (L)-5 (Cecchinato et al. 2016). The virus exhibits neither hemagglutinating nor neuraminidase activities. It is suggested that the G protein has immunological importance due to the presence of T-cell epitopes (Cecchinato et al. 2010). Additionally, it is used for performing molecular epidemiological studies and strain characterization.

Different antigenic subtypes (A–D) have been described (Brown et al. 2014), as well as two genetic lineages of AMPV-C (Toquin et al. 2006). Subtyping is based on the reactivity against monoclonal antibodies, cross-reactivity in the ELISA and neutralization tests, and nucleotide sequence analysis. The A, B, and D subtypes are more closely related to each other than to subtype C.

In 2019, two novel divergent AMPVs that do not belong to these four subtypes were recently isolated from wild birds (Retallack et al. 2019; Canuti et al. 2019). Molecular analysis of the L gene proved that they are closer to the subgroup C viruses (Retallack et al. 2019; Canuti et al. 2019). Subtype C exhibits the highest nucleotide homology with human Metapneumovirus.

4.2 History and Distribution

AMPV, or turkey rhinotracheitis (TRT) or avian rhinotracheitis virus (ARTV), was first reported in South Africa by Buys and DuPreez in 1980. Later, the virus was identified in Europe, Israel, South America, the Middle and Far East, the United States, and Brazil. Turkeys and chickens are the natural hosts for AMPV, while wild birds are the natural reservoirs (Rautenschlein 2018).

Subtypes A and B are distributed worldwide, including Europe, Africa, Asia, and the United States in chickens and turkeys, rarely identified in other species. Subtype C stains have not been identified in chickens. However, it is endemic in the United States and detected sporadically in Muscovy ducks and pheasants in Korea, France, and China (Bäyon-Auboyer et al. 2000; Senne et al. 1997). In 1985, subtype D was isolated from turkeys in France, which has never been reported again.

The virus is highly sensitive to chemical disinfectants such as H_2O_2, formic acid, and formalin (Hafez and Arns 1991). It is stable at a pH of 3.0–9.0 and inactivated at 56 °C after 30 min. The virus was isolated from the autoclaved litter at −12 °C for up to 60 days. From the non-autoclaved litter, viral RNA was detected for up to 60 days, and the virus was isolated for up to 14 days (Velayudhan et al. 2003).

4.3 Virus Transmission

Migratory birds are considered a natural reservoir of infection (Chacón et al. 2011; Shin et al. 2002). The disease spreads via direct and indirect contact. Direct contact between infected and susceptible birds could spread the virus; however, the transmission failed if the susceptible birds were housed in the same room but in different pens (Cook et al. 1991). Contaminated water and movement of diseased or recovered poults, personnel, equipment, and feed trucks have all been implicated in outbreaks.

Vertical egg transmission is suspected (Shin et al. 2002); indeed, high virus titers can be detected in the reproductive tract of laying birds (Jones et al. 1988). However, further studies are still needed to confirm. It was suggested that chickens could play a role in maintaining and circulating AMPV in a region with chicken and turkey production (Cecchinato et al. 2016). To date, the airborne spread cannot be entirely excluded under certain conditions. Inoculation with mucus, nasal washings, or other materials from the respiratory tract of affected birds established AMPV infection (Nagaraja et al. 2000). There is no experimental evidence of long-term AMPV persistence in chickens or turkeys. However, both vaccines and field strains may cocirculate in poultry-dense regions (Lupini et al. 2011).

4.4 Clinical Signs and Postmortem Lesions

Clinical signs are sneezing, nasal discharge, conjunctivitis, tracheal rales, sinusitis, and submaxillary edema. These signs have been seen in birds as young as 14 days old but are more commonly observed in birds between 3 and 9 weeks of age. The morbidity approaches 100% within 24–48 h.

The mortality rate is highly variable, from negligible to over 40%. In susceptible turkey breeder flocks, a drop in egg production that approaches 50% or more for 2–4 weeks mainly accompanies clinical signs. Increased numbers of thin and white shelled eggs and the number of un-settable hatching eggs. This effect may be more significant if infection occurs early in the lay period. Turkeys are more susceptible to AMPV (Hartmann et al. 2015).

AMPV is associated with swollen head syndrome (SHS), respiratory disease complexes, and a drop in egg production in chickens. The clinical signs in chickens are less clear and normally more difficult to detect than the disease in turkeys. In roosters, AMPV causes respiratory signs followed by a drop in fertility due to orchitis, reviewed by Rautenschlein (2018).

Gross lesions include rhinitis, tracheitis, and sinusitis. In many cases, pericarditis, airsacculitis with congestion of the lungs, and fibrinous exudate in the pleural cavity have been observed. Several reports showed synergism between TRT and *Bordetella avium*, *Mycoplasma gallisepticum*, *Ornithobacterium rhinotracheale* (ORT), and *E. coli* (Marien et al. 2005; Turpin et al. 2002; Cook et al. 1991). After natural exposure to TRT infection, there was a significant increase in TRT antibody levels, which was usually accompanied by an increase in the number of positive sera

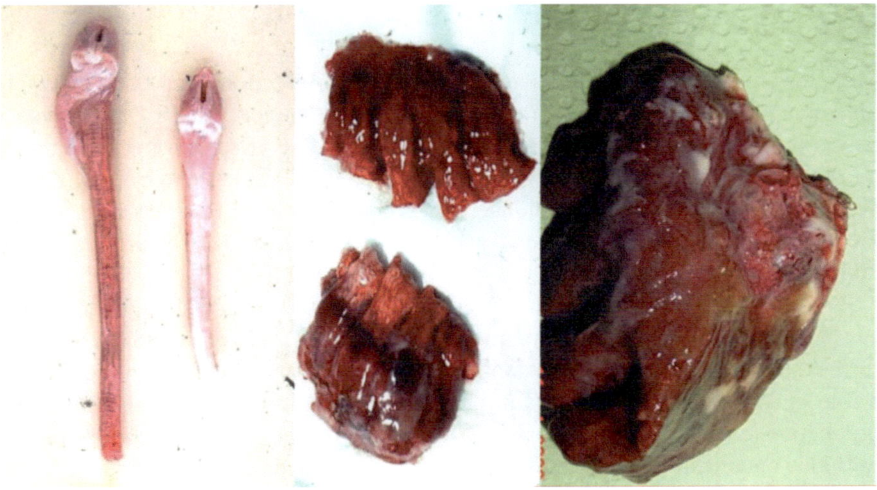

Fig. 4.1 Postmortem lesions of avian metapneumovirus in turkeys showing tracheitis and pneumonia ©Hafez, Fu-Berlin

to avian adenovirus, FAV 1 (Hafez 1990), and an increase in antibody levels against ORT (Jirjis et al. 2004) and against *Chlamydia psittaci* (Hafez et al. 1998) (Figs. 4.1 and 4.2).

4.5 Diagnosis

Diagnosis of AMPV based on clinical features and pathological lesions is mostly difficult since they may be confused with other respiratory infectious conditions. Accurate diagnosis based on laboratory methods. The virus can be isolated from choanal cleft, sinuses, nasal exudate, larynx, trachea, and lung samples. Tracheal organ cultures and cell cultures such as the monkey kidney cell line (Vero cells) and chicken embryo rough cell line (CER cells) can be used for virus isolation. The virus can also be isolated in 6- to 8-day-old embryonated chicken or turkey eggs free from maternal antibodies, via the yolk sac route.

PCR is widely used in many laboratories for detection and typing. However, virus isolation is difficult for the following reasons: (i) the virus shedding time is very short, and (ii) isolation of the virus is less successful from birds showing severe signs due to secondary bacterial infections. Recently, a single five-plex digital droplet RT-PCR targeting the conserved viral polymerase gene of AMPV to identify each of the four AMPV was developed (Lemaitre et al. 2022).

Serologically, serum neutralization test (SN), indirect immunofluorescence assay (IFA), and ELISA can be used to detect antibodies. However, the ELISA test is widely used since it has been developed in many laboratories and is available commercially. This method has a cost advantage and provides more rapid results than the neutralization test. Differentiation of AMPV subtypes can be done based on the

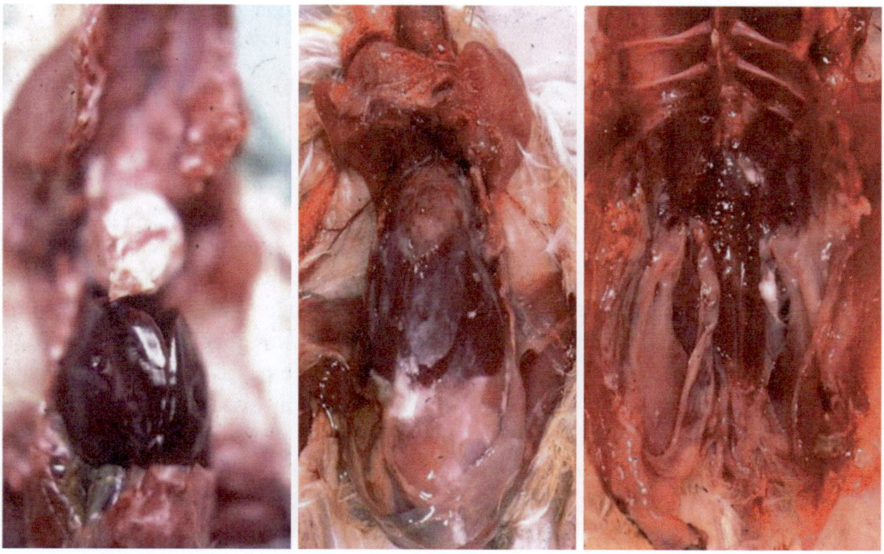

Fig. 4.2 Postmortem lesions of avian metapneumovirus in turkeys showing severely congested liver and airsacculitis. Gross lesions include rhinitis, tracheitis, and sinusitis. In many cases, pericarditis, airsacculitis with congestion of the lungs, and fibrinous exudate in the pleural cavity ©Hafez, Fu-Berlin

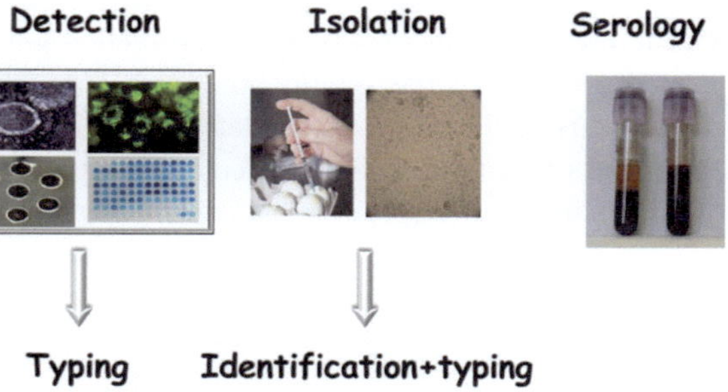

Fig. 4.3 Laboratory diagnosis of AMPV

nucleotide sequence analysis of the G protein and other genes as well using monoclonal antibodies.

AMPV should be differentiated from diseases causing respiratory manifestations and/or drop in egg production. Newcastle disease, PMV-3, avian influenza, *E. coli*, and *Mycoplasma* should be considered. Frequently, bacterial infections might be secondary infections to AMPV, which complicated the diagnosis (Fig. 4.3).

4.6 Prevention and Control

- Treatment of secondary bacterial infections using different antibiotics showed different results. However, treatment alone is of little value unless improvements in all aspects of management and biosecurity accompany it.
- Both live attenuated and inactivated vaccines are available. Several live AMPV and inactivated vaccines based on subtype A or B are available worldwide. However, live vaccines in turkeys provide better protection with low antibody responses. For optimal protection, it is recommended to use the live attenuated vaccine for priming, followed by injection with the inactivated vaccine.
- The maternally derived antibodies do not interfere with vaccine efficacy (Cook 2000). Therefore, the early application of AMPV vaccines is possible, especially in endemic areas. In addition, using live vaccines in ovo vaccination showed promising results (Worthington et al. 2003).
- For breeders, it is recommended to be boosted with inactivated vaccine at the 26th to 27th week of age. Effective monovalent inactivated AMPV vaccines or bivalent with TRT are available to protect laying and breeding turkeys against the drop in egg production.
- Cross-protection afforded by subtype A and B vaccines against a heterologous challenge by subtype A or B has been documented in turkeys (Cook et al. 1999; Eterradossi et al. 1995; Naylor et al. 1997). Recently, immunization of day-old broiler chickens with AMPV subtype B can protect against homologous (subtype B) and heterologous (subtype A) challenges (Ball et al. 2022).
- The level of antibodies is poorly correlated with protection. Virulent or attenuated AMPV in turkey poults produced virus-specific IgA (in tears) and IgG (in the tears and serum). Complete protection against challenge was obtained 3 weeks post-vaccination. Protection is attributed to the virus-neutralizing antibodies in tears, suggesting that cell-mediated immunity (CMI), either instead of or as well as antibody responses, may be necessary for immunity to APV infections (Khehra and Jones 1999).

References

Ball C, Manswr B, Herrmann A, Lemiere S, Ganapathy K (2022) Avian metapneumovirus subtype B vaccination in commercial broiler chicks: heterologous protection and selected host transcription responses to subtype A or B challenge. Avian Pathol 51(2):181–196. https://doi.org/10.1080/03079457.2022.2036697

Bäyon-Auboyer M-H, Arnauld C, Toquin D, Eterradossi N (2000) Nucleotide sequences of the F, L and G protein genes of two non-A/non-B avian pneumoviruses (APV) reveal a novel APV subgroup. J Gen Virol 81(11):2723–2733. https://doi.org/10.1099/0022-1317-81-11-2723

Brown PA, Lemaitre E, Briand F-X, Courtillon C, Guionie O, Allée C et al (2014) Molecular comparisons of full-length metapneumovirus (MPV) genomes, including newly determined French AMPV-C and –D isolates, further supports possible subclassification within the MPV genus. PLoS One 9(7):e102740. https://doi.org/10.1371/journal.pone.0102740

Canuti M, Kroyer ANK, Ojkic D, Whitney HG, Robertson GJ, Lang AS (2019) Discovery and characterization of novel RNA viruses in aquatic north American wild birds. Viruses 11(9):768. https://doi.org/10.3390/v11090768

Cecchinato M, Catelli E, Lupini C, Ricchizzi E, Clubbe J, Battilani M et al (2010) Avian metapneumovirus (AMPV) attachment protein involvement in probable virus evolution concurrent with mass live vaccine introduction. Vet Microbiol 146(1–2):24–34. https://doi.org/10.1016/j.vetmic.2010.04.014

Cecchinato M, Ferreira HL, Munir M, Catelli E (2016) Avian metapneumovirus. In: Munir M (ed) Mononegaviruses of veterinary importance. Volume 2: molecular epidemiology and control, 1st edn. CABI, pp 127–143

Chacón JL, Mizuma M, Vejarano MP, Toquín D, Eterradossi N, Patnayak DP et al (2011) Avian metapneumovirus subtypes circulating in Brazilian vaccinated and nonvaccinated chicken and Turkey farms. Avian Dis 55(1):82–89. https://doi.org/10.1637/9501-081310-Reg.1

Cook JKA (2000) Avian rhinotracheitis. Rev Sci Tech OIE 19(2):602–613. https://doi.org/10.20506/rst.19.2.1233

Cook JK, Ellis MM, Huggins MB (1991) The pathogenesis of Turkey rhinotracheitis virus in Turkey poults inoculated with the virus alone or together with two strains of bacteria. Avian Pathol 20(1):155–166. https://doi.org/10.1080/03079459108418750

Cook JKA, Huggins MB, Orbell SJ, Senne DA (1999) Preliminary antigenic characterization of an avian pneumovirus isolated from commercial turkeys in Colorado, USA. Avian Pathol 28(6):607–617. https://doi.org/10.1080/03079459994407

Eterradossi N, Toquin D, Guittet M, Bennejean G (1995) Evaluation of different Turkey rhinotracheitis viruses used as antigens for serological testing following live vaccination and challenge. J Vet Med Series B 42(1–10):175–186. https://doi.org/10.1111/j.1439-0450.1995.tb00698.x

Hafez HM (1990) Serological surveillance for antibodies against different avian infectious agents on Turkey flocks naturally infected with Turkey rhinotracheitis. J Vet Med Series B 37(1–10):369–376. https://doi.org/10.1111/j.1439-0450.1990.tb01071.x

Hafez HM, Arns CW (1991) Disinfection trials on turkey rhinotracheitis Proceedings of the XXIV World Veterinary Congress Rio de Janeiro. Brazil. Abst. No. 10.13

Hafez HM, Sting R, Jodas S, Stadler A (1998) Investigations on the interaction between Chlamydia psittaci und avian Pneumovirus infections. In: Kaleta FE, Heffels-Redmann U (eds) Proceedings of the international symposium on infectious bronchitis and Pneumovirus infections in poultry. Rauischholzhausen, pp 126–135

Hartmann S, Sid H, Rautenschlein S (2015) Avian metapneumovirus infection of chicken and Turkey tracheal organ cultures: comparison of virus-host interactions. Avian Pathol 44(6):480–489. https://doi.org/10.1080/03079457.2015.1086974

Jirjis FF, Noll SL, Halvorson DA, Nagaraja KV, Martin F, Shaw DP (2004) Effects of bacterial coinfection on the pathogenesis of avian pneumovirus infection in turkeys. Avian Dis 48(1):34–49. https://doi.org/10.1637/7017

Jones RC, Williams RA, Baxter-Jones C, Savage CE, Wilding GP (1988) Experimental infection of laying turkeys with rhinotracheitis virus: distribution of virus in the tissues and serological response. Avian Pathol 17(4):841–850. https://doi.org/10.1080/03079458808436506

Khehra RS, Jones RC (1999) Local and systemic class-specific antibody responses to avian pneumovirus: comparison of the chicken and Turkey. In: Hafez HM, Mazaheri A (eds) Proceedings of the 2nd international symposium on Turkey diseases, Berlin. DVG, Gießen

Lemaitre E, Bougeard S, Allée C, Eterradossi N, Courtillon C, Brown PA (2022) Avian metapneumovirus: a five-plex digital droplet RT-PCR method for identification of subgroups A, B, C, and D. Front Vet Sci 9:1058294. https://doi.org/10.3389/fvets.2022.1058294

Lupini C, Cecchinato M, Ricchizzi E, Naylor CJ, Catelli E (2011) A Turkey rhinotracheitis outbreak caused by the environmental spread of a vaccine-derived avian metapneumovirus. Avian Pathol 40(5):525–530. https://doi.org/10.1080/03079457.2011.607428

Marien M, Decostere A, Martel A, Chiers K, Froyman R, Nauwynck H (2005) Synergy between avian pneumovirus and Ornithobacterium rhinotracheale in turkeys. Avian Pathol 34(3):204–211. https://doi.org/10.1080/03079450500096414

Nagaraja KV, Shin HJ, Halvorson DA (2000) Avian pneumovirus of turkeys and its host range. In: Hafez HM (ed) 3rd International symposium on Turkey diseases. German Veterinary Medical Society, Berlin, Germany, pp 208–213

Naylor C, Shaw K, Britton P, Cavanagh D (1997) Appearance of type B avian Pneumovirus in Great Britain. Avian Pathol 26(2):327–338. https://doi.org/10.1080/03079459708419215

Rautenschlein (2018) Avian metapneumovirus. In: Williams SM, Dufour-Zavala L, Jackwood MW, Lee MD, Lupiani B, Reed WM, Speckman E, Woolcock PR (eds) A laboratory manual for the isolation, identification, and characterization of avian pathogens, 6th edn. American Association of Avian Pathologists, Jacksonville, FL, pp 173–178

Retallack H, Clubb S, DeRisi JL (2019) Genome sequence of a divergent avian metapneumovirus from a monk parakeet (*Myiopsitta monachus*). Microbiol Resour Announc 8(16):e00284–e00219. https://doi.org/10.1128/MRA.00284-19

Senne DA, Edson RK, Pederson JC, Panigrahy B (1997) Avian pneumovirus update, p. 190. In: Proceedings of the 134th annual convention of the American Veterinary Medical Association, Reno, Nevada. American Veterinary Medical Association, Washington, DC

Shin H-J, Cameron KT, Jacobs JA, Turpin EA, Halvorson DA, Goyal SM et al (2002) Molecular epidemiology of subgroup C avian Pneumoviruses isolated in the United States and comparison with subgroup A and B viruses. J Clin Microbiol 40(5):1687–1693. https://doi.org/10.1128/JCM.40.5.1687-1693.2002

Toquin D, Guionie O, Jestin V, Zwingelstein F, Allee C, Eterradossi N (2006) European and American subgroup C isolates of avian metapneumovirus belong to different genetic lineages. Virus Genes 32(1):97–103. https://doi.org/10.1007/s11262-005-5850-3

Turpin EA, Perkins LEL, Swayne DE (2002) Experimental infection of turkeys with avian pneumovirus and either Newcastle disease virus or Escherichia coli. Avian Dis 46(2):412–422. https://doi.org/10.1637/0005-2086(2002)046[0412:EIOTWA]2.0.CO;2

Velayudhan BT, Lopes VC, Noll SL, Halvorson DA, Nagaraja KV (2003) Avian pneumovirus and its survival in poultry litter. Avian Dis 47(3):764–768. https://doi.org/10.1637/7042

Worthington KJ, Sargent BA, Davelaar FG, Jones RC (2003) Immunity to avian pneumovirus infection in turkeys following in ovo vaccination with an attenuated vaccine. Vaccine 21(13–14):1355–1362. https://doi.org/10.1016/s0264-410x(02)00689-8

Poxvirus Infection

5

Dörte Lüschow and Hafez M. Hafez

Abstract

Avian pox is a globally distributed viral disease in a broad range of wild, captive, and domestic birds, including chickens and turkeys, caused by large DNA viruses of *Poxviridae* family. The slow-spreading disease is characterized by the formation of proliferative lesions of the unfeathered skin areas (cutaneous form or dry pox) or the upper respiratory and digestive tract (diphtheritic form or wet pox). In commercial poultry, avian poxvirus infection is associated with a significant economic impact, constrained by a drop in egg production, reduced growth rates in young birds, and increased mortality. Among other avian species, turkeys are also broadly affected. The disease can be controlled by vaccinating turkeys using live attenuated fowlpox vaccine by thigh application at the age of about 8 weeks. In breeder flocks, a booster may be necessary before the start of production. After a week of vaccination, birds should be checked for a "vaccinal take," such as skin swelling or scab, at the application site to assess the success of the vaccination. In this chapter, we will discuss in more detail the etiology, clinical signs, diagnosis, and vaccination of poxvirus in turkeys.

Keywords

Pox · Cutaneous form · Diphtheritic form · Diagnosis · Vaccinal take · Fowlpox vaccine

D. Lüschow (✉) · H. M. Hafez
Institute of Poultry Diseases, Faculty of Veterinary Medicine, Free University of Berlin, Berlin, Germany
e-mail: doerte.lueschow@fu-berlin.de; hafez.mohamed@fu-berlin.de

H. M. Hafez, A. A. Shehata (eds.), *Turkey Diseases and Disorders Volume 2*,
https://doi.org/10.1007/978-3-031-63322-5_5

5.1 Etiology

Avian poxviruses are classified according to the International Committee on Taxonomy of Viruses (ICTV) (https://ictv.global/taxonomy) into the genus *Avipoxvirus* within the family *Poxviridae*. This genus comprises seven species, including the species *Avipoxvirus fowlpox*, formerly designated as fowlpox virus (FWPV), and *Avipoxvirus turkeypox*, formerly designated as turkeypox virus (TKPV). In this chapter the previous designations will be used. FWPV was the type species of the *Avipoxvirus* genus, and the main focus of research studies in the past was based on this virus. The morphology of avian poxviruses is similar to other members of the *Poxviridae* family (Tripathy and Reed 2013). Virus particles are brick-shaped and measure about 330 × 280 × 200 nm. They consist of an electron-dense biconcave core flanked by two lateral bodies and are surrounded by an envelope (Fig. 5.1). Unusual for most DNA viruses, avian poxviruses multiply in the cytoplasm of infected cells and induce the formation of inclusion bodies (Bollinger bodies) (Woodruff and Goodpasture 1929; Moss 1990).

The viral genome contains a linear double-stranded DNA of about 300 kbp (Coupar et al. 1990). In the case of FWPV, the 288 kbp genome encodes 260 putative genes (Afonso et al. 2000). Three major genetic clades are distinguished by phylogenetic analysis of different avipoxvirus species based on selected genomic regions (Jarmin et al. 2006; Gyuranecz et al. 2013). In this context, most sequence data obtained from different countries revealed a close relationship between TKPV and FWPV, and in some cases, FWPV was identified in diseased turkeys (Lüschow et al. 2004, 2006; Hess et al. 2011; Ferreira et al. 2018). In addition, a novel avipoxvirus strain with a very compact genomic organization was detected in a flock of turkeys previously vaccinated against avipoxvirus (Bányai et al. 2015).

FWPV is highly resistant in the poultry environment and remains infectious in dried scabs for months or years (Tripathy and Reed 1997). A photolyase coding gene in the genome of FWPV appears to contribute to a high environmental tenacity (Srinivasan et al. 2001). Turkey poxviruses are resistant to ether treatment. Heating at 60 °C for 30 min and treatment with 0.25% trypsin, 0.5% phenol, and 5% chloroform inactivate the virus (Singh et al. 2003).

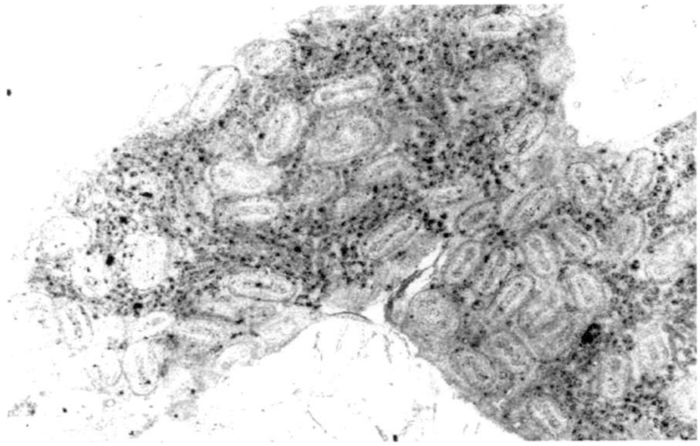

Fig. 5.1 EM of fowlpox virus showing brick-shaped virus particles (©Hafez, FU-Berlin)

5.2 Transmission

Infections are frequently found in wild turkeys but have also been reported in commercial as well as backyard turkey flocks (Davidson et al. 1985; Prukner-Radovčić et al. 2006; Camacho-Escobar et al. 2008; Hess et al. 2011; Lüschow et al. 2012; Elsmo et al. 2016; Ferreira et al. 2018). Virus transmission occurs after contact of the virus with damaged skin, for example, as a result of fighting and pecking (Tripathy 1993). Infections also appear when mucous membranes of the oral cavity and upper respiratory tract come in contact with aerosols containing poxviruses generated by feathers and dried scabs in the poultry house environment (Tripathy and Reed 1997). Arthropods like mites, mosquitoes, or other biting insects may serve as mechanical vectors (Kligler et al. 1929; Huong et al. 2014). In addition, artificial insemination has been described as facilitating the mechanical transmission of poxvirus in turkey breeder flocks (Metz et al. 1985).

5.3 Clinical Signs and Pathology

The incubation period after natural infection varied from 4 to 10 days (Tripathy and Reed 2013). The progression of the disease is influenced by the strain involved, the route of infection, and the distribution of the lesions (Tripathy 1993). Comparable to the clinical feature in chickens, the disease in turkeys is usually manifested in cutaneous and diphtheritic forms and occasionally a mixed form where both clinical manifestations occur in the same bird (Tripathy and Reed 2013). The more common cutaneous form (dry pox) is characterized by developing epithelial and proliferative skin lesions of unfeathered areas, especially the head and upper neck (Fig. 5.2).

They are frequently located on wattle, snood, and around the beak and eyelids. Legs, feet, and regions around the cloaca may also be affected. Lesions evolve from small-blanched nodules to finally thick, dark, crusty scabs (Mayr and Wittmann 1957). After a couple of weeks, scabs drop off, usually leaving a scar. In more severe forms, generalized lesions may also spread to feathered parts of the skin. The cutaneous form is mostly associated with low mortality and usually heals within 3 weeks. The economic impact might be induced by reduced growth rates and a drop in egg production in breeder flocks (Tripathy and Reed 2013). Secondary bacterial infections may complicate this form (Hess et al. 2011; Weli and Tryland 2011). In addition, eye lesions may lead to partial or complete closing of the eye and blindness, accompanied by the bird's inability to find food and water (Tripathy and Hanson 1978).

The more uncommon diphtheritic form (wet pox) involves proliferative lesions on mucous membranes of the oral cavity (Fig. 5.3), the esophagus, and the upper respiratory tract (larynx and trachea). Their fusion leads to yellowish necrotic pseudomembranes that leave a bleeding surface when removed (Tripathy 1993). This form is frequently associated with reduced feed intake, drinking problems, and breathing difficulties, resulting in a greater mortality rate (Tripathy and Reed 2013). An uncommon outbreak of venereal pox infection was identified in turkey breeders

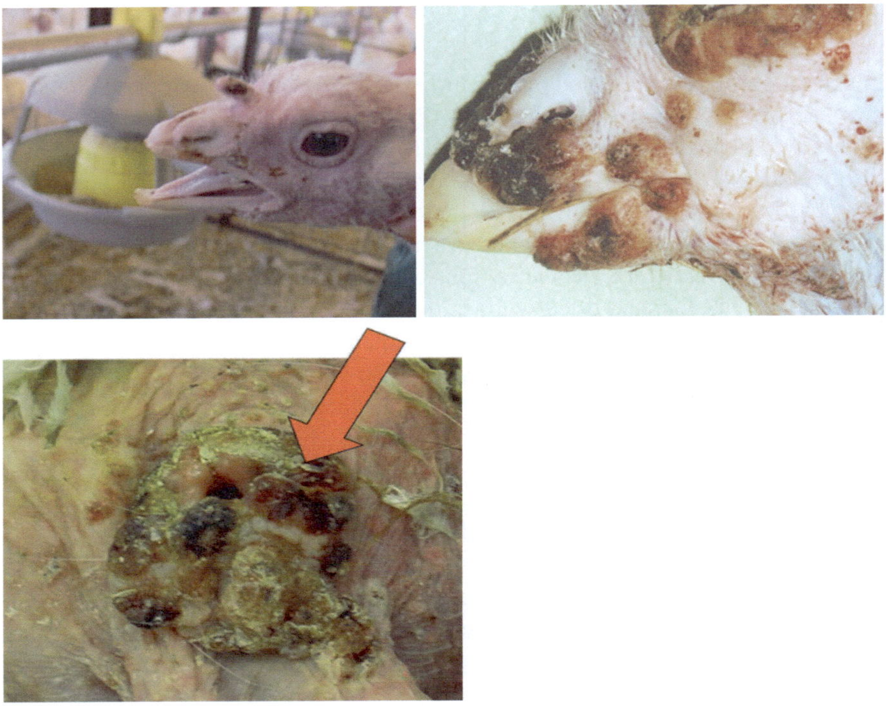

Fig. 5.2 Cutaneous form of pox showing proliferative skin lesions on unfeathered areas (©Hafez, FU-Berlin)

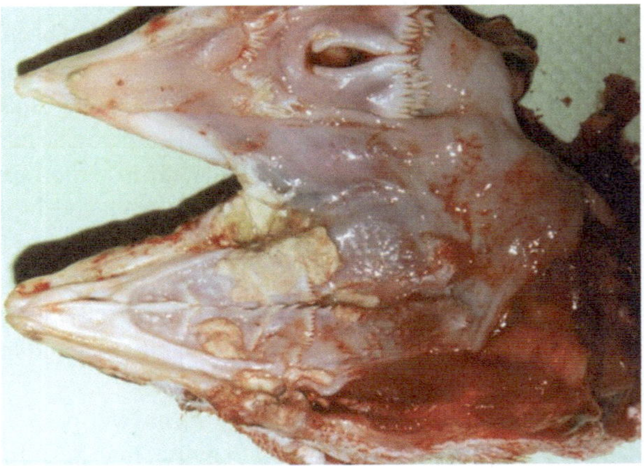

Fig. 5.3 Diphtheritic form (wet pox) of pox showing proliferative lesions on mucous membranes of the oral cavity with the presence of yellowish necrotic pseudomembranes © Hafez, FU-Berlin

after artificial insemination. Lesions were limited to the cloacal mucosa, lips of the vent, and the oviduct mucosa (Metz et al. 1985).

Microscopically, epithelial hyperplasia and enlargement of cells accompanied by inflammatory changes and the characteristic Bollinger bodies are observed (Tanizaki et al. 1987).

5.4 Diagnosis

Clinical signs and gross lesions enable a presumptive diagnosis of avian pox, which has to be confirmed by laboratory methods. In histological sections of cutaneous or diphtheritic lesions, characteristic eosinophilic intracytoplasmic inclusion bodies (Bollinger bodies) can be demonstrated after hematoxylin and eosin (H&E) staining (Tripathy and Hanson 1978). Electron microscopy can be used to identify the typical poxvirus morphology by negative staining or in ultrathin sections of infected tissues (McFerran et al. 1971; Catroxo et al. 2009).

Virus isolation can be attempted by inoculating suspected material onto chorioallantoic membranes (CAMs) of 9- to 12-day-old embryonated chicken eggs (Tripathy 2008). After 5–7 days of incubation, CAMs are investigated for poxvirus-specific lesions like focal white pocks or a generalized thickening. Confirmation can be acquired by histopathological or electron microscopy investigation or by detecting specific poxvirus antigens or DNA using different methods. Primary cell cultures like chicken embryo fibroblasts or permanent cell lines like QT–35 or LMH are also suitable for virus propagation attempts (Mayr and Kalcher 1961; Schnitzlin et al. 1988; Tripathy 2008).

Amplification of viral-specific DNA sequences by polymerase chain reaction (PCR) allows a fast and accurate diagnosis. A PCR assay targeting the p4b protein gene has been used successfully for the molecular detection of different avian poxviruses, including poxviruses recovered from turkeys (Lee and Lee 1997; Lüschow et al. 2004). Additional sequence analysis of amplified fragments enables further genetic differentiation (Lüschow et al. 2004; Jarmin et al. 2006).

To investigate the immune response after infection and/or vaccination, several serological tests like agar gel immunodiffusion (AGID), mixed antigen agar gel enzyme assay (AGEA), or enzyme-linked immunosorbent assay (ELISA) have been described (Buscaglia et al. 1985; Tadese et al. 2003; Hauck et al. 2009; Tripathy 2018).

5.5 Control

There are no effective treatments known for poxvirus infection in turkeys. Prophylactic measures to prevent infection include good hygienic measures, control of cannibalism, arthropod vector control, and vaccine administration.

Vaccination of commercial poultry flocks is applied routinely in areas with a widespread occurrence of the disease (Tripathy and Reed 2013). Different vaccine types have been developed in recent decades (Giotis and Skinner 2019). Currently, live attenuated vaccines propagated in embryonated chicken eggs or avian cell culture and recombinant vector vaccines are commercially available. In general, for a successful immunization, the directions for the use of the manufacturers have to be followed strictly.

Turkeys are usually vaccinated by live attenuated vaccines based on the fowlpox virus. Typically, FPWV-based vaccines are applied by the wing-web method; however, due to the behavior of turkeys, the virus might spread and infect the skin of the head (Hinshaw 1947). As an alternate site, the vaccine should be administered by thigh application, usually at the age of about 8 weeks. In breeder flocks, a repeated vaccination might be needed before the start of production. About 1 week after vaccination, birds should be investigated for a "vaccinal take" consisting of a skin swelling or scab at the application site to evaluate the vaccination's success (Tripathy and Reed 2013). In turkeys, vaccination is usually effective. Nevertheless, outbreaks have also been described in previously vaccinated flocks (Bickford et al. 1971; Odoya et al. 2006; Ferreira et al. 2018).

References

Afonso CL, Tulman ER, Lu Z, Zsak L, Kutish GF, Rock DL (2000) The genome of fowlpox virus. J Virol 74(8):3815–3831. https://doi.org/10.1128/JVI.74.8.3815-3831.2000

Bányai K, Palya V, Dénes B, Glávits R, Ivanics É, Horváth B et al (2015) Unique genomic organization of a novel Avipoxvirus detected in Turkey (Meleagris gallopavo). Infect Genet Evol 35:221–229. https://doi.org/10.1016/j.meegid.2015.08.001

Bickford AA, Gallina AM, Winterfield RW, Bolte H (1971) Case report: studies of an unusual pox infection in turkeys. Avian Dis 15(3):614. https://doi.org/10.2307/1588741

Buscaglia C, Bankowski RA, Miers L (1985) Cell-culture virus-neutralization test and enzyme-linked immunosorbent assay for evaluation of immunity in chickens against fowlpox. Avian Dis 29(3):672–680

Camacho-Escobar MA, Arroyo-Ledezma J, Ramirez-Cancino L (2008) Diseases of backyard turkeys in the Mexican tropics. Ann N Y Acad Sci 1149:368–370. https://doi.org/10.1196/annals.1428.004

Catroxo MHB, Pongiluppi T, Melo NA, Milanelo L, Petrella S, Martins AMCPF et al (2009) Identification of poxvirus under transmission electron microscopy during outbreak period in wild birds, in São Paulo, Brazil. Int J Morphol 27(2). https://doi.org/10.4067/S0717-95022009000200043

Coupar BEH, Teo T, Boyle DB (1990) Restriction endonuclease mapping of the fowlpox virus genome. Virology 179(1):159–167. https://doi.org/10.1016/0042-6822(90)90285-Y

Davidson WR, Nettles VF, Couvillion CE et al (1985) Diseases diagnosed in wild turkeys (Meleagris gallopavo) of the southeastern united states. J Wildl Dis 21:386–390

Elsmo EJ, Allison AB, Brown JD (2016) A retrospective study of causes of skin lesions in wild turkeys (meleagris gallopavo) in the eastern USA, 1975–2013. J Wildl Dis 52:582–591

Ferreira BC, Ecco R, Couto RM, Coelho HE, Rossi DA, Beletti ME et al (2018) Outbreak of cutaneous form of avian poxvirus disease in previously pox-vaccinated commercial turkeys. Pesqui Veterinária Bras 38(3):417–424. https://doi.org/10.1590/1678-5150-pvb-4463

Giotis ES, Skinner MA (2019) Spotlight on avian pathology: Fowlpox virus. Avian Pathol 48(2):87–90. https://doi.org/10.1080/03079457.2018.1554893

Gyuranecz M, Foster JT, Dán Á, Ip HS, Egstad KF, Parker PG et al (2013) Worldwide phylogenetic relationship of avian poxviruses. J Virol 87(9):4938–4951. https://doi.org/10.1128/JVI.03183-12

Hauck R, Prusas C, Hafez HM, Lüschow D (2009) Serologic response against fowl poxvirus and reticuloendotheliosis virus after experimental and natural infections of chickens with fowl poxvirus. Avian Dis 53(2):205–210. https://doi.org/10.1637/8451-081908-Reg.1

Hess C, Maegdefrau-Pollan B, Bilic I, Liebhart D, Richter S, Mitsch P et al (2011) Outbreak of cutaneous form of poxvirus on a commercial Turkey farm caused by the species fowlpox. Avian Dis 55(4):714–718. https://doi.org/10.1637/9771-050511-Case.1

Hinshaw WR (1947) Fowl pox in Turkey breeding flocks. Vet Med 42(3):113

Huong CTT, Murano T, Uno Y, Usui T, Yamaguchi T (2014) Molecular detection of avian pathogens in poultry red mite (*Dermanyssus gallinae*) collected in chicken farms. J Vet Med Sci 76(12):1583–1587. https://doi.org/10.1292/jvms.14-0253

Jarmin S, Manvell R, Gough RE, Laidlaw SM, Skinner MA (2006) Avipoxvirus phylogenetics: identification of a PCR length polymorphism that discriminates between the two major clades. J Gen Virol 87(8):2191–2201. https://doi.org/10.1099/vir.0.81738-0

Kligler IJ, Muckenfuss RS, Rivers TM (1929) Transmission of fowl-pox by mosquitoes. J Exp Med 49(4):649–660. https://doi.org/10.1084/jem.49.4.649

Lee LH, Lee KH (1997) Application of the polymerase chain reaction for the diagnosis of fowl poxvirus infection. J Virol Methods 63(1–2):113–119. https://doi.org/10.1016/S0166-0934(96)02119-2

Lüschow D, Hoffmann T, Hafez HM (2004) Differentiation of avian poxvirus strains on the basis of nucleotide sequences of 4b Gene Fragment. Avian Dis 48(3):453–462. https://doi.org/10.1637/7111

Lüschow D, Hoffmann T, Hauck R, Hafez HM (2006) Investigations on poxviruses in Turkey flocks. In: Proceedings of the 6th international symposium on Turkey diseases. Berlin, pp 184–187

Lüschow D, Hoffmann T, Hafez HM (2012) Detection of fowlpox in chickens and turkeys in Germany. Berl Munch Tierarztl Wochenschr 125(1–2):60–66

Mayr A, Kalcher K (1961) Vergleichende Studien über die Züchtung von Geflügelpockenviren in der Zellkultur. Arch Für Gesamte Virusforschung 10(1):72–102. https://doi.org/10.1007/BF01258768

Mayr A, Wittmann G (1957) Zur Pathogenese der Hühnerpockeninfektion. Mh Tierheilkd 9:44–53

McFerran JB, Clarke JK, Curran WL (1971) The application of negative contrast electron microscopy to routine veterinary virus diagnosis. Res Vet Sci 12(3):253–257

Metz AL, Hatcher L, Newman JA, Halvorson DA (1985) Venereal pox in breeder turkeys in Minnesota. Avian Dis 29(3):850. https://doi.org/10.2307/1590679

Moss B (1990) Poxviridae and their replication. In: Fields BN, Knipe DM, Chanock RM, Melnick JL, Roizman B, Shope RE (eds) Virology, 2nd edn. Raven Press, New York

Odoya E, Abegunde A, Agyogbo B, Omatainse S, Gwankat E, Okpara U (2006) Outbreak of Turkey pox disease in fowl pox vaccinated poults in Vom Plateau State of Nigeria. Afr J Clin Exp Microbiol 7(2):136–138. https://doi.org/10.4314/ajcem.v7i2.7443

Prukner-Radovčić E, Lüschow D, Grozdanić IC, Tišljar M, Mazija H, Vranešić Ł et al (2006) Isolation and molecular biological investigations of avian poxviruses from chickens, Turkey, and pigeon in Croatia. Avian Dis 50(3):440–444. https://doi.org/10.1637/7506-012006R.1

Schnitzlin WM, Ghildyal N, Tripathy DN (1988) Genomic and antigenic characterization of avipoxviruses. Virus Res 10(1):65–75. https://doi.org/10.1016/0168-1702(88)90058-5

Singh A, Dash BB, Kataria JM, Dandapat S, Dhama K (2003) Characterisation of an Indian isolate of Turkey pox virus. Indian J Comp Microbiol Immunol Infect Dis 24(2):149–152

Srinivasan V, Schnitzlein WM, Tripathy DN (2001) Fowlpox virus encodes a novel DNA repair enzyme, CPD-photolyase, That restores infectivity of UV light-damaged Virus. J Virol 75(4):1681–1688. https://doi.org/10.1128/JVI.75.4.1681-1688.2001

Tadese T, Potter EA, Reed WM (2003) Development of a mixed antigen agar gel enzyme assay (AGEA) for the detection of antibodies to poxvirus in chicken and Turkey sera. J Vet Med Sci 65(2):255–258. https://doi.org/10.1292/jvms.65.255

Tanizaki E, Kotani T, Odagiri Y (1987) Pathological changes of tracheal mucosa in chickens infected with fowl pox virus. Avian Dis 31(1):169. https://doi.org/10.2307/1590791

Tripathy DN (1993) Avipox viruses. In: McFerran JB, McNulty MS (eds) Virus infections of birds. Elsevier

Tripathy DN (2008) Pox. In: Dufour-Zavala L, Swayne DE, Glisson JR et al (eds) A laboratory manual for the isolation, identification, and characterization of avian pathogens, 5th edn. American Association of Avian Pathologists, New Bolton Center, Kennet Square, PA

Tripathy DN (2018) Fowlpox. In: Manual of diagnostic tests and vaccines for terrestrial animals, 8th edn. OIE World Organisation for Animal Health. ISBN: 978-92-95108-18-9

Tripathy DN, Hanson LE (1978) Pathogenesis of fowlpox in laying hens. Avian Dis 22(2):259. https://doi.org/10.2307/1589537

Tripathy DN, Reed WM (1997) Pox. In: Calnek BW, Barnes HJ, Beard CW, McDougald LR, Saif YM (eds) Diseases of poultry, 10th edn. Iowa State University Press, Ames, IA

Tripathy DN, Reed WM (2013) Pox. In: Swayne DE (ed) Diseases of poultry, 1st edn. Wiley, pp 333–349

Weli SC, Tryland M (2011) Avipoxviruses: infection biology and their use as vaccine vectors. Virol J 8(1):49. https://doi.org/10.1186/1743-422X-8-49

Woodruff CE, Goodpasture EW (1929) The infectivity of isolated inclusion bodies of fowl-pox. Am J Pathol. 5:1–10

Haemorrhagic Enteritis (Siadenovirus) 6

Awad A. Shehata and Hafez M. Hafez

Abstract

Turkey hemorrhagic enteritis (HE) was first recognized in turkey poults in 1937 by Pomeroy and Fenster macher in the United States. Afterward, the disease has been reported in several countries worldwide, that is, Canada, England, Germany, Australia, India, Japan, Israel, and the United States. The disease is caused by turkey adenovirus-3 (TAV-3), the genus *Siadenovirus*, family *Adenoviridae*. The susceptible age is 4 weeks and older turkeys mainly 6–11 weeks old). The disease is characterized by general clinical signs such as bloody dropping and sudden death. The main postmortem lesions are swollen intestines, filled with bloody contents, splenomegaly, and hepatomegaly. The mortality rates ranged from 10 to 15%; however, it can reach 60%. Sometimes, the intestinal mucosa is covered by a yellowish fibrinonecrotic membrane. Petechial hemorrhages can also be seen in several tissues. Histologically, the presence of intranuclear inclusion bodies in the macrophages and lymphocytes of the spleen is a characteristic of HE. The diagnosis is based on the clinical signs, postmortem lesions, and histopathological examination. AGID, PCR, and ELISA can be used for virus identification. The disease can be prevented by vaccination with avirulent live vaccines at 3–6 weeks, accompanied by hygienic measures.

A. A. Shehata (✉)
TUM School of Natural Sciences, Bavarian NMR Center (BNMRZ), Structural Membrane Biochemistry, Technical University of Munich, Garching, Germany
e-mail: awad.shehata@tum.de

H. M. Hafez
Faculty of Veterinary Medicine, Institute of Poultry Diseases, Free University of Berlin, Berlin, Germany
e-mail: hafez.mohamed@fu-berlin.de

H. M. Hafez, A. A. Shehata (eds.), *Turkey Diseases and Disorders Volume 2*,
https://doi.org/10.1007/978-3-031-63322-5_6

Keywords
Hemorrhagic enteritis · *Adenoviridae* · Fibrinonecrotic membrane · Diagnosis · Vaccination

6.1 Etiology

Adenoviruses, belonging to *Adenoviridae*, are DNA viruses with an icosahedral capsid and a double-stranded, linear genome. Adenoviruses are described in many species of vertebrate animals, including mammals, birds, reptiles, amphibians, and fish. Five genera of adenoviruses are known: *Mastadenovirus*, *Aviadenovirus*, *Siadenovirus*, *Atadenovirus*, and *Ichtdenovirus*. Three of them, *Aviadenovirus (Group I)*, *Siadenovirus (Group II)*, and *Atadenovirus (Group III)*, can infect poultry (Shehata et al. 2021).

The *Aviadenovirus* comprises the fowl aviadenovirus (FAdV) and the turkey aviadenoviruses (TAdV) species (Fig. 6.1). FAdVs are grouped into five species (FAdV A to FAdV-E), 12 serotypes (FAdV-1 to FAdV-8a and FAdV-8b to FAdV-11), and 12 genotypes [75]. Three TAdV species, namely, TAdV-B (type TAdV-1), TAdV-C (type TAdV-4), and TAdV-D (type TAdV-5), were isolated from respiratory disease and poult enteritis and mortality syndrome (PEMS) (Kaján et al. 2010; Kleine et al. 2017; Marek et al. 2014). These viruses also cause inclusion body hepatitis in turkey poults and may be responsible for lower hatchability rates in breeder flocks (Guy et al. 1988; Guy 1998; Shivaprasad et al. 2001). Generally,

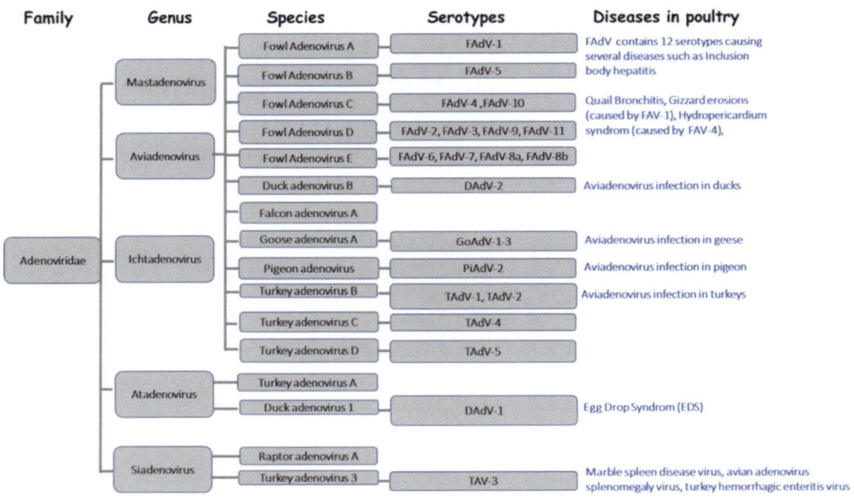

Fig. 6.1 Classification of avian adenoviruses. Turkey hemorrhagic enteritis (HE) is caused by turkey adenovirus-3 (TAV-3), the genus *Siadenovirus*, family *Adenoviridae (The information collected from* https://talk.ictvonline.org/taxonomy/*)*. The genus Aviadenovirus comprises 14 species (Fowl AdV A-D, Duck AdV-B, Falcon AdV-A, Goose AdV-A, Pigeon AdV-A/ B, TAdV-B/C/D, and Psittacine AdV-B)

further studies are required to understand the pathogenicity of aviadenoviruses in turkeys.

The *Aviadenovirus* comprises the fowl aviadenovirus (FAdV) and the turkey aviadenovirus (TAdV) species. FAdVs are grouped into five species (FAdV A to FAdV-E), 12 serotypes (FAdV-1 to FAdV-8a and FAdV-8b to FAdV-11), and 12 genotypes [75]. Three TAdV species, namely, TAdV-B (type TAdV-1), TAdV-C (type TAdV-4), and TAdV-D (type TAdV-5), were isolated from respiratory disease and poult enteritis and mortality syndrome (PEMS) (Kaján et al. 2010; Kleine et al. 2017; Marek et al. 2014). These viruses also cause inclusion body hepatitis in turkey poults and may be responsible for lower hatchability rates in breeder flocks (Guy et al. 1988; Shivaprasad et al. 2001; Guy 1998). Generally, further studies are required to understand the pathogenicity of aviadenoviruses in turkeys; hence, all aviadenoviruses were identified.

Turkey HE virus was first recognized in turkey poults in 1937 by Pomeroy and Fenstermacher in the United States (Pomeroy and Fenstermacher 1937). HE virus replicates in the endothelial cells but does not replicate in the intestinal epithelium, causing vascular damage and ischemic necrosis of intestinal villi. Since adenoviruses are non-enveloped, they are resistant to lipid solvents and pH 3–9 but are sensitive to formaldehyde. HEV remains infective for 6 months and 4 years at 48 °C and −208 °C, respectively (Domermuth and Gross 1984).

Due to the close relationship between the pheasant marble spleen disease virus and HE, the pheasant marble spleen virus can infect turkeys, and the turkey HE virus may infect pheasants (Dhama et al. 2017). The virus causes severe immunosuppression, which is accompanied by infections with opportunistic bacteria, leading to growth retardation and reducing feed conversion rate. The disease is accompanied by economic losses due to the acute intestinal disorder caused by virulent strains. The disease typically lasts after 7–10 days in the flock, although losses could last for an extra 2–3 weeks in the case of mixed or secondary infections.

6.2 Clinical Signs

The clinical disease is commonly reported in four to 12-week-old turkey poults. The main signs include depression, bloody droppings, sudden death, and bloody droppings. The mortality rates ranged from 10 to 15%; however, it can reach 60% (Fig. 6.2).

6.3 Postmortem Lesions

Postmortem lesions are characterized by an enlarged mottled spleen and distended and congested intestines, more prominent in the proximal small intestine. The intestine might be filled with bloody exudate. Sometimes, the intestinal mucosa is covered by a yellowish fibrinonecrotic membrane. Moreover, severely congested

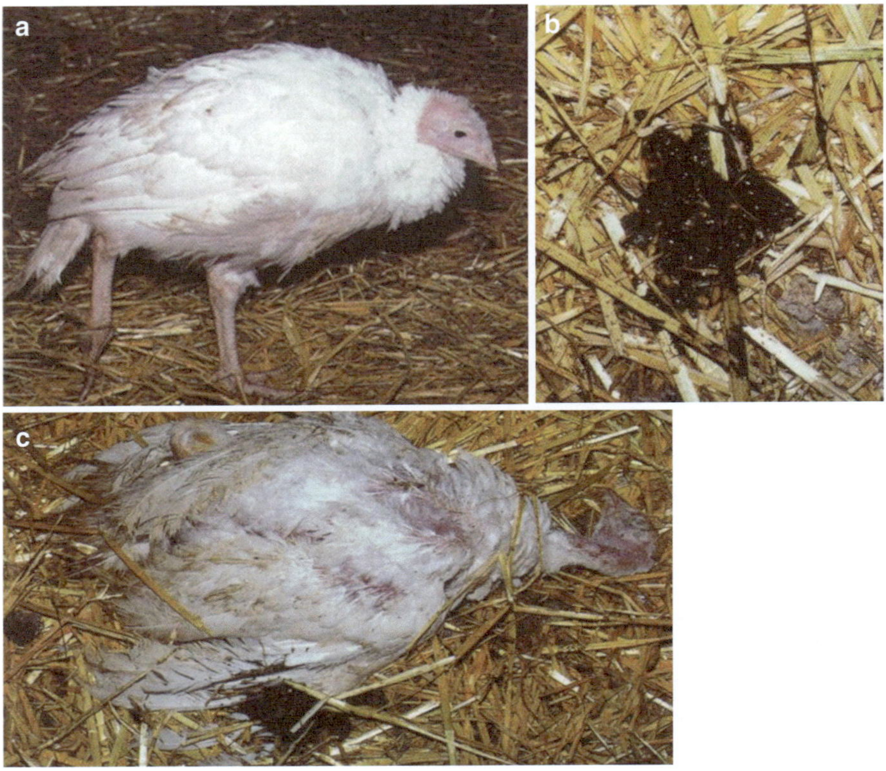

Fig. 6.2 Clinical signs: (**a**) depression, (**b**) bloody dropping, and (**c**) death (©Hafez, FU-Berlin)

mucosa, degeneration and sloughing of villus epithelium, and hemorrhage at villus tips were reported (Figs. 6.3 and 6.4).

6.4 Histopathological Examination

Microscopic examination revealed characteristic lesions, including hyperplasia of the white pulp and lymphoid necrosis in the spleen. Petechial hemorrhages can also be seen in several tissues (Shehata et al. 2021).

6.5 Diagnosis

The clinical signs, postmortem lesions, and histopathology are the presumptive diagnosis of HE. A definitive diagnosis depends on virus isolation and molecular identification. Adenovirus isolation can be done using cell cultures derived from the homologous species (Hess et al. 1999). However, adenoviruses could be successfully isolated from turkeys using chicken embryo liver (CEL) cells from SPF

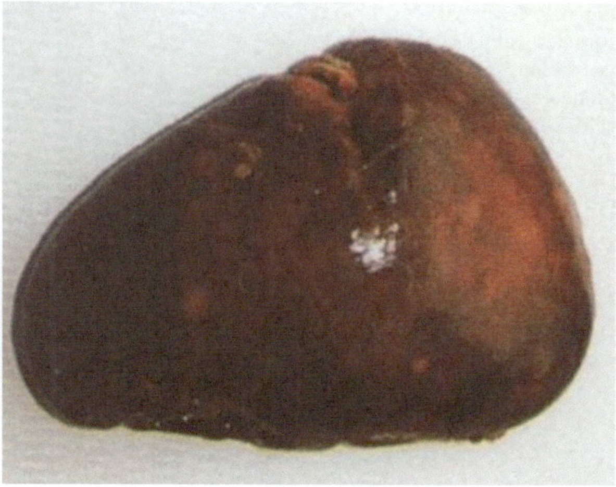

Fig. 6.3 Hemorrhagic enteritis in turkeys, showing enlarged mottled spleen

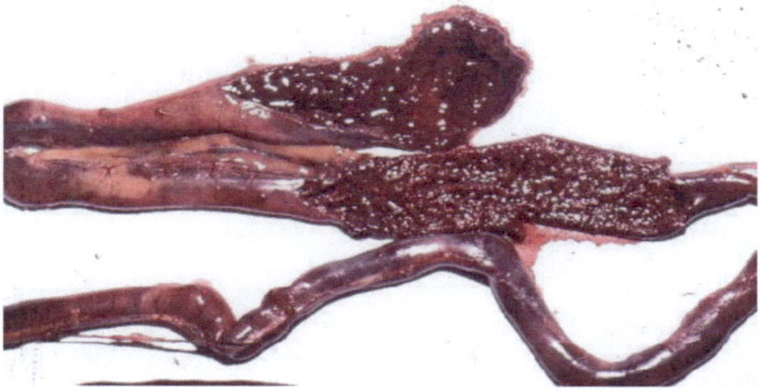

Fig. 6.4 Hemorrhagic enteritis in turkeys showing hemorrhage of the intestine and a velvety appearance (©Hafez, FU-Berlin)

chickens. The main cytopathic effects on CEL cells are rounded cell degeneration after the first and fourth passages (Kleine et al. 2017). Electron microscopy can also be used to detect the virus in turkey poults (Fig. 6.5). The agar gel immunodiffusion test is the most common test to detect the virus in the splenic extracts. The highest titer of the HEV is present in the spleen.

Molecular typing can be done based on conventional PCR targeting the L1 region of the hexon gene (Meulemans et al. 2001; Kleine et al. 2017). Amplification and sequence analysis of polymerase genes can be used to distinguish between TAdV-B, TAdV-C, and TAdV-D (Ye et al. 2012).

Differential diagnosis from pathogens causing splenic and intestinal lesions such as *Erysipelothrix rhusiopathiae*, *Pasteurella multocida*, *E. coli*, *Salmonella* sp., reticuloendotheliosis, lymphoproliferative disease, avian influenza, and ND should be considered (Dhama et al. 2017). Aortic rupture should also be considered in the differential diagnosis. However, aortic rupture is characterized by blood coming from the nostrils and mouth before performing the postmortem lesions.

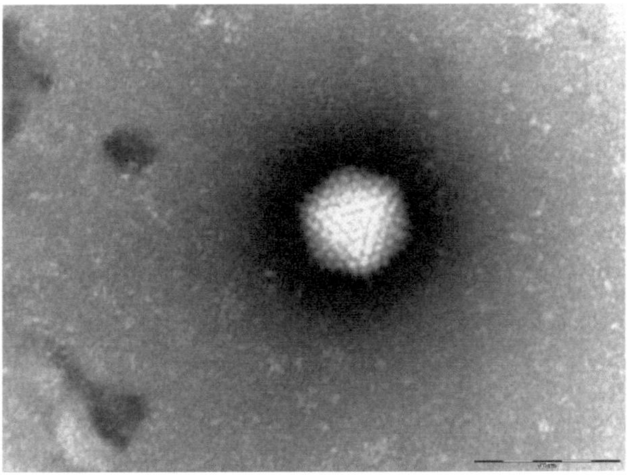

Fig. 6.5 Adenovirus particles using electron microscopy (CVUA-Stuttgart 2010) (Shehata et al. 2021)

6.6 Prevention and Control

Live HE vaccines prepared from naturally avirulent strains are effective against HEV. These vaccines can be prepared in cell culture using RP19 cells or peripheral blood leukocytes (van den Hurk 1992), commercially available, or from spleen homogenate obtained from 6-week-old poults vaccinated with the avirulent HEV vaccine (Fadly et al. 1985). However, these vaccines can cause transient immunosuppression.

By designing a vaccination program against HE, it is recommended to consider several factors: (1) To avoid exposure to virulent field strains, it is suggested to immunize poults against HEV before the maternal antibody begins to decline. Typically, vaccinations are given to poults between 3 and 6 weeks old. (2) Seroconversion in response to vaccination developed slowly. It was suggested that a seroconversion rate of 60% or higher using AGID or ELISA 3 weeks after vaccination with splenic homogenate is a good indication of 100% protection. (3) when cell culture-based vaccinations are used, a booster dose can be given one week following the initial vaccination. (4) Presence of immunosuppressive agents such as avian metapneumoviruses (Chary et al. 2002) or residual water sanitizers in the pipeline can negatively impact the vaccine efficacy. (5) If less than 100% of poults get vaccinated, lateral transmission within 2–3 weeks can be expected due to fecal excretion of the virus (Pierson and Fitzgerald 2013). (6) The application of *E. coli* autogenous vaccines reduces mortality during the most critical period of HEV infection (Dhama et al. 2017).

References

Chary P, Rautenschlein S, Sharma JM (2002) Reduced efficacy of hemorrhagic enteritis virus vaccine in turkeys exposed to avian pneumovirus. Avian Dis 46(2):353–359. https://doi.org/10.1637/0005-2086(2002)046[0353:REOHEV]2.0.CO;2

Dhama K, Gowthaman V, Karthik K, Tiwari R, Sachan S, Kumar MA et al (2017) Haemorrhagic enteritis of turkeys—current knowledge. Vet Q 37(1):31–42. https://doi.org/10.1080/01652176.2016.1277281

Domermuth CH, Gross WB (1984) Hemorrhagic enteritis and related infections. In: Hoftad MS, Barnes HJ, Calnek BW, Reid WM, Yoder HW Jr (eds) Diseases of poultry, 8th edn. Iowa State University Press, Ames, pp 511–516

Fadly AM, Nazerian K, Nagaraja K, Below G (1985) Field vaccination against hemorrhagic enteritis of turkeys by a cell-culture live-virus vaccine. Avian Dis 29(3):768. https://doi.org/10.2307/1590669

Guy JS (1998) Virus infections of the gastrointestinal tract of poultry. Poult Sci 77(8):1166–1175

Guy JS, Schaeffer JL, Barnes HJ (1988) Inclusion-body hepatitis in day-old turkeys. Avian Dis 32(3):587–590

Hess M, Raue R, Hafez HM (1999) PCR for specific detection of haemorrhagic enteritis virus of turkeys, an avian adenovirus. J Virol Methods 81(1–2):199–203. https://doi.org/10.1016/s0166-0934(99)00067-1

Kaján GL, Stefancsik R, Ursu K, Palya V, Benko M (2010) The first complete genome sequence of a non-chicken aviadenovirus, proposed to be Turkey adenovirus 1. Virus Res 153(2):226–233. https://doi.org/10.1016/j.virusres.2010.08.006

Kleine A, Hafez HM, Lüschow D (2017) Investigations on aviadenoviruses isolated from Turkey flocks in Germany. Avian Pathol 46(2):181–187. https://doi.org/10.1080/03079457.2016.1237013

Marek A, Ballmann MZ, Kosiol C, Harrach B, Schlötterer C, Hess M (2014) Whole-genome sequences of two Turkey adenovirus types reveal the existence of two unknown lineages that merit the establishment of novel species within the genus Aviadenovirus. J Gen Virol 95(Pt 1):156–170. https://doi.org/10.1099/vir.0.057711-0

Meulemans G, Boschmans M, Berg TP, Decaesstecker M (2001) Polymerase chain reaction combined with restriction enzyme analysis for detection and differentiation of fowl adenoviruses. Avian Pathol 30(6):655–660. https://doi.org/10.1080/03079450120092143

Pierson FW, Fitzgerald SD (2013) Hemorrhagic enteritis and related infections. In: Swayne DE, Glisson JR, McDougald LR, Nolan LK, Suarez DL, Nair V (eds) Diseases of poultry, 13th edn. Iowa State University Press, Ames, pp 237–247

Pomeroy BS, Fenstermacher R (1937) Hemorrhagic enteritis in turkeys. Poult Sci 16:378–383

Shehata AA, Basiouni S, Sting R, Akimkin V, Hoferer M, Hafez HM (2021) Poult enteritis and mortality syndrome in Turkey poults: causes, diagnosis and preventive measures. Animals 11(7):2063. https://doi.org/10.3390/ani11072063

Shivaprasad HL, Woolcock PR, McFarland MD (2001) Group I avian adenovirus and avian adeno-associated virus in Turkey poults with inclusion body hepatitis. Avian Pathol 30(6):661–666. https://doi.org/10.1080/03079450120092152

van den Hurk JV (1992) Characterization of the structural proteins of hemorrhagic enteritis virus. Arch Virol 126(1–4):195–213. https://doi.org/10.1007/BF01309695

Ye J, Coulouris G, Zaretskaya I, Cutcutache I, Rozen S, Madden TL (2012) Primer-BLAST: a tool to design target-specific primers for polymerase chain reaction. BMC Bioinformatics 13:134. https://doi.org/10.1186/1471-2105-13-134

Turkey Arthritis Reovirus

7

Hafez M. Hafez and Awad A. Shehata

Abstract

Turkey reovirus is a progressive condition affecting mainly tom turkeys at 12th–16th weeks of age, causing enteritis, immunosuppression, arthritis/tenosynovitis, and myocarditis. The disease is characterized by lameness, stunted and uneven growth, poor feed conversion, and increased mortality. The primary postmortem lesions are unilateral or bilateral swelling of the hock joints and the presence of clear yellow to serosanguinous synovial fluid, which may even reach the sheaths of the gastrocnemius and digital flexor tendons. Rupture of gastrocnemius and/or digital flexor tendon may occur, resulting in hemorrhages followed by a fibrotic lump above the hock joint. Reoviruses can be isolated in embryonated chicken eggs via either a yolk sac or on chorioallantoic membrane routes. PCR, electron microscopy, and immunohistochemistry can be used for virus identification. ELISA and AGID can be used for serological diagnosis and/or monitoring of vaccinated flocks. Although live attenuated and inactivated vaccines against chicken reovirus are available; however, no commercial vaccine is available against turkey reovirus. In some countries, the autogenous inactivated turkey reovirus vaccine can be used.

H. M. Hafez
Faculty of Veterinary Medicine, Institute of Poultry Diseases, Free University of Berlin, Berlin, Germany
e-mail: hafez.mohamed@fu-berlin.de

A. A. Shehata (✉)
TUM School of Natural Sciences, Bavarian NMR Center (BNMRZ), Structural Membrane Biochemistry, Technical University of Munich, Garching, Germany
e-mail: awad.shehata@tum.de

H. M. Hafez, A. A. Shehata (eds.), *Turkey Diseases and Disorders Volume 2*,
https://doi.org/10.1007/978-3-031-63322-5_7

Keywords
Reovirus · Arthritis · Tenosynovitis · Chorioallantoic membrane · Immunohistochemistry · ELISA · AGID

7.1 Etiology

The family *Reoviridae* comprises 15 genera and is classified into two subfamilies, namely, *Spinareovirinae* and *Sedoreovirinae*. Avian reoviruses are members of the genus *Orthoreoviruses* and can infect domestic and wild birds. Like *Aquareoviruses*, *Orthoreoviruses* are non-enveloped, icosahedral virions comprising a double-stranded RNA of ten segments (Tang et al. 2016), designated large (L1, L2, and L3), medium (M1, M2, and M3), and small (S1, S2, S3, and S4).

The genome encodes eight structural proteins (λA, λB, λC, μA, μB, σA, σB, and σC) and four nonstructural proteins (μNS, P10, P17, and σNS) (Varela et al. 1996). The σC protein is encoded by the S1 segment, which is responsible for cell attachment and immunogenicity (Martínez-Costas et al. 1997).

Based on the molecular differences between avian reoviruses, these viruses are classified into five species-specific, for example, turkey reovirus, duck reovirus, goose reovirus, psittacines reovirus, and chicken reovirus (Jones et al. 1989). It is believed that turkey reovirus might have emerged as an evolution of avian reoviruses due to genetic reassortment (Day et al. 2007; Sharafeldin et al. 2014).

7.2 History and Epidemiology

The turkey reovirus is ubiquitous and distributed worldwide, suggesting that it may be part of the virosome of healthy turkeys. The first isolation of reoviruses from turkeys suffering from tenosynovitis/arthritis was in the 1980s. However, the isolated virus could not fulfill Koch's postulates when inoculated into footpads of one-day-old poults (Afaleq and Jones 1989). Basically, the pathogenicity of avian reoviruses depends on the age, pathogenicity, reovirus strain, dose, route of infection, presence of maternal antibodies, and immune system. Young birds free from maternally derived antibodies can be infected with reoviruses (van der Heide 2000), highlighting the importance of vaccination of breeders against reovirus to provide protection conferred by maternal-derived antibodies.

Reoviruses are resistant to heat and environmental stress. The virus survives in litter and drinking water for about a week but is inactivated within 10 min or less when exposed to most commercial disinfectants (Mor et al. 2014). It was found that oxidizing agents and quaternary ammonium compounds plus aldehyde were more effective against turkey rotaviruses.

The clinical disease is more common in Tom turkeys at 12–16 weeks of age. However, it was found that poults aged 28 days or less are more susceptible to infection following oral challenge (Kumar et al. 2021).

Although chicken reoviruses can be transmitted vertically via hatching eggs and horizontally through fecal and contaminated litter (Al-Muffarej et al. 1996). The vertical transmission of turkey reovirus in turkeys has not yet been approved. Recently, it was also found that turkey reovirus transmitted the infection to naïve sentinel turkeys of the same age, and this highlights the potential vertical transmission (Kumar et al. 2021).

7.3 Pathogenic Conditions Associated with Turkey Reovirus

7.3.1 Turkey Reovirus-Associated Poult Enteritis and Mortality Syndrome (PEMS) and Immunosuppression

Turkey reovirus grows well in the intestine (enterocytes), causing mild diarrhea and depression 7 days post-oral inoculation in commercial poults. The virus can cross the intestinal barrier and reach the blood and then the gastrocnemius tendon to produce tenosynovitis (Sharafeldin et al. 2015, 2016) A mild increase in crypt depth due to crypt hyperplasia has been observed with granulocyte and lymphocyte infiltration in the lamina propria and submucosa 2 weeks after oral challenge.

Turkey reovirus was also detected in the bursa Fabricius and caused lymphoid depletion that caused bursal atrophy and fibroplasia in poults at a young age, leading to permanent immunosuppression and stimulating opportunistic pathogens (Heggen-Peay et al. 2002; Spackman et al. 2005b; Day et al. 2008).

Turkey reoviruses are also associated with poult enteritis and mortality syndrome (PRMS) (Shehata et al. 2021). However, experimentally infected SPF poults showed mild clinical signs and no postmortem lesions, highlighting that turkey reovirus might not be the primary cause of PEMS (Spackman et al. 2005a).

7.3.2 Turkey Reovirus-Associated Arthritis/Tenosynovitis

Infection with turkey reovirus is associated with uni- or bilateral lameness, with swelling of tarsal and metatarsal tendons. Sometimes, swelling of the hock joint can be seen. Birds are reluctant to move, and infected flocks exhibit uneven growth, low feed conversion ratio, and reduced performance. Postmortem lesions reveal unilateral or bilateral swelling of the hock joints and the presence of clear yellow to serosanguinous synovial fluid, which may even reach the sheaths of the gastrocnemius and digital flexor tendons. A gastrocnemius and/or digital flexor tendon rupture may occur, resulting in bleeding followed by a fibrotic lump above the hock joint. In chronic cases, there is bruising of the skin of the hock with prominent periarticular fibrosis, edema, and occasional large flecks of fibrin within the subcutis and tendon sheaths.

Microscopically, the main lesions are lymphocytic infiltration in the subsynovium of tendons and synoviocyte hypertrophy. Fibrosis of the tendon and tendon sheath that promotes stiffness (reduced tendon elasticity) and ultimately leads to gastrocnemius tendon rupture in some birds was documented (Jones et al. 2013; Sharafeldin et al. 2016).

7.3.3 Turkey Reovirus-Associated Myocarditis

Turkey reovirus was detected in 19-day-old turkey poults with myocarditis with a history of diarrhea and increased mortality, suggesting that this virus is probably the etiology of myocarditis in turkeys. The virus was isolated from enlarged hearts with mild to severe ventricular dilation, whitish discoloration, hydropericardium, pale liver areas, congestion, lung edema, and enteritis with pale serosa and watery contents. The virus was detected from the heart, spleen, intestine, liver, lungs, intestine, and bursa of Fabricius samples using electron microscopy and immunohistochemistry using polyclonal antibodies for turkey reovirus. Microscopically, severe degeneration and necrosis of myocytes and lymphocytic infiltration in the myocardium and epicardium were documented (Shivaprasad et al. 2009; França et al. 2010).

7.4 Diagnosis

Avian reoviruses can be isolated in embryonated chicken eggs at 9 days using yolk sac inoculation of 6-day-old or chorioallantoic membrane route. Virus identification can be carried out using PCR, transmission electron microscopy (85–88 μm in diameter), and immunohistochemistry.

7.5 Vaccination

In some countries such as the United States, turkey breeders have been vaccinated with the autogenous vaccine against turkey reovirus for two purposes: (1) prevention of the suspected vertical transmission to the turkey poults and (2) providing maternally derived antibodies to the progeny during the initial days of their life. However, these vaccines provide different efficacy, especially against the variant strains (Porter 2018). A recombinant Pichinde virus-vectored vaccine expressing S1 and S3 antigens was developed recently. However, further studies are still required to produce such vaccines commercially.

References

Afaleq AA, Jones RC (1989) Pathogenicity of three Turkey and three chicken reoviruses for poults and chicks with particular reference to arthritis/tenosynovitis. Avian Pathol 18(3):433–440. https://doi.org/10.1080/03079458908418616

Al-Muffarej SI, Savage CE, Jones RC (1996) Egg transmission of avian reoviruses in chickens: comparison of a trypsin-sensitive and a trypsin-resistant strain. Avian Pathol 25(3):469–480. https://doi.org/10.1080/03079459608419156

Day JM, Pantin-Jackwood MJ, Spackman E (2007) Sequence and phylogenetic analysis of the S1 genome segment of Turkey-origin reoviruses. Virus Genes 35(2):235–242. https://doi.org/10.1007/s11262-006-0044-1

Day JM, Spackman E, Pantin-Jackwood MJ (2008) Turkey origin reovirus-induced immune dysfunction in specific pathogen free and commercial Turkey poults. Avian Dis 52(3):387–391. https://doi.org/10.1637/8190-120607-Reg

França M, Crespo R, Chin R, Woolcock P, Shivaprasad HL (2010) Retrospective study of myocarditis associated with reovirus in turkeys. Avian Dis 54(3):1026–1031. https://doi.org/10.1637/9262-020110-Reg.1

Heggen-Peay CL, Qureshi MA, Edens FW, Sherry B, Wakenell PS, O'Connell PH et al (2002) Isolation of a reovirus from poult enteritis and mortality syndrome and its pathogenicity in Turkey poults. Avian Dis 46(1):32–47. https://doi.org/10.1637/0005-2086(2002)046[0032:IOARFP]2.0.CO;2

Jones RC, Islam MR, Kelly DF (1989) Early pathogenesis of experimental reovirus infection in chickens. Avian Pathol 18(2):239–253. https://doi.org/10.1080/03079458908418599

Jones RC, Swayne DE, Glisson JR, McDougald LR, Nolan LK, Suarez DL (2013) Reovirus infections. In: Diseases of poultry, 13th edn. Wiley, Somerset, pp 351–373

Kumar R, Sharafeldin T, Goyal S, Mor S, Porter R (2021) Infection and transmission dynamics of Turkey arthritis reovirus in different age Turkeys. SSRN J. https://doi.org/10.2139/ssrn.3997874

Martínez-Costas J, Grande A, Varela R, García-Martínez C, Benavente J (1997) Protein architecture of avian reovirus S1133 and identification of the cell attachment protein. J Virol 71(1):59–64. https://doi.org/10.1128/JVI.71.1.59-64.1997

Mor SK, Bekele AZ, Sharafeldin TA, Porter RE, Goyal SM (2014) Efficacy of five commonly used disinfectants against turkey arthritis reovirus. Avian Dis 59(1):71–73

Porter R (2018) Turkey reoviral arthritis update; proceedings of the MPF convention Turkey health workshop, Midwest Poultry Federation of the Conference; Minneapolis, MN, USA

Sharafeldin TA, Mor SK, Bekele AZ, Verma H, Goyal SM, Porter RE (2014) The role of avian reoviruses in Turkey tenosynovitis/arthritis. Avian Pathol 43(4):371–378. https://doi.org/10.1080/03079457.2014.940496

Sharafeldin TA, Mor SK, Sobhy NM, Xing Z, Reed KM, Goyal SM et al (2015) A newly emergent Turkey arthritis reovirus shows dominant enteric tropism and induces significantly elevated innate antiviral and T-helper-1 cytokine responses. PLoS One 10(12):e0144085. https://doi.org/10.1371/journal.pone.0144085

Sharafeldin TA, Chen Q, Mor SK, Goyal SM, Porter RE (2016) Altered biomechanical properties of gastrocnemius tendons of turkeys infected with Turkey arthritis reovirus. Vet Med Int 2016:7829138. https://doi.org/10.1155/2016/7829138

Shehata AA, Basiouni S, Sting R, Akimkin V, Hoferer M, Hafez HM (2021) Poult enteritis and mortality syndrome in Turkey poults: causes, diagnosis and preventive measures. Animals 11(7):2063. https://doi.org/10.3390/ani11072063

Shivaprasad HL, Franca M, Woolcock PR, Nordhausen R, Day JM, Pantin-Jackwood M (2009) Myocarditis associated with reovirus in Turkey poults. Avian Dis 53(4):523–532. https://doi.org/10.1637/8916-050309-Reg.1

Spackman E, Kapczynski D, Sellers H (2005a) Multiplex real-time reverse transcription-polymerase chain reaction for the detection of three viruses associated with poult enteritis complex: Turkey astrovirus, Turkey coronavirus, and Turkey reovirus. Avian Dis 49(1):86–91. https://doi.org/10.1637/7265-082304R

Spackman E, Pantin-Jackwood M, Michael Day J, Sellers H (2005b) The pathogenesis of Turkey origin reoviruses in turkeys and chickens. Avian Pathol 34(4):291–296. https://doi.org/10.1080/03079450500178501

Tang Y, Lin L, Sebastian A, Lu H (2016) Detection and characterization of two co-infection variant strains of avian orthoreovirus (ARV) in young layer chickens using next-generation sequencing (NGS). Sci Rep 6(1):24519. https://doi.org/10.1038/srep24519

van der Heide L (2000) The history of avian reovirus. Avian Dis 44(3):638–641

Varela R, Martínez-Costas J, Mallo M, Benavente J (1996) Intracellular posttranslational modifications of S1133 avian reovirus proteins. J Virol 70(5):2974–2981. https://doi.org/10.1128/jvi.70.5.2974-2981.1996

Avian Encephalomyelitis

8

Awad A. Shehata and Hafez M. Hafez

Abstract

Avian encephalomyelitis (AE), or epidemic tremor, is a worldwide infectious viral disease affecting the central nervous system of young birds (during the first 2 weeks) such as chickens, turkeys, quail, pheasants, and pigeons, causing ataxia and rapid tremors with milder clinical signs in turkeys than in chickens. AE virus, belonging to the genus *Tremovirus* of the family *Picornaviridae*, is enterotropic and can be transmitted horizontally and/or vertically. In layers and breeders, the virus caused a drop in egg production and reduced hatchability. The diagnosis depends on the clinical signs, histopathological lesions of the brain, pancreas, and duodenum, and PCR. The virus can be isolated in 6-day-old embryonated chicken eggs. One vaccination of breeder hens before egg production with inactivated or live vaccines prevents disease in breeders and progeny. Commercial ELISA is available for monitoring vaccination status.

Keywords

Avian encephalomyelitis · *Picornaviridae* · Epidemic tremor histopathological · Natural field strains · Embryo-adapted strains

A. A. Shehata (✉)
TUM School of Natural Sciences, Bavarian NMR Center (BNMRZ), Structural Membrane Biochemistry, Technical University of Munich, Garching, Germany
e-mail: awad.shehata@tum.de

H. M. Hafez
Faculty of Veterinary Medicine, Institute of Poultry Diseases, Free University of Berlin, Berlin, Germany
e-mail: hafez.mohamed@fu-berlin.de

H. M. Hafez, A. A. Shehata (eds.), *Turkey Diseases and Disorders Volume 2*,
https://doi.org/10.1007/978-3-031-63322-5_8

8.1 Etiology

AEV of the *Tremovirus* genus, family *Picornaviridae* is non-enveloped, with a size of 24–32 nm in diameter, and contains a single-stranded RNA genome of 7032 nucleotides (nt), not including the poly-A tail. The AEV genome comprises a single open reading frame (ORF) that encodes 11 proteins, including four viral structural proteins (vp1, vp2, vp3, and vp4) and seven nonstructural proteins. The VP1 is the most immunogenic part of the virus (Liu et al. 2014; Marvil et al. 1999). Because of the absence of an envelope, the AEV is highly resistant in the environment and to chloroform, acids, trypsin, pepsin, and DNase and is protected against the effects of heat by divalent magnesium ions. However, it is susceptible to formaldehyde fumigation and beta-propiolactone.

8.2 Epidemiology

Jones first described AEV in 1930 in the United States (Jones 1932). Turkey poults, chickens, pheasants, and quails can also be infected naturally. Baby chicks (1–3 weeks old) obtained from non-vaccinated breeder flocks are susceptible to AE. Duckling, pigeons, and guinea fowl can be experimentally infected.

The virus is relatively nonpathogenic to adult birds but can cause a temporary drop in egg production (5–10%). However, in 2016, sporadic cases with neurological signs of AE in pullets at 1–2 were reported (Sentíes-Cué et al. 2016). AE can be transmitted vertically through eggs from the infected non-vaccinated breeders. Infection in incubators at hatching or brooder by contact with infected chicks.

The virus can be transmitted through direct contact. Ingestion of contaminated food and water by the droppings of infected birds. Infected birds harbor the virus in the intestine and shed it in the feces for 10–15 days, as the virus can survive for at least 3 weeks in droppings.

The incubation period after vertical transmission is 1–7 days. While in horizontal transmission, the incubation period is 11 days.

8.3 Pathotypes

Two Distinct Pathotypes of AEV Are Known

1. **Natural field strains**. These viruses are enterotropic; however, some show tropism to the nervous system. The virus is shed in the feces for several days (1–2 weeks) and can be transmitted orally (Westbury and Sinkovic 1978; Shafren and Tannock 1988). The virus can remain infectious for a long time as it is quietly resistant to environmental conditions. Natural strains do not induce gross lesions in embryos.
2. **Embryo-adapted strains**. These viruses are subjected to multiple passages in antibody-free chicken embryos. They are neurotropic viruses causing severe neurological signs following intracerebral, intramuscular, or subcutaneous injec-

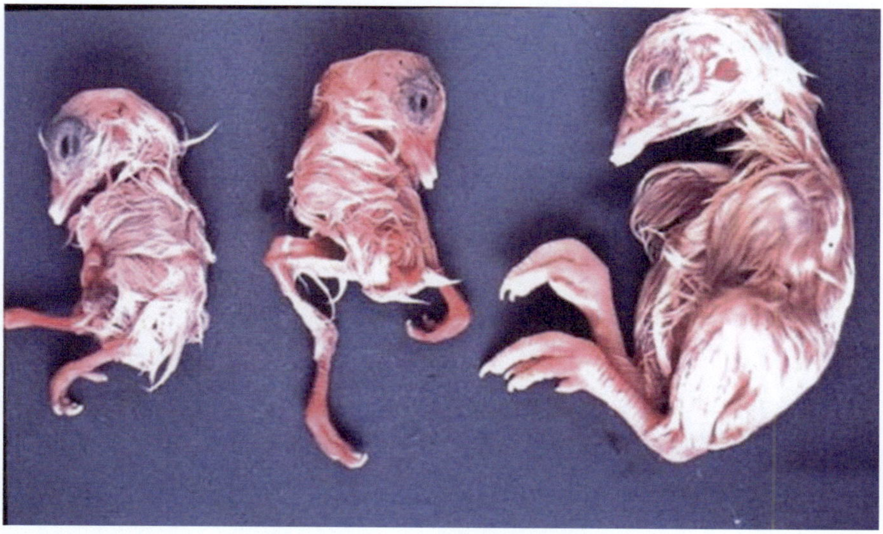

Fig. 8.1 Avian encephalitis in chicken embryos, showing severe stunting growth at 5 days post-inoculation. The embryo on the right is a normal control (© Hafez)

tion. Embryo-adapted viruses are not spread horizontally because they are not infectious if given orally, except with high doses (Shafren and Tannock 1991). In contrast to natural field strains, these viruses are pathogenic for embryos, causing muscular dystrophy and immobilization of skeletal muscles. A severe stunting growth at 5 days post-inoculation is shown in Fig. 8.1.

Maternal antibodies effectively protect chicks and/or poults against AEV. Maternal antibodies waned between the second and third week post-hatch (Szabó 2012). Therefore, detecting high antibodies against AEV in **chickens** at 21 days of age or older indicates field infection.

8.4 Clinical Signs

The main clinical signs of AE at an early age (1–3 weeks old) are inappetence, weakness, and the development of various neurological signs such as ataxia, paralysis, and opisthotonus, leading to prostration and death (Fig. 8.2). Head and neck tremors were also reported. A tremor frequently precedes ataxia; in some cases, only tremors have been seen. Usually, ataxia worsens until the chick cannot move, leading to inanition, prostration, and ultimately death.

The disease is characterized by high morbidity (40–60%) and mortality (25–50%) rates, depending on the immune status of breeders. Infection at an early age can lead to the development of cataracts and opacity, which gives a bluish discoloration to

Fig. 8.2 Avian encephalomyelitis in chicks showing ataxia and inability to stand (© Hafez)

the lens, in some survivors, and blindness. In mature birds, AE is characterized by a temporary drop in egg production (5–10%) for 1–2 weeks (Meroz et al. 1990).

8.5 Postmortem Lesions

There are no specific gross lesions for AE, only pale areas in the muscular layer of the gizzard.

8.6 Histopathology

Non-suppurative encephalomyelitis is an apathogenomic microscopic lesion of AE. The main lesions are multifocal severe perivascular cuffing by lymphocytes, gliosis, severe local edema, vascular proliferation, and pyknosis (Fig. 8.3). Swelling and chromatolysis of neurons in nuclei of the (nucleus rotundus and nucleus ovoidolis) in the midbrain and cerebellum, in combination with infiltration and

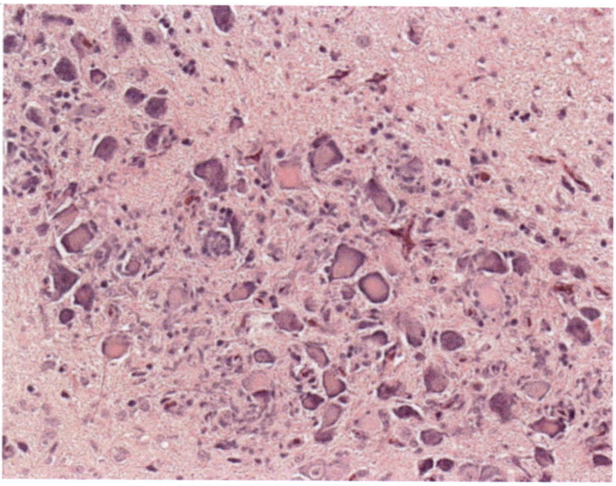

Fig. 8.3 Histopathological lesions of Medulla oblongata of a chick affected with avian encephalomyelitis showing central chromatolysis of neurons, neuronal necrosis, and satellitosis, neuronophagia, and gliosis. Avian Encephalomyelitis—Poultry—MSD Veterinary Manual. (©Dr. Tahseen Abdul-Aziz)

aggregation of lymphocytes in the muscular layers of the proventriculus, have been reported. Lymphocytic inflammation of the pancreas, myocardium, skeletal muscles, nerves, and muscular layers of the gizzard, crop, and esophagus is also documented. Muscular changes consisted primarily of eosinophilic swelling and necrosis, fragmentation, and loss of striations of affected fibers with rare sarcolemma proliferation and heterophil infiltration (Hishida et al. 1986).

8.7 Diagnosis

- The presumptive diagnosis of AE depends on the neurological signs in baby chicks and poults at 1–3 weeks. Biphasic mortality is a feature of vertical infection with AE. Some recovered birds develop cataracts in one or both eyes. In older ages (after 6 weeks), the main signs are a transient drop in egg production (5–10% for 1–2 weeks) and decreased hatchability due to embryonic deaths during the last 3 days of incubation. Poults hatched from these eggs develop ataxia and leg paralysis.
- Egg or embryo susceptibility test. It is used to determine the immunity of a flock. Fertile eggs from the flock will be tested, and control eggs from a known susceptible flock will be incubated. Each embryonated egg will be inoculated at 6 days old via the yolk sac with 100 EID_{50} of the egg-adapted virus. Embryos are examined 10–12 days PI for characteristic lesions. The flock is considered susceptible if 100% of embryos are affected; less than 50% affected indicates an immune flock.
- Histopathological examination is pathognomonic for AE, particularly: (1) central chromatolysis of neurons in the brain and spinal cord and (2) clusters of lymphocytes in the proventriculus muscles, gizzard, and pancreas.
- Virus isolation from a suspension of the brain, spleen, and duodenum can be inoculated in embryonated chicken eggs via the yolk sac route at 5- to 6-day-old embryos. The field viruses do not exhibit any embryonic lesions. However, the

hatched chicks exhibit typical clinical signs during the first 7–10 days of life. Embryonic fibroblast, kidney, and neuroglial cell cultures and pancreatic cells from young chicks can be used for virus propagation and isolation. Replication in cell cultures can be detected by inoculation of embryos (adapted strains only) or by IFA or ELISA (Nicholas et al. 1986, 1987).

- The virus can be identified using RT-PCR (Xie et al. 2005; Liu et al. 2014). The brain is an excellent source of virus for RT-PCR or isolation, although other tissues and organs induce the disease when injected into chicks.
- Serological monitoring can be done using ELISA. However, commercial AE ELISA kits are designed for chickens, making implementing these tests in the serological monitoring of turkey flocks difficult. Differences in the sensitivity of the ELISAs for testing specific anti-AE antibody levels in turkey serum samples were reported (Śmiałek et al. 2022).

8.8 Differential Diagnosis

AE should be differentiated from other neurological diseases, including vitamin E deficiency: its diagnosis depends on the therapeutic diagnosis, negative virus isolation, and the microscopic examination (degenerative changes).

The diagnosis of vitamin A or B2 deficiency can be confirmed based on the therapeutic diagnosis, negative virus isolation, microscopic examination (degenerative changes), and the bilateral thickening of the sciatic nerves. Vitamin B2 deficiency: poults are reluctant to move or walk on their hocks with the aid of wings. In some birds, the toes are bent inward and downward.

Newcastle disease, Marek's disease, should also be considered in diagnosing AE.

8.9 Prevention and Control

No treatment is effective against avian encephalomyelitis.

The control of AE is achieved by vaccination of breeder flocks during the growing period. The maternal antibodies protect baby chicks during the first 2–3 weeks, the critical susceptible age (Hauck et al. 2017). Commercial layers can also be vaccinated to prevent the egg production drop associated with AE. Vaccines used to control AE in chickens have also been effective in turkeys (Deshmukh et al. 1973).

Live vaccines, embryo-propagated but not embryo-adapted viruses, such as strain 1143, can be applied in drinking water wing-web inoculation, or intracutaneous injection (Smyth et al. 1994; Yu et al. 2015; Sentíes-Cué et al. 2016). Vaccination of breeders (growing period) protects against drop in egg production and their progeny through maternal antibodies (Hauck et al. 2017).

It is crucial to avoid the use of embryo-adapted vaccine for two reasons: (1) adapted virus loses its ability to infect via the intestinal tract and is no longer effective when administered orally. (2) Egg-adapted viruses are similar to field viruses and can cause clinical disease.

Generally, vaccination with live vaccines is done after eight-week-of age and at least 4 weeks before egg production. Too early vaccination may cause clinical disease in young birds, and too late vaccination can cause a drop in egg production and shedding of the virus in the eggs, causing clinical disease in hatching birds.

References

Deshmukh DR, Larsen CT, Rude TA, Pomeroy BS (1973) Evaluation of live-virus vaccine against avian encephalomyelitis in Turkey breeder hens. Am J Vet Res 34(7):863–867

Hauck R, Sentíes-Cué CG, Wang Y, Kern C, Shivaprasad HL, Zhou H et al (2017) Evolution of avian encephalomyelitis virus during embryo-adaptation. Vet Microbiol 204:1–7. https://doi.org/10.1016/j.vetmic.2017.04.005

Hishida N, Odagiri Y, Kotani T, Horiuchi T (1986) Morphological changes of neurons in experimental avian encephalomyelitis. Nihon Juigaku Zasshi 48:169–172

Jones EE (1932) An encephalomyelitis in the chicken. Science 76(1971):331–332. https://doi.org/10.1126/science.76.1971.331

Liu Q, Yang Z, Hao H, Cheng S, Fan W, Du E et al (2014) Development of a SYBR green real-time RT-PCR assay for the detection of avian encephalomyelitis virus. J Virol Methods 206:46–50. https://doi.org/10.1016/j.jviromet.2014.05.015

Marvil P, Knowles NJ, Mockett AP, Britton P, Brown TD, Cavanagh D (1999) Avian encephalomyelitis virus is a picornavirus and is most closely related to hepatitis A virus. J Gen Virol 80(Pt 3):653–662. https://doi.org/10.1099/0022-1317-80-3-653

Meroz M, Elkin N, Hadash D, Abrams M (1990) Egg drop associated with avian encephalomyelitis virus. Vet Rec 127(21):532

Nicholas RA, Hopkins IG, Southern SJ, Thornton DH (1986) A comparison of titration methods for live avian encephalomyelitis virus vaccines. Dev Biol Stand 64:207–212

Nicholas RA, Ream AJ, Thornton DH (1987) Replication of avian encephalomyelitis virus in chick embryo neuroglial cell cultures. Brief report. Arch Virol 96(3–4):283–287. https://doi.org/10.1007/BF01320969

Sentíes-Cué CG, Gallardo RA, Reimers N, Bickford AA, Charlton BR, Shivaprasad HL (2016) Avian encephalomyelitis in layer pullets associated with vaccination. Avian Dis 60(2):511–515. https://doi.org/10.1637/11306-102115-Case

Shafren DR, Tannock GA (1988) An enzyme-linked immunosorbent assay for the detection of avian encephalomyelitis virus antigens. Avian Dis 32(2):209. https://doi.org/10.2307/1590806

Shafren DR, Tannock GA (1991) Pathogenesis of avian encephalomyelitis viruses. J Gen Virol 72(Pt 11):2713–2719. https://doi.org/10.1099/0022-1317-72-11-2713

Śmiałek M, Kowalczyk J, Ogonowska-Woźniak B, Koncicki A (2022) Comparison of two commercial ELISA kits for serological monitoring of avian encephalomyelitis in a reproductive Turkey flock. Pol J Vet Sci 25(4):621–624. https://doi.org/10.24425/pjvs.2022.143542

Smyth JA, McNeilly F, Reilly GA, McKillop ER, Cassidy JP (1994) Avian encephalomyelitis following oral vaccination. Avian Pathol 23(3):435–445. https://doi.org/10.1080/03079459408419014

Szabó C (2012) Transport of IGY from egg-yolk to the chicken embryo. J Microbiol Biotechnol Food Sci 2(3):612–620

Westbury HA, Sinkovic B (1978) The pathogenesis of infectious avian encephalomyelitis. 4. The effect of maternal antibody on the development of the disease. Aust Vet J 54(2):81–85. https://doi.org/10.1111/j.1751-0813.1978.tb00352.x

Xie Z, Khan MI, Girshick T, Xie Z (2005) Reverse transcriptase-polymerase chain reaction to detect avian encephalomyelitis virus. Avian Dis 49(2):227–230. https://doi.org/10.1637/7307-111804R

Yu X, Zhao J, Qin X, Zhang G (2015) Serological evidence of avian encephalomyelitis virus infection associated with vertical transmission in chicks. Biologicals 43(6):512–514. https://doi.org/10.1016/j.biologicals.2015.09.003

Rotaviruses

9

Hafez M. Hafez and Awad A. Shehata

Abstract

Rotaviruses were first identified in 1973 by children at 4 and 31 months, showing symptoms of acute gastroenteritis. Rotaviruses were detected in birds for the first time by Bergland in 1977. They, belonging to the family *Reoviridae*, cause enteric diseases in several avian and mammalian species. These viruses are classified into nine distinct groups (A–D and F–J) based on the antigenicity and genetic characteristics of capsid protein VP6. Avian Rotavirus (ARV) was first reported in turkey poults in the United States in 1977, and since then, the following groups, RVA, ARVD, RVF, and RVG, have been identified worldwide. The pathogenicity of rotaviruses in turkeys depends on several factors, including virulence of involved strains, co-infections with other pathogens, and management. The main clinical signs of rotavirus infections are enteric disorders such as diarrhea, depression, and runting-stunting syndrome. The diagnosis involves identifying the virus using electron microscopy and/or PCR. Commercially available ELISAs for identifying RVA in feces are also available. Currently, no commercial vaccines are available. Implementing biosecurity and strict sanitation and hygienic measures could minimize the negative impacts of rotaviruses.

H. M. Hafez
Institute of Poultry Diseases, Faculty of Veterinary Medicine, Free University of Berlin, Berlin, Germany
e-mail: hafez.mohamed@fu-berlin.de

A. A. Shehata (✉)
TUM School of Natural Sciences, Bavarian NMR Center (BNMRZ), Structural Membrane Biochemistry, Technical University of Munich, Garching, Germany
e-mail: awad.shehata@tum.de

H. M. Hafez, A. A. Shehata (eds.), *Turkey Diseases and Disorders Volume 2*, https://doi.org/10.1007/978-3-031-63322-5_9

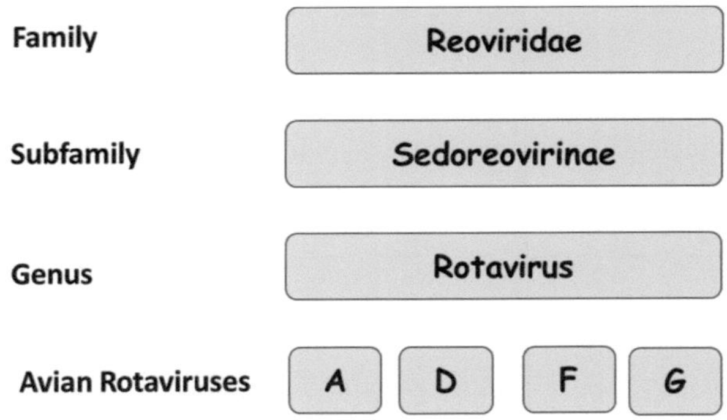

Fig. 9.1 Classification of avian rotaviruses

Keywords
Reoviridae · Capsid protein VP6 · Enteric disorders · Electron microscopy · Runting-stunting syndrome · Hygienic measures

9.1 Etiology

Rotaviruses belong to the genus Rotavirus, one among 15 genera within the *Reoviridae* family. *Reoviridae* is further subdivided into two subfamilies, *Sedoreoviridae* and *Spinareovirinae*. Rotaviruses are non-enveloped icosahedral particles and contain double-stranded RNA of 11 segments that encode six structural (VP1–VP4, VP6, and VP7) and six nonstructural proteins (NSP1–NSP6). According to the VP6, rotaviruses are classified into ten serogroups (RVA–RVJ) (Bányai et al. 2017).

The RVA is of greatest epidemiological importance worldwide. The classification of avian rotaviruses is shown in Fig. 9.1. Rotaviruses are extremely environmentally resistant. The antigenicity of the virus is determined by two surface proteins (VP4 and VP7), which are also used to classify the serogroup into different serotypes (genotypes). There are 16 VP7 ("G-") and 27 VP4 ("P-") types.

Recently, it has been found that avian rotaviruses can exchange genetic material with mammalian rotaviruses, leading to the emergence of new types of rotaviruses (BFR 2020).

9.2 Epidemiology

Rotaviruses have been associated with enteritis in mammalian and poultry species (Matthijnssens and Van Ranst 2012; Martella et al. 2010; Estes and Greenberg 2013). In poultry, several RVA, RVD, RVF, and RVG have been identified worldwide, such as in the United States, Italy, Argentina, Belgium, Brazil, China, Cuba, Germany, India, Bangladesh, the United Kingdom, and Russia (Day 2020). The global prevalence of avian Rotavirus in poultry ranges from 19 to 70% in turkeys and 10 to 47% in chicken flocks.

The pathogenicity of the disease depends on several factors such as (1) Rotavirus strain, (2) mixed infections, (3) maternal antibodies (with the protection of up to 4 weeks), and (4) hygienic measures. Usually, natural infection is more common in young poults (<6 weeks of age). Younger poults at 1–2 weeks old are more susceptible (Shehata et al. 2021).

Natural co-infections with Rotavirus and other avian enteric viruses, such as astrovirus, reovirus, and secondary bacterial pathogens, are common (Pantin-Jackwood et al. 2007; Otto et al. 2006). The avian rotaviruses are classified into genogroups/serologic groups: RVA, RVD, RVF, and RVG. The RVD, RVF and RVG are exclusive to avian species. RVA and RVD are the most prevalent, and interspecies transmission is known to occur within RVA. Broiler and turkey flocks often have a simultaneous and sequential disease with different avian rotavirus groups.

Avian rotaviruses transmit horizontally (Pantin-Jackwood et al. 2007). No evidence of vertical transmission via hatching eggs. However, Rotavirus was detected in a 3-day-old turkey (Theil and Saif 1987), raising the question of whether the transmission is mechanical or vertical. There is also evidence that larvae of the darkling beetle can act as a mechanical vector for TRV (Despins et al. 1994).

9.3 Clinical Signs

Experimentally, the incubation period is 3 days. The clinical signs of rotavirus infections vary from subclinical infections to severe diarrhea, dehydration, poor growth, and increased mortality. In the first week of age, infected poults show mild diarrhea (Horrox 1980). In the second and fifth weeks of life, poults show diarrhea associated with the runting-stunting syndrome and increased mortality. In the third week, the symptoms are characterized by restlessness, litter eating, and watery droppings. The mortality rate reaches 4–7% (Bergeland et al. 1977). No mortality occurred in experimentally infected turkeys or chickens (Spackman et al. 2010). Infection of turkeys with crude intestinal homogenates that contained Rotavirus showed the same clinical signs (Jindal et al. 2009).

9.4 Postmortem

The most common postmortem lesions are the presence of abnormal amounts of fluid and gas in the intestinal tract and caeca. The intestinal tract's pallor, accompanied by tonicity loss, may be evident. Secondary findings include dehydration, stunting of growth, pasted and inflamed vents, anemia due to vent pecking, litter in the gizzard, and inflammation of the feet (Bergeland et al. 1977; Yason et al. 1987; Shawky et al. 1993). Discrete, multifocal, superficial, brownish-red erosions were found in the duodenum and jejunum of turkeys orally infected at 84 and 112 days of age (Yason et al. 1987).

9.5 Diagnosis

The diagnosis of rotavirus depends on identifying the virus in feces or intestinal contents by direct electron microscopy (McNulty et al. 1979). The immune-EM can be used to identify and differentiate the serogroups, albeit the method requires the availability of particular reference antisera (Theil et al. 1986). Commercially available ELISAs are commonly used to identify RVA in mammalian and avian feces. However, no commercial ELISAs are available to detect RVD, RVF, and RVG.

Only avian RVA viruses can be isolated in cell culture (Devitt and Reynolds 1993). These viruses can be isolated in primary chicken, turkey, and chicken embryonic liver cells. The MA104 cell line (Fig. 9.2), derived from the rhesus monkey kidney, is commonly used for avian RVA. Trypsin activation of rotaviruses is essential for isolation on cell culture. Other rotavirus serogroups have shown to be particularly challenging to be isolated.

PCR is highly sensitive in determining the rotaviruses' genotype and can be used as an alternative to EM (Fig. 9.3) or virus antigen ELISA (Kirkwood 2010). Highly conserved primers of the NSP4 gene can be used (Pantin-Jackwood et al. 2007; Day et al. 2007) to identify ARV.

A multiplex RT-PCR test was developed and validated to identify concurrent enteric viruses, such as avian astroviruses and avian rotaviruses (Day et al. 2007). Akimkin et al. (2011) developed a one-step RT-qPCR with an internal control system for detecting turkey rotaviruses in fecal samples.

Because of the high incidence of antibodies, it is challenging to interpret the results of serologic diagnosis of rotavirus infections (McNulty et al. 1984; Minamoto et al. 1988). Additionally, antigens are unavailable due to the failure to adapt some avian serogroups to cell culture. Indirect immunofluorescence or ELISA (Myers et al. 1989) can be used for monitoring the status of specific pathogen-free flocks.

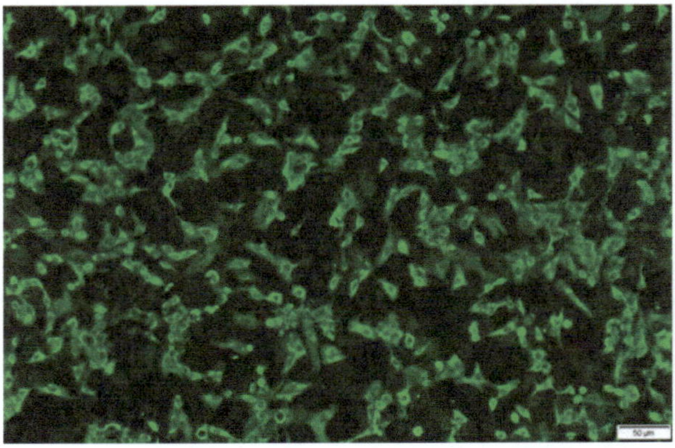

Fig. 9.2 Immunofluorescent assay of Rotavirus on MA104 cells © Awad Shehata

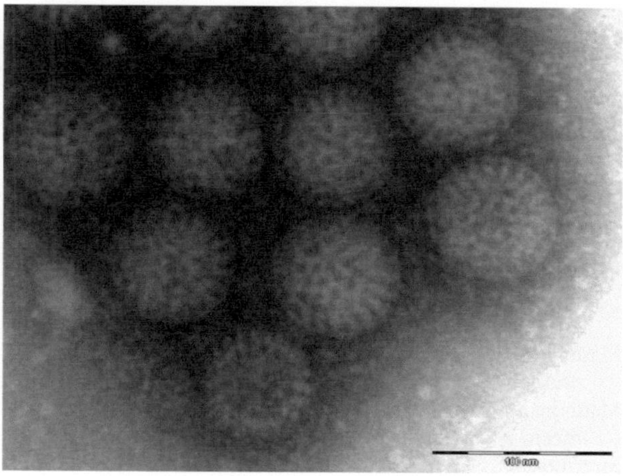

Fig. 9.3 Rotavirus particles using electron microscopy (Shehata et al. 2021)

9.6 Prevention and Control

There is no specific treatment for avian rotaviruses. Good hygienic measures, management, ventilation, and adding fresh litter are advisable. It is also recommended to avoid the reuse of litter for each cycle of birds. Currently, there are no available commercial vaccines. Trials to develop a vaccine against group A Rotavirus in turkeys indicated that inactivated vaccines administered to the breeders are unlikely to protect the progeny against challenge for more than the first week of life (Shawky et al. 1994).

References

Akimkin V, Bindel F, Hoferer M, Sting R, Polley B, Hänel A, Hafez HM (2011) One-step RT-qPCR with an internal control system for the detection of turkey rotaviruses in faecal samples. J Virol Methods 177:112–117

Bányai K, Kemenesi G, Budinski I, Földes F, Zana B, Marton S, Varga-Kugler R, Oldal M, Kurucz K, Jakab F (2017) Candidate new rotavirus species in Schreiber's bats, Serbia. Infect Genet Evol 48:19–26. https://doi.org/10.1016/j.meegid.2016.12.002

Bergeland ME, McAdaragh JP, Stotz I (1977) Rotaviral enteritis in Turkey poults. In: Presented at the 26th Western poultry disease conference, pp 129–130

BFR (2020) Rotaviruses from poultry flocks can exchange genes with rotaviruses from mammals—the risk of infection for humans is low. https://www.bfr.bund.de/cm/349/rotaviruses-from-poultry-flocks-can-exchange-genes-with-rotaviruses-from-mammals-the-risk-of-infection-for-humans-is-low.pdf

Day DM (2020) Rotavirus infections. In: Diseases of poultry. Wiley-Blackwell, Hoboken

Day JM, Spackman E, Pantin-Jackwood M (2007) A multiplex RT-PCR test for the differential identification of turkey astrovirus type 1, turkey astrovirus type 2, chicken astrovirus, avian nephritis virus, and avian rotavirus. Avian Dis 51:681–684. https://doi.org/10.1637/0005-2086(2007)51[681:AMRTFT]2.0.CO;2

Despins JL, Axtell RC, Rives DV, Guy JS, Ficken MD (1994) Transmission of enteric pathogens of turkeys by darkling beetle larva (*Alphitobius diaperinus*). J Appl Poult Res 3:61–65

Devitt CM, Reynolds DL (1993) Characterization of a group D rotavirus. Avian Dis 37:749. https://doi.org/10.2307/1592024

Estes M, Greenberg H (2013) Rotaviruses. In: Knipe DM, Howley PM (eds) Fields virology, 6th edn. Lippincott Williams & Wilkins, Philadelphia

Horrox NE (1980) Some observations and comments on rotaviruses in Turkey poults. In: Presented at the 29th Western poultry disease conference, pp 162–164

Jindal N, Patnayak DP, Ziegler AF, Lago A, Goyal SM (2009) Experimental reproduction of poult enteritis syndrome: clinical findings, growth response, and microbiology. Poult Sci 88:949–958. https://doi.org/10.3382/ps.2008-00490

Kirkwood C (2010) Rotavirus. In: Schuller M, Sloots TP, James GS, Halliday CL, Und CIWJ (eds) PCR for clinical microbiology: an Australian and international perspective. Springer, Dordrecht

Martella V, Bányai K, Matthijnssens J, Buonavoglia C, Ciarlet M (2010) Zoonotic aspects of rotaviruses. Vet Microbiol 140:246–255. https://doi.org/10.1016/j.vetmic.2009.08.028

Matthijnssens J, Van Ranst M (2012) Genotype constellation and evolution of group A rotaviruses infecting humans. Curr Opin Virol 2:426–433. https://doi.org/10.1016/j.coviro.2012.04.007

McNulty MS, Curran WL, Todd D, McFerran JB (1979) Detection of viruses in avian faeces by direct electron microscopy. Avian Pathol 8:239–247. https://doi.org/10.1080/03079457908418349

McNulty MS, Allan GM, McFerran JB (1984) Prevalence of antibody to conventional and atypical rotaviruses in chickens. Vet Rec 114:219. https://doi.org/10.1136/vr.114.9.219

Minamoto N, Oki K, Tomita M, Kinjo T, Suzuki Y (1988) Isolation and characterization of rotavirus from feral pigeon in mammalian cell cultures. Epidemiol Infect 100:481–492. https://doi.org/10.1017/S0950268800067212

Myers TJ, Schat KA, Mockett AP (1989) Development of immunoglobulin class-specific enzyme-linked immunosorbent assays for measuring antibodies against avian rotavirus. Avian Dis 33:53–59

Otto P, Liebler-Tenorio EM, Elschner M, Reetz J, Löhren U, Diller R (2006) Detection of rotaviruses and intestinal lesions in broiler chicks from flocks with runting and stunting syndrome (RSS). Avian Dis 50:411–418. https://doi.org/10.1637/7511-020106R.1

Pantin-Jackwood MJ, Spackman E, Day JM, Rives D (2007) Periodic monitoring of commercial turkeys for enteric viruses indicates continuous presence of astrovirus and rotavirus on the farms. Avian Dis 51:674–680. https://doi.org/10.1637/0005-2086(2007)51[674:PMOCTF]2.0.CO;2

Shawky SA, Saif YM, Swayne DE (1993) Role of circulating maternal anti-rotavirus IgG in protection of intestinal mucosal surface in Turkey poults. Avian Dis 37:1041–1050

Shawky SA, Saif YM, McCormick J (1994) Transfer of maternal anti-rotavirus IgG to the mucosal surfaces and bile of Turkey Poults. Avian Dis 38:409. https://doi.org/10.2307/1592060

Shehata AA, Basiouni S, Sting R, Akimkin V, Hoferer M, Hafez HM (2021) Poult enteritis and mortality syndrome in turkey poults: causes. Diagn Prev Meas Anim 11:2063. https://doi.org/10.3390/ani11072063

Spackman E, Day JM, Pantin-Jackwood MJ (2010) Astrovirus, reovirus, and rotavirus concomitant infection causes decreased weight gain in broad-breasted white poults. Avian Dis 54:16–21. https://doi.org/10.1637/8986-070909-Reg.1

Theil KW, Saif YM (1987) Age-related infections with rotavirus, rotaviruslike virus, and atypical rotavirus in turkey flocks. J Clin Microbiol 25:333–337. https://doi.org/10.1128/jcm.25.2.333-337.1987

Theil KW, Reynolds DL, Saif YM (1986) Comparison of immune electron microscopy and genome electropherotyping techniques for detection of turkey rotaviruses and rotavirus-like viruses in intestinal contents. J Clin Microbiol 23:695–699. https://doi.org/10.1128/JCM.23.4.695-699.1986

Yason CV, Summers BA, Schat KA (1987) Pathogenesis of rotavirus infection in various age groups of chickens and turkeys: pathology. Am J Vet Res 48:927–938

Turkey Coronavirus

10

Awad A. Shehata and Hafez M. Hafez

Abstract

Turkey coronavirus (TCoV), a Gammacoronavirus, causes acute contagious enteritis in turkey poults, leading to impaired growth rate, low feed conversion, increased mortality, and decreased egg production in turkey breeder hens. The TCoV infections, in association/combination with other enteropathogenic viruses, bacteria, and protozoa, are associated with poult enteritis-mortality syndrome (PEMS) in turkey poults of 1–4 weeks of age. The diagnosis of TCoV is based on the isolation of the virus, detecting the antigen or RNA in the intestines or bursa of Fabricius. Serological detection of TCoV-specific antibodies can be done using immunofluorescence procedures or enzyme-linked immunosorbent assays. Currently, no licensed vaccine is available, and the control is based on eliminating TCoV from contaminated premises by depopulation, cleaning, and/or disinfection.

Keywords

Coronaviruses · Enteropathogenic · PEMS · IFA · ELISA

A. A. Shehata (✉)
TUM School of Natural Sciences, Bavarian NMR Center (BNMRZ), Structural Membrane Biochemistry, Technical University of Munich, Garching, Germany
e-mail: awad.shehata@tum.de

H. M. Hafez
Institute of Poultry Diseases, Faculty of Veterinary Medicine, Free University of Berlin, Berlin, Germany
e-mail: hafez.mohamed@fu-berlin.de

H. M. Hafez, A. A. Shehata (eds.), *Turkey Diseases and Disorders Volume 2*,
https://doi.org/10.1007/978-3-031-63322-5_10

10.1 Etiology

TCoV is a member of the family Coronaviridae, which is divided into two *subfamilies*, namely, *Letovirinae* and *Orthocoronavirinae*. The subfamily *Letovirinae* includes the genus *Alphaletovirus*, while the subfamily *Orthocoronavirinae* has four genera based on molecular analysis, namely, *alphacoronavirus* (αCoV), *betacoronavirus* (βCoV), *gammacoronavirus* (γCoV), and *deltacoronavirus* (δCoV) (ICTV 2020). Both γCoV and δCoV infect birds, but some can also infect mammals (de Wit and Cook 2020), Fig. 10.1.

The avian coronaviruses (ACoVs), such as infectious bronchitis virus (IBV) in chickens and guinea fowl coronavirus (GfCoV), and TcoV are members of the genus γCoV and subgenus *Igacovirus* (Houta et al. 2021). The γCoV contains three subgenera, namely, *Igacovirus* and *Brangacovirus*. Both were identified in birds, and Cegacovirus was reported in a mammal (beluga whale, SW1 virus).

TCoV is an enveloped virus with a single-stranded positive-sense, non-segmented RNA of 28 kb (Jackwood et al. 2010). It consists of four structural proteins, namely, spike (S), envelope (E), membrane (M), and nucleocapsid (N), encoded by the open reading frame (ORF) at the 3′-end. The other ORF 1a/b at the 5′-end encodes 15 nonstructural proteins (Ducatez et al. 2015). ACoVs exhibit high phylogenetic relationships, genomic structures, and close nucleotide identities. The IBV, TCoV, and GfCoV showed nucleotide identities of 90% for the replicase, E, M, and N genes. However, the S gene of ACoVs shares at most 36% of identity (Miłek and Blicharz-Domańska 2018).

10.2 History and Distribution

TCoV was isolated for the first time in the United States by Peterson and Hymas (1951). Later, the virus was reported in several countries, including Australia, Brazil, Italy, the United Kingdom, France, and Poland (Villarreal et al. 2006). In 2008, TCoV was isolated from turkey poults that had enteritis (Maurel et al. 2011).

Three distinct genetic groups of TCoV isolates were identified in the United States: North Carolina isolates formed Group I, Texas isolates formed Group II, and Minnesota isolates formed Group III, suggesting the endemic circulation of distinct TCoV genotypes in different geographic states (Chen et al. 2015).

10.3 Susceptibility

Turkeys of all ages are susceptible to TCV infection. However, the clinical disease is more common in young turkeys during the first few weeks of age. Chickens, pheasants, seagulls, coturnix quail, and hamsters are refractory susceptible to infection.

Recently, it was found that younger turkeys are more susceptible to infection than older birds after infection with TCoV NC1743 isolate at a dose of 10^6 EID50/

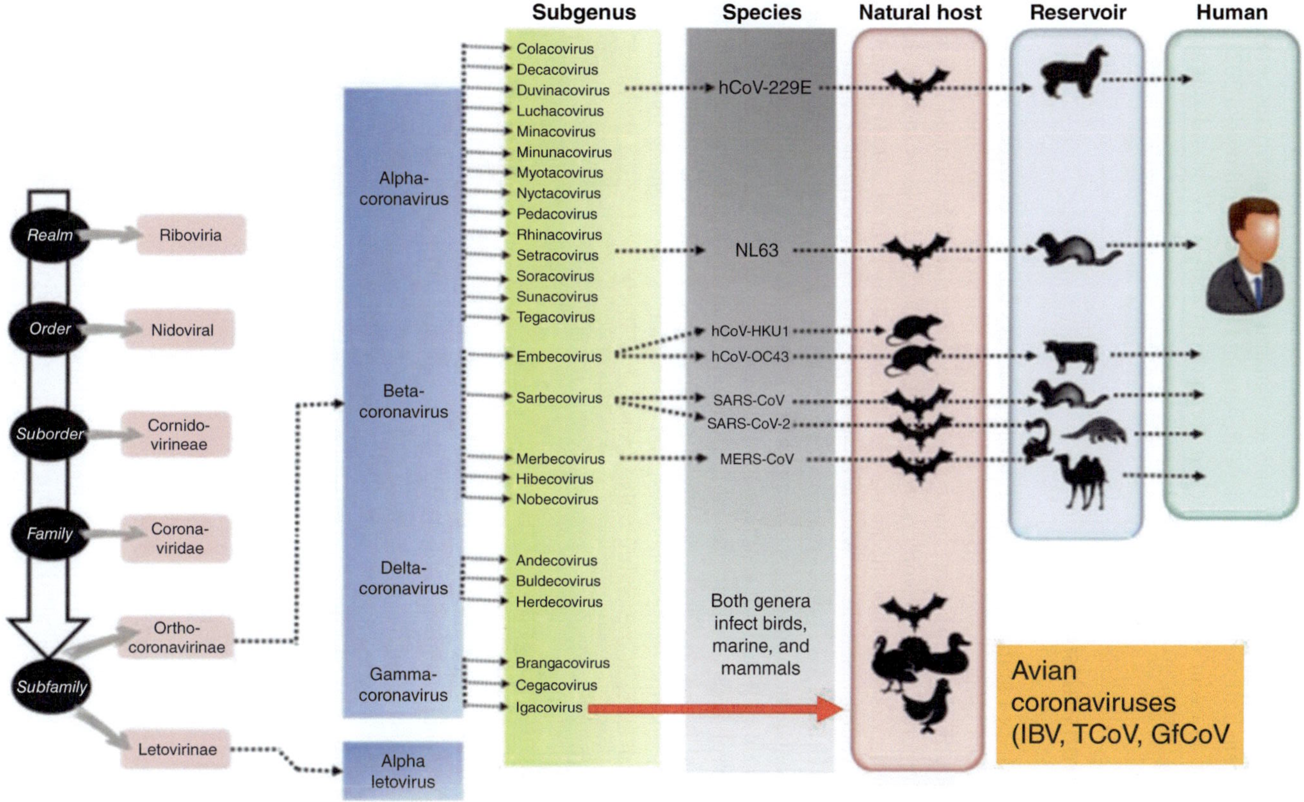

Fig. 10.1 Classification of coronaviruses. The avian coronaviruses (ACoVs), such as infectious bronchitis virus (IBV) and guinea fowl coronavirus (GfCoV), and TcoV, are members of the genus γCoV and subgenus *Igacovirus*. [*Modified after* (Shehata et al. 2022)]

bird. Therefore, 1-day-old turkeys could be used as a TCoV disease model to study the disease pathogenesis (Kang et al. 2021).

Molecular analysis revealed recombination between different γCoV genomic backbones, suggesting potential interspecies transmission of coronaviruses between other different bird species (Domanska-Blicharz and Sajewicz-Krukowska 2021). A recombination event between a chicken coronavirus and TCoV was documented (Wang et al. 2020).

10.4 Transmission

The virus is shed in the droppings of infected turkeys for several weeks, even after recovery from clinical signs (Breslin et al. 2000). Therefore, the virus transmits horizontally by ingesting feces and contaminated materials. Mechanical transmission was reported through workers, equipment, vehicles, and insects such as darkling beetle larvae and domestic house flies (Calibeo-Hayes et al. 2003) as well as wild birds, rodents, and dogs (Watson et al. 2000); there is no evidence for vertical transmission; however, poults may become infected in the hatchery via contaminated personnel and fomites such as egg boxes from infected farms. The tropism of TCoV is primarily in the intestine, namely, enterocytes in the jejunum and ileum. The virus replicates in enterocytes at the apical portion of the intestinal villi of the jejunum and ileum and bursa of Fabricius (Guy et al. 2002). The virus was also detected in dendritic cells, monocytes, and macrophages, highlighting its potential replication in antigen-presenting cells (Brown et al. 2019).

10.5 Clinical Signs

The incubation period after natural infection is about 15 days; experimentally, 2–3 days. TCoV causes high morbidity rates that may reach 100% and a sudden increase in mortality rates ranging between 10 and 50% in turkey poults during the first 4 weeks of age. It is associated with PEMS, a multifactorial syndrome (Day et al. 2010). In older ages, the infection results in stunting growth with low mortality rates (Guy 2003). The disease is more common and severe during the summer months (May–August), with sporadic occurrence in autumn. Once turkeys are infected with the virus, they remain lifelong shedders (Guy and Barnes 1996).

10.6 Postmortem Lesions

The postmortem lesions are mainly found in the gastrointestinal tract (GIT) and in the bursa of Fabricius. Pale duodenum and jejunum distended with watery gaseous contents were reported. In addition, the caeca were distended and filled with watery contents. Atrophy of the bursa of Fabricius can also be observed.

Microscopic lesions include villus atrophy, infiltration with mononuclear inflammatory cells in the *Lamina propria*, and decreased numbers of goblet cells on villous tips. Lymphoid atrophy of follicular cells of the bursa of Fabricius and heterophilic infiltration are also reported (Guy 2020). Experimentally, TCoV shedding persists for 14 weeks post-inoculation (Gomaa et al. 2009). However, poults infected with TCoV NC95 isolate shed the virus up to 7 weeks post-infection (Breslin et al. 1999). Additionally, latent infection with TCoV without clinical signs is also reported (Larsen 1998).

10.7 Diagnosis

TCoV can be isolated from samples collected from the intestine and/or bursa of Fabricius in >16-day-old embryonated turkey eggs (preferred) or in >15-day-old embryonated chicken eggs and via amniotic cavity route (Guy 2015). Electron microscopy (EM) can also be used for the diagnosis of TCoV. TcoV was determined using EM in fecal samples collected from turkey poults (Fig. 10.2). It is crucial to consider that cell membrane fragments frequently resemble TCoV particles, making it challenging to assess electron microscopic specimens. Immuno-EM is recommended for definitive identification; however, a virus-specific antiserum is needed (Shehata et al. 2021).

PCR can also be used for the detection of TcoV-RNA. The highly conserved noncoding region (3′UTR) at the 3′-end of the RNA strand is particularly suitable for the design of PCR primers (Jonassen et al. 2005). The 3′UTR sequence fragments showed high homologies between the TCoV strains and IBV (Cavanagh et al. 2001); therefore, infectious bronchitis virus (e.g., H120-strain) can also be used as a positive control for identifying TCoV based on 3′UTR-PCR (Villarreal et al. 2006; Culver et al. 2006). The N and M genes are also highly conserved and can be used

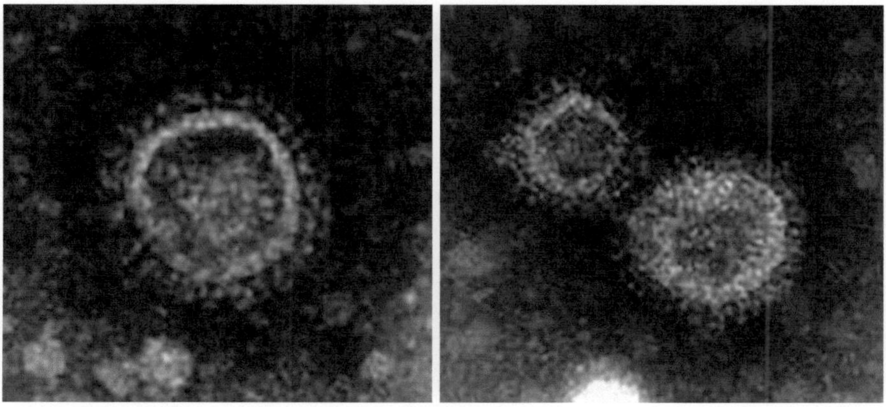

Fig. 10.2 Corona-like particles electron microscopy (EM). Negative staining was done using phosphotungstic acid. TCoV is enveloped particles, roughly spherical, with a diameter of 100–200 nm (Shehata et al. 2021)

for PCR targeting (Sellers et al. 2004; Cavanagh et al. 2001). PCR is highly sensitive and detects even minimal amounts of virus particles. It was found that TCoV can be detected in cloacal swabs just 24 h after oral infection of turkey poults (Spackman et al. 2005). The use of the 3′UTR primers and the N-gene primers had identical results. Even though no coronavirus genome was found in the Bursa Fabricii samples, the detection was successful in at least 27% of the cloacal swabs. These results highlight that feces and intestinal samples are the best samples suited for PCR detection of TCoV (Teixeira et al. 2007).

Serological tests using ELISA and/or immunofluorescent assay (IFA) can be used to detect antibodies. Commercially available ELISA plates coated with IBV antigens could successfully detect antibodies to TCoV in antibody-capture ELISA (Loa et al. 2000). The recombinant S1 spike polypeptide was also used to develop a TCoV-specific antibody ELISA (Gomaa et al. 2009). Also, Abdelwahab et al. (2015) developed a recombinant ELISA based on the N protein of TCoV expressed in a prokaryotic system for detecting the antibody of TCoV. The relative sensitivity and specificity of the developed recombinant ELISA compared with IFA were 86% and 96%, respectively (Abdelwahab et al. 2015).

10.8 Treatment and Vaccination

Antibiotic treatment can be used in case of secondary bacterial infections. On the other hand, no beneficial effect was observed when glucose, electrolytes, or calf milk replacer was added to drinking water (Dea et al. 1989). Generally, there is no specific treatment for TCoV, and there is no licensed vaccine available. Several attempts for the development of vaccines have been reviewed in detail (Houta et al. 2021). These trials include the development of live attenuated, recombinant, DNA, and nucleocapsid-based DNA vaccines. However, efforts to develop effective TCoV vaccines failed to induce early and protective humoral and cellular immune responses. Further investigations on dosages, frequency, and adjuvants are still required to improve the vaccine efficacy against TCoVs in turkeys.

References

Abdelwahab M, Loa CC, Wu CC, Lin TL (2015) Recombinant nucleocapsid protein-based enzyme-linked immunosorbent assay for detection of antibody to turkey coronavirus. J Virol Methods 217:36–41. https://doi.org/10.1016/j.jviromet.2015.02.024

Breslin JJ, Smith LG, Fuller FJ, Guy JS (1999) Sequence analysis of the turkey coronavirus nucleocapsid protein gene and 3′ untranslated region identifies the virus as a close relative of infectious bronchitis virus. Virus Res 65(2):187–193. https://doi.org/10.1016/S0168-1702(99)00117-3

Breslin JJ, Smith LG, Barnes HJ, Guy JS (2000) Comparison of virus isolation, immunohistochemistry, and reverse transcriptase-polymerase chain reaction procedures for detection of turkey coronavirus. Avian Dis 44(3):624. https://doi.org/10.2307/1593102

Brown PA, Courtillon C, Weerts EAWS, Andraud M, Allée C, Vendembeuche A et al (2019) Transmission kinetics and histopathology induced by European turkey coronavirus during experimental infection of specific pathogen free turkeys. Transbound Emerg Dis 66(1):234–242. https://doi.org/10.1111/tbed.13006

Calibeo-Hayes D, Denning SS, Stringham SM, Guy JS, Smith LG, Watson DW (2003) Mechanical transmission of turkey coronavirus by domestic houseflies (*Musca domestica* Linnaeaus). Avian Dis 47(1):149–153. https://doi.org/10.1637/0005-2086(2003)047[0149:MTOTCB]2.0.CO;2

Cavanagh D, Mawditt K, Sharma M, Drury SE, Ainsworth HL, Britton P et al (2001) Detection of a coronavirus from turkey poults in Europe genetically related to infectious bronchitis virus of chickens. Avian Pathol 30(4):355–368. https://doi.org/10.1080/03079450120066368

Chen Y-N, Loa CC, Ababneh MM-K, Wu CC, Lin TL (2015) Genotyping of turkey coronavirus field isolates from various geographic locations in the Unites States based on the spike gene. Arch Virol 160(11):2719–2726. https://doi.org/10.1007/s00705-015-2556-2

Culver F, Dziva F, Cavanagh D, Stevens MP (2006) Poult enteritis and mortality syndrome in turkeys in Great Britain. Vet Rec 159(7):209–210. https://doi.org/10.1136/vr.159.7.209

Day JM, Ballard LL, Duke MV, Scheffler BE, Zsak L (2010) Metagenomic analysis of the turkey gut RNA virus community. Virol J 7:313. https://doi.org/10.1186/1743-422X-7-313

de Wit JJ, Cook JKA (2020) Spotlight on avian coronaviruses. Avian Pathol 49(4):313–316. https://doi.org/10.1080/03079457.2020.1761010

Dea S, Garzon S, Tijssen P (1989) Isolation and trypsin-enhanced propagation of turkey enteric (bluecomb) coronaviruses in a continuous human rectal adenocarcinoma cell line. Am J Vet Res 50(8):1310–1318

Domanska-Blicharz K, Sajewicz-Krukowska J (2021) Recombinant turkey coronavirus: are some S gene structures of gammacoronaviruses especially prone to exchange? Poultry Sci 100(4):101018. https://doi.org/10.1016/j.psj.2021.101018

Ducatez MF, Liais E, Croville G, Guérin J-L (2015) Full genome sequence of Guinea fowl coronavirus associated with fulminating disease. Virus Genes 50(3):514–517. https://doi.org/10.1007/s11262-015-1183-z

Gomaa MH, Yoo D, Ojkic D, Barta JR (2009) Use of recombinant S1 spike polypeptide to develop a TCoV-specific antibody ELISA. Vet Microbiol 138(3–4):281–288. https://doi.org/10.1016/j.vetmic.2009.04.010

Guy JS (2003) Turkey coronavirus. In Diseases of poultry. Ed Saif, Y. M., Barnes, H. J., Glisson, J. R., Fadly, A. M., McDougald, L. R. And D. E. Swayne. Iowa State University Press, Ames, pp. 300–307

Guy JS (2015) Isolation and propagation of coronaviruses in embryonated eggs. In: Maier HJ, Bickerton E, Britton P (eds) Coronaviruses. Springer, New York, pp 63–71

Guy J (2020) Turkey Coronavirus enteritis. In: Swayne DE, Boulianne M, McDougald LR, Nair V, Suarez DL (eds) , vol 4. Wiley-Blackwell, Hoboken, pp 402–408

Guy J, Barnes H (1996) Poults enteritis and mortality syndrome ("spiking mortality"): an acute, transmissible diseases unknown etiology. 31–34

Guy JS, Smith LG, Breslin JJ, Pakpinyo S (2002) Development of a competitive enzyme-linked immunosorbent assay for detection of Turkey coronavirus antibodies. Avian Dis 46(2):334–341. https://doi.org/10.1637/0005-2086(2002)046[0334:DOACEL]2.0.CO;2

Houta MH, Awe OO, Ali A (2021) Infection with avian coronaviruses: a recurring problem in turkeys. Ger J Vet Res 1(3):19–27. https://doi.org/10.51585/gjvr.2021.3.0016

ICTV (2020) International committee on taxonomy of viruses. Master Species List 2019 V1

Jackwood MW, Boynton TO, Hilt DA, McKinley ET, Kissinger JC, Paterson AH et al (2010) Emergence of a group 3 coronavirus through recombination. Virology 398(1):98–108. https://doi.org/10.1016/j.virol.2009.11.044

Jonassen CM, Kofstad T, Larsen I-L, Løvland A, Handeland K, Follestad A et al (2005) Molecular identification and characterization of novel coronaviruses infecting graylag geese (*Anser anser*), feral pigeons (*Columbia livia*) and mallards (*Anas platyrhynchos*). J Gen Virol 86(Pt 6):1597–1607. https://doi.org/10.1099/vir.0.80927-0

Kang K-I, Day JM, Eldemery F, Yu Q (2021) Pathogenic evaluation of a turkey coronavirus isolate (TCoV NC1743) in turkey poults for establishing a TCoV disease model. Vet Microbiol 259:109155. https://doi.org/10.1016/j.vetmic.2021.109155

Larsen C (1998) Turkey coronavirus (TCV): cleanup and prevention of TCV enteritis. In: Clark SR (ed) Proceedings and technical supplement of Roche animal nutrition and health-turkey coronavirus workshop, pp 181–183

Loa CC, Lin TL, Wu CC, Bryan TA, Thacker HL, Hooper T, Schrader D (2000) Detection of antibody to turkey coronavirus by antibody-capture enzyme-linked immunosorbent assay utilizing infectious bronchitis virus antigen. Avian Dis 44(3):498–506

Maurel S, Toquin D, Briand FX, Quéguiner M, Allée C, Bertin J et al (2011) First full-length sequences of the S gene of European isolates reveal further diversity among turkey coronaviruses. Avian Pathol 40(2):179–189. https://doi.org/10.1080/03079457.2011.551936

Miłek J, Blicharz-Domańska K (2018) Coronaviruses in avian species—review with focus on epidemiology and diagnosis in wild birds. J Vet Res 62(3):249–255. https://doi.org/10.2478/jvetres-2018-0035

Peterson EH, Hymas TA (1951) Antibiotics in the treatment of an unfamiliar turkey disease. Poult Sci 30(3):466–468. https://doi.org/10.3382/ps.0300466

Sellers HS, Koci MD, Linnemann E, Kelley LA, Schultz-Cherry S (2004) Development of a multiplex reverse transcription–polymerase chain reaction diagnostic test specific for turkey astrovirus and coronavirus. Avian Dis 48(3):531–538. https://doi.org/10.1637/7128

Shehata AA, Basiouni S, Sting R, Akimkin V, Hoferer M, Hafez HM (2021) Poult enteritis and mortality syndrome in turkey poults: causes, diagnosis and preventive measures. Animals 11(7):2063. https://doi.org/10.3390/ani11072063

Shehata AA, Attia YA, Rahman MT, Basiouni S, El-Seedi HR, Azhar EI et al (2022) Diversity of coronaviruses with particular attention to the interspecies transmission of SARS-CoV-2. Animals 12(3):378. https://doi.org/10.3390/ani12030378

Spackman E, Kapczynski D, Sellers H (2005) Multiplex real-time reverse transcription–polymerase chain reaction for the detection of three viruses associated with poult enteritis complex: turkey astrovirus, turkey coronavirus, and turkey reovirus. Avian Dis 49(1):86–91. https://doi.org/10.1637/7265-082304R

Teixeira MCB, Luvizotto MCR, Ferrari HF, Mendes AR, da Silva SEL, Cardoso TC (2007) Detection of turkey coronavirus in commercial turkey poults in Brazil. Avian Pathol 36(1):29–33. https://doi.org/10.1080/03079450601102939

Villarreal LYB, Assayag MS, Brandão PE, Chacón JLV, Bunger AND, Astolfi-Ferreira CS et al (2006) Identification of turkey Astrovirus and turkey coronavirus in an outbreak of poult enteritis and mortality syndrome. Bra J Poult Sci 8(2):131–135

Wang Y, Cui X, Chen X, Yang S, Ling Y, Song Q et al (2020) A recombinant infectious bronchitis virus from a chicken with a spike gene closely related to that of a turkey coronavirus. Arch Virol 165(3):703–707. https://doi.org/10.1007/s00705-019-04488-3

Watson DW, Guy JS, Stringham SM (2000) Limited transmission of turkey coronavirus in young turkeys by adult *Alphitobius diaperinus* (Coleoptera: *Tenebrionidae*). J Med Entomol 37(3):480–483. https://doi.org/10.1093/jmedent/37.3.480

Turkey Viral Hepatitis

11

Awad A. Shehata and Hafez M. Hafez

Abstract

Turkey viral hepatitis (TVH) is an acute to subacute highly contagious viral disease of young turkeys younger 6 weeks old, characterized by hepatitis and pancreatitis. The turkey hepatitis virus of the family *Picornaviridae* causes the disease. The main clinical signs are stunted growth, high morbidity rate (100%), and mortality rates of 25%. In breeders, the disease causes a drop in egg production. On necropsy, TVH is characterized by pale white necrotic foci in the liver, hepatomegaly, necrosis in the pancreas, and enteritis. Diagnosis is based on the microscopic lesions in the liver and pancreas, including multifocal necrosis and mononuclear inflammatory cell infiltrations. Definitive diagnosis can be done using RT-PCR and virus isolation in embryonated chicken eggs. To date, there is no treatment or vaccination commercially available.

Keywords

TVH · *Picornaviridae* · Necrotic foci · Multifocal necrosis · Multifocal necrosis

A. A. Shehata (✉)
TUM School of Natural Sciences, Bavarian NMR Center (BNMRZ), Structural Membrane Biochemistry, Technical University of Munich, Garching, Germany
e-mail: awad.shehata@tum.de

H. M. Hafez
Institute of Poultry Diseases, Faculty of Veterinary Medicine, Free University of Berlin, Berlin, Germany
e-mail: hafez.mohamed@fu-berlin.de

H. M. Hafez, A. A. Shehata (eds.), *Turkey Diseases and Disorders Volume 2*,
https://doi.org/10.1007/978-3-031-63322-5_11

11.1 Etiology

TVH is caused by the turkey hepatitis virus (THV), belonging to the family *Picornaviridae* (Honkavuori et al. 2011). The virus is a non-enveloped capsid that encloses an ssRNA genome of more than 9040 nucleotides and 2813 amino acids. It is thermostable and resistant to ether, phenol, and creolin but not formalin. THV survives 6 h at 60 °C, 14 h at 56 °C, and 4 weeks at 37 °C in yolk. It survived for 1 h at pH 2 but not at pH 12 (Tzianabos and Snoeyenbos 1965; Guy 2018).

11.2 Susceptibility

Turkeys are the only natural host, particularly in young turkeys under 6 weeks old. It was found that chickens, pheasants, ducks, quails, mice, and rabbits are refractory to infection (Tzianabos and Snoeyenbos 1965).

11.3 Transmission

THV is suggested to be transmitted by direct and indirect contact with infected birds. Vertical transmission has also been suggested based on field observations and by virus isolation from an ovarian follicle of experimentally infected turkey hens (Snoeyenbos and Basch 1960).

11.4 Incubation Period

The incubation period ranged from 2 to 7 days in intraperitoneally inoculated and in-contact poults.

11.5 Clinical Signs

The disease has several forms:

1. **Subclinical form.** Infected birds exhibited no clinical signs. However, due to stress, the disease may become apparent. Infected breeders may exhibit a drop in egg production, decreased fertility, and hatchability. Mortality in turkeys older than 6 weeks of age has not been reported.
2. **Subacute form.** This form is characterized by the sudden death of apparently normal birds, common in young poults under 6 weeks.
3. **Acute form.** This form is characterized by variable degrees of depression, common in young poults under 6 weeks. The morbidity and mortality rates may reach up to 100% and 25%, respectively (Guy 2018).

11.6 Postmortem Lesions

The main postmortem lesions are hepatomegaly, with few to multiple necrotic foci in the liver and pancreas. Occasionally, mottled pale or gray splenomegaly could be observed. Distended enteritis with the presence of watery contents has been reported.

11.7 Microscopic Examination

The main microscopic lesions are the presence of pale foci representing multifocal acute coagulative necrosis with dense infiltration by mononuclear leukocytes and lymphocytes. The proliferation of reticuloendothelial cells with giant cells could be observed late in the course of infection. Pancreas may exhibit lesions similar to those observed in livers. Acinar cell degeneration and necrosis are observed with infiltration of macrophages and lymphocytes. Enteritis is characterized by increased cellularity of the *Lamina propria* (Guy 2018).

11.8 Diagnosis

The diagnosis is based on histopathology, virus isolation, and RT-PCR on feces, cloacal swabs, and liver and pancreas tissues (Honkavuori et al. 2011). THV can be propagated in embryonated chicken eggs (5–7 days old) and turkey eggs (10 days old) via the yolk sac. However, maternal antibodies in turkey eggs might negatively impact virus replication. Transmission electron microscopy of the liver and/or pancreas can be used to demonstrate the virus particles.

11.9 Treatment and Control

To date, there is no known treatment or vaccine against turkey viral hepatitis. Proper sanitation and biosecurity are recommended to prevent virus spread.

References

Guy JS (2018) Turkey viral hepatitis. In: Swayne DE et al (eds) In "disease of poultry", 14th edn. Blackwell Publishing, Iowa, pp 516–520

Honkavuori KS, Shivaprasad HL, Briese T, Street C, Hirschberg DL, Hutchison SK, Lipkin WI (2011) Novel picornavirus in turkey poults with hepatitis, California, USA. Emerg Infect Dis 17(3):480–487. https://doi.org/10.3201/eid1703.101410

Snoeyenbos GH, Basch HI (1960) Further studies of virus hepatitis of turkeys. Avian Dis 4(4):477. https://doi.org/10.2307/1587700

Tzianabos T, Snoeyenbos GH (1965) Some physicochemical properties of turkey hepatitis virus. Avian Dis 9(1):152–156

Marek's Disease 12

John Dunn

Abstract

Marek's disease (MD) is a contagious and oncogenic herpesvirus that causes immunosuppression and tumors in chickens. While there have been several reports of lymphomas (MD-like conditions) in turkeys, this species has received less attention in relation to the disease. Recent studies have shown that Marek's disease virus (MDV) has been found in lymphomatous tumors in commercial turkeys in several countries. The pathogenesis of MDV infection in turkeys is still not fully understood, and further research is needed. Although the herpesvirus of turkey (HVT) vaccine did not protect turkeys from challenges with a virulent MDV, the Rispens strain proved effective, emphasizing the need for further evaluation of MDV vaccines. This review aims to describe the history and current situation of MD in turkeys, including its clinical presentation and diagnostic methods in turkeys.

Keywords
Oncogenic viruses · *Herpesviridae* · Herpesvirus of turkey · Rispens strain

12.1 Introduction

While common in chickens, Marek's disease (MD) in turkeys is a relatively infrequent diagnosis but was recently reviewed (Shehata et al. 2021). Visceral T-cell lymphomas and peripheral nerve enlargement, leading to paralysis, are hallmarks of MD caused by Marek's disease virus (MDV). There are several early reports of

J. Dunn (✉)
Southeast Poultry Research Laboratory, US National Poultry Research Center, Agricultural Research Service, United States Department of Agriculture, Athens, GA, USA
e-mail: John.dunn@usda.gov

H. M. Hafez, A. A. Shehata (eds.), *Turkey Diseases and Disorders Volume 2*,
https://doi.org/10.1007/978-3-031-63322-5_12

possible MD in turkeys, such as neurolymphomatosis in a turkey in the UK (Andrewes and Glover 1939) and MD-like lesions from wild turkeys in Florida (USA) (Busch and Lovett 1970) and table turkeys in the Netherlands (Voute and Wagenaar-Schaafsma 1974), but neither virus isolation nor serology was attempted in most early reports. More recent cases have included Poland (Gawel et al. 1996), Israel (Davidson et al. 2002; Malkinson et al. 1996), France (Coudert et al. 1997), Germany (Voelckel et al. 1999), the United Kingdom (Demkin 2012; Deuchande et al. 2012; Pennycott and Venugopal 2002), the United States (Hauck et al. 2020), Italy (Mescolini et al. 2020), and Turkey (Ongor et al. 2022). While some severe cases have been reported in commercial turkeys, most of these recent cases are associated with small noncommercial flocks, and many times, the affected turkeys were raised in close proximity to chickens.

12.2 Etiology

Marek's disease virus (MDV) is a cell-associated herpesvirus in the subfamily *Alphaherpesvirinae*, genus Mardivirus (King et al. 2011). The three MDV serotypes are now designated as three species, which are *Gallid alphaherpesvirus* 2 (serotype 1) (Churchill and Biggs 1967; Nazerian et al. 1968; Solomon et al. 1968), *Gallid alphaherpesvirus* 3 (serotype 2) (Biggs and Milne 1972; Cho and Kenzy 1972; Schat and Calnek 1978), and *Meleagrid alphaherpesvirus* 1 (turkey herpesvirus (HVT) (serotype 3) (Kawamura et al. 1969; Witter et al. 1970). As the species names imply, serotypes 1 and 2 are of chicken origin, and serotype 3 is of turkey origin. *Gallid alphaherpesvirus* 2 is the only species capable of causing disease and will be synonymous with the designation MDV for the remainder of the chapter. It is particularly interesting to note that MDV (originating from chickens) is capable of causing MD in both chickens and turkeys, whereas HVT (originating from turkeys) is non-oncogenic in both chickens and turkeys (Witter 1972). MDV is further divided into pathotypes based on virulence in vaccinated chickens, designated as mild (m), virulent (v), very virulent (vv), and very virulent plus (vv+) (Witter 1997; Witter et al. 2005).

MDV is a linear double-stranded DNA virus, approximately 180 kb in length (Lee et al. 1971; Ross 1985). The virus is highly cell-associated, and while lymphotropic properties are more similar to a gammaherpesvirus, its genomic similarity to other alphaherpesviruses resulted in its designation within the subfamily Alphaherpesvirinae (Buckmaster et al. 1988; Lee et al. 2000; Tulman et al. 2000). Structurally, the virus is composed of a uniquely long and unique short sequence, each of which is flanked by inverted repeat sequences.

12.3 Transmission

MDV is highly contagious and readily transmitted by inhalation of infectious feather dander. Horizontal transmission between turkeys and between turkeys and chickens was established (Coudert et al. 1997). Feather follicle epithelial cells are the source of infectious viruses that slough off and are shed as feather dander (Calnek et al. 1970), which can remain viable in a poultry house for months to years, depending on environmental conditions. While chickens are commonly exposed to residual viruses in the poultry house from previous flocks, this is likely uncommon for turkey flocks, given the low prevalence. In many cases of MD in turkeys, it was reported that the affected turkeys were in close proximity to chickens, which may be the most common source of infection. Examples of presumed crossover from chickens to turkeys include a research flock (Pennycott and Venugopal 2002), commercial farms currently raising both species or recently raising chickens (Blake-Dyke and Baigent 2013; Voelckel et al. 1999), or backyard flocks raising both chickens and turkeys (Deuchande et al. 2012; Hauck et al. 2020).

12.4 Clinical Signs and Gross Lesions

While turkeys are more naturally resistant to MD compared to chickens, MDV can cause similar clinical signs and gross lesions in turkeys as in chickens. An outbreak in France reported fatigue, weight loss, dehydration, diarrhea, and occasional paralysis, such as shown in Fig. 12.1 (Coudert et al. 1997). This outbreak manifested with a high level of morbidity and mortality, with mortality reaching a rate of 80–100% in some flocks between 12 and 30 weeks of age. Earlier studies in a research setting have demonstrated experimental infection of turkeys with MDV can result in mortality up to 70% (Paul et al. 1977; Witter et al. 1974). Figure 12.2 illustrates an example of lymphoma from a turkey housed at a zoo (Hauck et al. 2020). Tumors are most commonly found in the liver and spleen but may also be found in kidneys, heart, gonads, or skin (Elmubarak et al. 1981) (Fig. 12.3).

Fig. 12.1 MD in turkeys showing paralysis © Hafez M. Hafez, Berlin

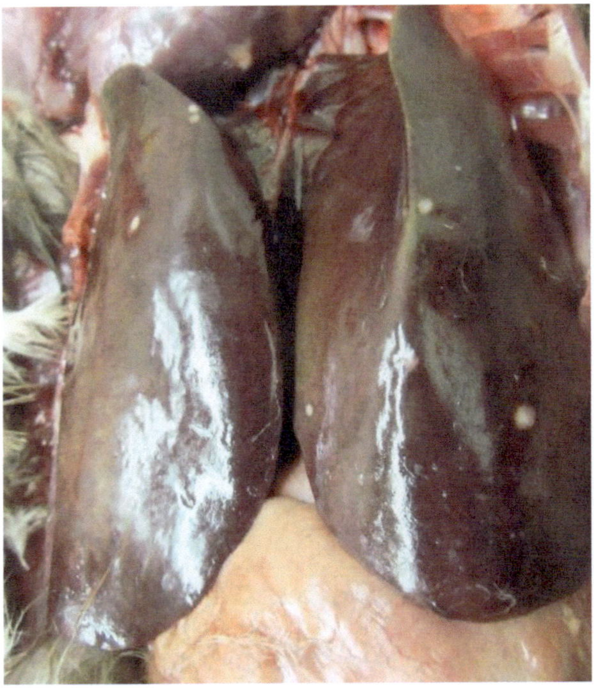

Fig. 12.2 Lymphoma in the liver of 3-year-old Bourbon Red turkey tom © AAAP

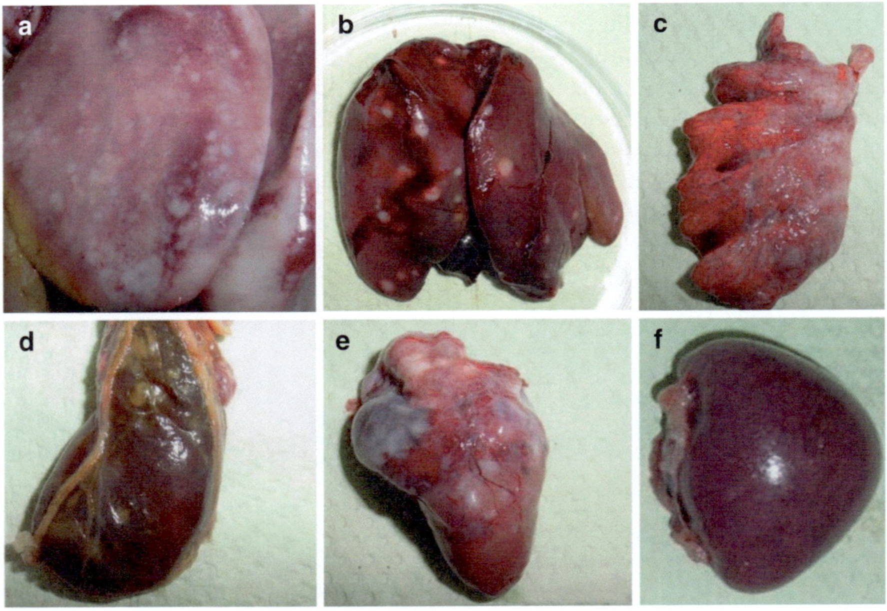

Fig. 12.3 Postmortem lesions of MD in small turkey flocks at 9 weeks of age. On necropsy, pale white nodules were found primarily in the liver (**a**, **b**), lung (**c**), gallbladder (**d**), heart (**e**), and spleen (**f**). On this farm, turkeys and broiler chickens were kept close to each other (Shehata et al. 2021)

Peripheral nerve enlargement is variable with some studies noting enlargement and others reporting rare or no enlargements.

12.5 Diagnosis

Diagnosis of MD in turkeys is in many ways similar to diagnosis in chickens, as detailed elsewhere (Witter et al. 2010). Standard diagnostic criteria include history, epidemiology, gross necropsy, and histopathology. Advanced criteria include tumor cell characterization and virological assays. In general, *diagnosis of the disease* is critical versus *diagnosis of infection* since vaccinated birds are commonly infected with MDV. However, in turkeys, vaccination is uncommon, significantly increasing the diagnostic value of detecting MDV infection. Reticuloendotheliosis virus is a much more common cause of lymphoid tumors in turkeys, so ruling out this differential diagnosis is an important early step. The leukosis/sarcoma group of viruses has never been isolated from turkeys as a natural host, although the disease can be induced experimentally with certain virus strains (Holmes 1964; Venugopal et al. 2000).

12.5.1 Diagnosis of the Disease

Evaluating the flock history and gross necropsy coupled with characteristics of the tumor cells will lead to a presumptive diagnosis. MD can occur as early as ~3 weeks of age up through maturity. Typical gross lesions include enlarged peripheral nerves and/or visceral lymphomas. Tumors and nerves should be fixed in formalin, and sections examined with hematoxylin and eosin staining, as shown in Fig. 12.4. Typical microscopic lesions include a pleomorphic population of lymphocytes, lymphoblasts, plasma cells, and macrophages (Gimeno et al. 1999, 2011a, b; Payne and Biggs 1967). Impression smears of tumors can also be used to evaluate cellular morphology.

Since MDV and REV induce lymphomas of different cell types, immunohistochemistry is of particular value for differentiating T-cell tumors (MDV) from B-cell tumors (REV). CD3 and CD4 T-cell markers are particularly useful (example of antibody labeling in Fig. 12.5), although some CD8 T cells may also be present (Schat et al. 1991). T-cell markers are commercially available, such as CT4 antibody, which labels CD4 cells from both chickens and turkeys. However, care should be taken in the selection of antibodies, as not all commercially available chicken leukocyte markers cross-react in the turkey (Meyerhoff et al. 2012). A step further in making a definitive diagnosis is by labeling viral antigens expressed in the tumor. Although methods have not been validated in turkeys, the viral antigen Meq is uniformly expressed by MD tumor cells, and pp38 is expressed in scattered tumor cells undergoing late lytic infection. Antibodies against these antigens can be used with immunohistochemistry, fluorescent antibody assays, or in situ hybridization (Ahmed et al. 2018; Naito et al. 1986; Ross et al. 1997).

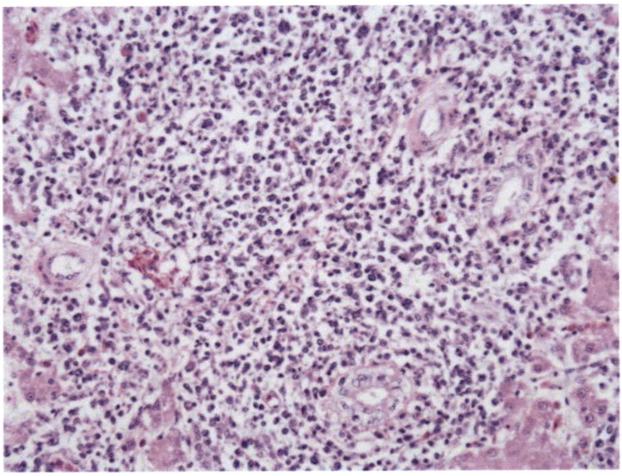

Fig. 12.4 Pleomorphic lymphocytic cell population in turkey liver. H&E © AAAP

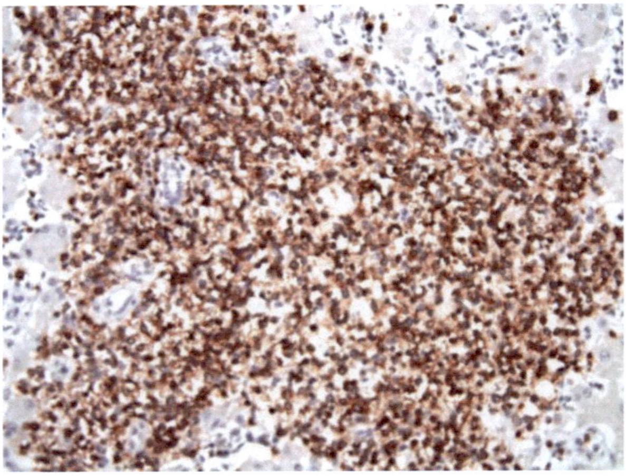

Fig. 12.5 IHC detecting CD3 in liver lymphoma from 3-year-old Bourbon Red turkey © AAAP

Quantitative PCR (qPCR) is also highly valuable in diagnosing MD. A variety of qPCR assays have been designed for measuring virus load from infected tissues of chickens (Baigent et al. 2005; Bumstead et al. 1997; Burgess and Davison 1999; Islam et al. 2004; Reddy et al. 2000). The qPCR assay has been further validated for the diagnosis of MD based on the criteria that tumorous tissue has approximately 10^2-fold higher virus load compared to latently infected tissue (Gimeno et al. 2005). Although more variable, this difference in viral load can also be detected from FFPE tissue sections (Ahmed et al. 2018). Assays have also been developed for quantitating viral load in both serotype 2 and serotype 3 MDV (Islam et al. 2006; Renz et al. 2006).

Because MD and MD vaccination are both uncommon in turkeys, principal methods of diagnosing infection have higher diagnostic value compared to chickens. These methods include virus isolation, detection of viral DNA by standard PCR assays, and detection of antibodies. All such methods would be identical methodology in both turkeys and chickens. Virus isolation is the gold standard for confirming the presence of infection and allowing further virus characterization. Techniques for isolating viruses have been described elsewhere (Nair and Dunn 2018). MDV is highly cell-associated, so the preferred sample for virus isolation is blood lymphocytes collected from heparinized whole blood, but other options include splenocytes or tumor cells. To maintain the viability of live cells, samples should be shipped overnight in cold packs, with care to avoid freezing. Chick kidney cells or duck embryo fibroblasts are the preferred substrates for primary isolation. Successful virus isolation results in the development of cell culture plaques, usually within 3–12 days. MDV plaques can be distinguished morphologically from HVT plaques (Schat 1985; Witter 1983), although confirmation should be made using either PCR or fluorescent antibody labeling with serotype-specific monoclonal antibodies (Lee et al. 1983).

PCR is an important tool for detecting the presence of viral DNA. PCR assays have been available to differentially detect MDV and HVT (Handberg et al. 2001) almost as soon as the full sequences of each MDV serotype were made available (Afonso et al. 2001; Lee et al. 2000; Spatz and Schat 2011). PCR tests for MD have since been developed using specific primer sets for a variety of different MDV genes, such as pp38 (Cao et al. 2013), gB (Gimeno et al. 2005) and Meq (Dunn et al. 2010).

Serology may also be of value in turkeys to diagnose the presence of antibodies to MDV. As most turkeys are not vaccinated against MD, the assay is not complicated by the presence of maternal antibodies in young poults. Serological tests currently in use include agar gel immunodiffusion, fluorescent antibody, enzyme-linked immunosorbent assays (Mohammadi et al. 2007; Zelnik et al. 2004), and virus neutralization (Sharma 2008). Detailed techniques for most of these assays are described in the OIE Terrestrial Manual (Nair and Dunn 2018).

12.6 Treatment and Control

There is no effective treatment for MD in turkeys or any other poultry species.

While vaccination against MD has proven to be a highly effective control method in chickens and can provide protection in turkeys, the low prevalence of MD in turkeys does not warrant routine vaccination. The first generation of MD vaccines, and most widely used in chickens, includes attenuated serotype 1 MDV (Rispens et al. 1972a, b), as well as the naturally non-oncogenic HVT (Okazaki et al. 1970), and serotype 2 viruses (Schat and Calnek 1978; Witter et al. 1987). Interestingly, while HVT was isolated and is endemic in domestic turkeys (Witter and Solomon 1971), it appears to provide no protection against MDV challenge (Elmubarak et al. 1982). An extensive review of HVT can be found in other sources (Nair et al. 2020) but is not included here, given the lack of protection provided to turkeys. CVI988/Rispens vaccine, on the other hand, appears to provide protection (Coudert et al. 1997), although another study failed to demonstrate protection (Davidson et al. 2002).

Because MDV is not endemic in commercial turkeys and vaccines are not commonly employed, biosecurity is currently the most significant form of control. Based on previous reports described above, one of the most important control methods is to avoid keeping turkeys in close proximity to chickens. In addition, all-in, all-out management allows for extensive cleaning and disinfection and downtime between flocks and also reduces mixed ages in close proximity.

References

Afonso CL, Tulman ER, Lu Z, Zsak L, Rock DL, Kutish GF (2001) The genome of turkey herpesvirus. J Virol 75:971–978

Ahmed H, Mays J, Kiupel M, Dunn JR (2018) Development of reliable techniques for the differential diagnosis of avian tumour viruses by immunohistochemistry and polymerase chain reaction from formalin-fixed paraffin-embedded tissue sections. Avian Pathol 47:364–374. https://doi.org/10.1080/03079457.2018.1451620

Andrewes CH, Glover RE (1939) A case of neurolymphomatosis in a turkey. Vet Rec 51:934–935

Baigent SJ, Petherbridge LJ, Howes K, Smith LP, Currie RJ, Nair VK (2005) Absolute quantitation of Marek's disease virus genome copy number in chicken feather and lymphocyte samples using real-time PCR. J Virol Methods 123:53–64

Biggs PM, Milne BS (1972) Biological properties of a number of Marek's disease virus isolates. In: Biggs PM, de The G, Payne LN (eds) Oncogenesis and herpesviruses. IARC, Lyon, pp 88–94

Blake-Dyke C, Baigent S (2013) Marek's disease in commercial turkey flocks. Vet Rec 173:376. https://doi.org/10.1136/vr.f6229

Buckmaster AE, Scott SD, Sanderson MJ, Boursnell MEG, Ross LJN, Binns MM (1988) Gene sequence and mapping data from Marek's disease virus and herpesvirus of turkeys implications for herpesvirus classification. J Gen Virol 69:2033–2042

Bumstead N, Sillibourne J, Rennie M, Ross N, Davison F (1997) Quantification of Marek's disease virus in chicken lymphocytes using the polymerase chain reaction with fluorescence detection. J Virol Methods 65:75–81

Burgess SC, Davison TF (1999) A quantitative duplex PCR technique for measuring amounts of cell-associated Marek's disease virus: differences in two populations of lymphoma cells. J Virol Methods 82:27–37

Busch RH, Lovett EW (1970) Case report—a Marek's disease-like condition in Florida turkeys. Avian Dis 14:550–554

Calnek BW, Adldinger HK, Kahn DE (1970) Feather follicle epithelium: a source of enveloped and infectious cell-free herpesvirus from Marek's disease. Avian Dis 14:219–233

Cao W, Mays J, Dunn J, Fulton R, Silva R, Fadly A (2013) Use of polymerase chain reaction in detection of Marek's disease and reticuloendotheliosis viruses in formalin-fixed, paraffin-embedded tumorous tissues. Avian Dis 57:785–789. https://doi.org/10.1637/10542-032713-ResNote.1

Cho BR, Kenzy SG (1972) Isolation and characterization of an isolate (HN) of Marek's disease virus with low pathogenicity. Appl Microbiol 24:299–306

Churchill AE, Biggs PM (1967) Agent of Marek's disease in tissue culture. Nature 215:528–530

Coudert F, Vuillaume A, Wyers M, Chauss AM (1997) Marek's disease in turkeys. World Poultry August:28–29

Davidson I, Malkinson M, Weisman Y (2002) Marek's disease in turkeys. I. A 7-year survey of commercial flocks and experimental infection using two field isolates. Avian Dis 46:314–321

Demkin GP (2012) Marek's disease in turkeys. Veterinariya (Moscow) 48–49

Deuchande R, Murphy A, Otter A, Baigent S, Wood A, Irvine RM (2012) Marek's disease in turkeys. Vet Rec 171:602. https://doi.org/10.1136/vr.e8274

Dunn JR, Southard T, Cooper C, Kiupel M, Witter RL (2010) Diagnosis of Marek's disease from a Japanese quail (*Coturnix japonica*) using paraffin-embedded liver [abstract]. In American Association of Avian Pathologists Symposium and Scientific Program, Atlanta

Elmubarak AK, Sharma BD, Witter RL, Nazerian K, Sanger VL (1981) Induction of lymphomas and tumor antigen by Marek's disease virus in turkeys. Avian Dis 25:911–926
Elmubarak AK, Sharma JM, Witter RL, Sanger VL (1982) Marek's disease in turkeys: lack of protection by vaccination. Am J Vet Res 43:740–742
Gawel A, Samorek-Salamonowicz E, Mazurkiewicz M, Kozdrun W (1996) Marek's disease in slaughter turkeys. Nowa Weterynaria 1:54–56
Gimeno IM, Witter RL, Reed WM (1999) Four distinct neurologic syndromes in Marek's disease: effect of viral strain and pathotype. Avian Dis 43:721–737
Gimeno IM, Witter RL, Fadly AM, Silva RF (2005) Novel criteria for the diagnosis of Marek's disease virus-induced lymphomas. Avian Pathol 34:332–340
Gimeno IM, Cortes AL, Montiel ER, Lemiere S, Pandiri AKR (2011a) Effect of diluting Marek's disease vaccines on the outcomes of Marek's disease virus infection when challenged with highly virulent Marek's disease viruses. Avian Dis 55:263–272
Gimeno IM, Witter RL, Cortes AL, Reed WM (2011b) Replication ability of three highly protective Marek's disease vaccines: implications in lymphoid organ atrophy and protection. Avian Pathol 40:573–579
Handberg KJ, Nielsen OL, Jorgensen PH (2001) The use of serotype 1- and serotype 3-specific polymerase chain reaction for the detection of Marek's disease virus in chickens. Avian Pathol 30:243–249
Hauck R, Mays J, Dunn JR, Shivaprasad HL (2020) Two cases of Marek's disease in backyard turkeys. Avian Dis 64:347–351. https://doi.org/10.1637/aviandiseases-D-19-00177
Holmes J (1964) Avian osteopetrosis. Natl Cancer Inst Monogr 63–79
Islam A, Harrison B, Cheetham BF, Mahony TJ, Young P, Walkden-Brown SW (2004) Differential amplification and quantitation of Marek's disease viruses using real-time polymerase chain reaction. J Virol Methods 119:103–113
Islam A, Cheetham BF, Mahony TJ, Young PL, Walkden-Brown SW (2006) Absolute quantitation of Marek's disease virus and herpesvirus of turkeys in chicken lymphocyte, feather tip and dust samples using real-time PCR. J Virol Methods 132:127–134
Kawamura H, King DJ, Anderson DP (1969) A herpesvirus isolated from kidney cell culture of normal turkeys. Avian Dis 13:853–886
King AMQ, Adams MJ, Carstens EB, Lefkowitz EJ (2011) Ninth report of the international committee on taxonomy of viruses. Elsevier Academic Press, San Diego
Lee LF, Kieff ED, Bachenheimer SL, Roizman B, Spear PG, Burmester BR et al (1971) Size and composition of Marek's disease virus deoxyribonucleic acid. J Virol 7:289–294
Lee LF, Liu X, Witter RL (1983) Monoclonal antibodies with specificity for three different serotypes of Marek's disease viruses in chickens. J Immunol 130:1003–1006
Lee LF, Wu P, Sui D, Ren D, Kamil J, Kung HJ et al (2000) The complete unique long sequence and the overall genomic organization of the GA strain of Marek's disease virus. Proc Natl Acad Sci 97:6091–6096
Malkinson M, Davidson I, Weisman Y (1996) Marek's disease in turkeys. Vet Rec 139:504–504
Mescolini G, Lupini C, Davidson I, Massi P, Tosi G, Fiorentini L et al (2020) Molecular characterization of a Marek's disease virus strain detected in tumour-bearing turkeys. Avian Pathol 49:202–207. https://doi.org/10.1080/03079457.2019.1691715
Meyerhoff RR, Ali RA, Liu K, Huang GQ, Koci MD (2012) Comprehensive analysis of commercially available mouse antichicken monoclonal antibodies for cross-reactivity with peripheral blood leukocytes from commercial turkeys. Poult Sci 91:383–392. https://doi.org/10.3382/ps.2011-01846
Mohammadi A, Spencer JL, Chan M, Ansari Lari M (2007) Antibody response of chickens to serotype 1, 2, or 3 Marek's disease vaccines based on ELISA with infected cells as antigen. Avian Dis 51:982–985
Nair V, Dunn JR (2018) Marek's disease. In: Manual of diagnostic tests and vaccines for terrestrial animals. World Organisation for Animal Health (OIE), Paris, pp 952–963
Nair V, Gimeno IM, Dunn JR, Zavala G, Williams SM, Reece RL et al (2020) Neoplastic diseases. In: Swayne DE, Boulianne M, Logue CM, McDougald LR, Nair V, Suarez DL, de Wit S,

Grimes T, Johnson D, Kromm M, Prajitno TY, Rubinoff I, Zavala G (eds) Diseases of poultry. Wiley, pp 548–715

Naito M, Nakajima K, Iwa N, Ono K, Yoshida I, Konobe T et al (1986) Demonstration of a Marek's disease virus-specific antigen in tumour lesions of chickens with Marek's disease using monoclonal antibody against a virus phosphorylated protein. Avian Pathol 15:503–510

Nazerian K, Solomon JJ, Witter RL, Burmester BR (1968) Studies on etiology of Marek's disease. II. Finding of a herpesvirus in cell culture. Proc Soc Exp Biol Med 127:177–182

Okazaki W, Purchase HG, Burmester BR (1970) Protection against Marek's disease by vaccination with a herpesvirus of turkeys. Avian Dis 14:413–429

Ongor H, Timurkaan N, Abayli H, Karabulut B, Kalender H, Tonbak S et al (2022) First report of serotype-1 Marek's disease virus (MDV-1) with oncogenic form in backyard turkeys in Turkey: a molecular analysis study. BMC Vet Res 18:30. https://doi.org/10.1186/s12917-021-03130-2

Paul PS, Sautter JH, Pomeroy BS (1977) Susceptibility of turkeys to Georgia strain of Marek's disease virus of chicken origin. Am J Vet Res 38:1653–1656

Payne LN, Biggs PM (1967) Studies on Marek's disease. II. Pathogenesis. J Natl Cancer Inst 39:281–302

Pennycott TW, Venugopal K (2002) Outbreak of Marek's disease in a flock of turkeys in Scotland. Vet Rec 150:277–279

Reddy SM, Witter RL, Gimeno I (2000) Development of a quantitative-competitive polymerase chain reaction assay for serotype 1 Marek's disease virus. Avian Dis 44:770–775

Renz KG, Islam A, Cheetham BF, Walkden-Brown SW (2006) Absolute quantification using real-time polymerase chain reaction of Marek's disease virus serotype 2 in field dust samples, feather tips and spleens. J Virol Methods 135:186–191

Rispens BH, van Vloten H, Mastenbroek N, Maas HJ, Schat KA (1972a) Control of Marek's disease in The Netherlands. I. Isolation of an avirulent Marek's disease virus (strain CVI 988) and its use in laboratory vaccination trials. Avian Dis 16:108–125

Rispens BH, van Vloten H, Mastenbroek N, Maas JL, Schat KA (1972b) Control of Marek's disease in The Netherlands. II. Field trials on vaccination with an avirulent strain (CVI 988) of Marek's disease virus. Avian Dis 16:126–138

Ross LJN (1985) Molecular biology of the virus. In: Payne LN (ed) Marek's disease. Martinus Nijhoff, Boston, pp 113–150

Ross N, O'Sullivan G, Rothwell C, Smith G, Burgess SC, Rennie M et al (1997) Marek's disease virus EcoRI-Q gene (meq) and a small RNA antisense to ICP4 are abundantly expressed in CD4+ cells and cells carrying a novel lymphoid marker, AV37, in Marek's disease lymphomas. J Gen Virol 78:2191–2198

Schat KA (1985) Characteristics of the virus. In: Payne LN (ed) Marek's disease. Martinus Nijhoff, Boston, pp 77–112

Schat KA, Calnek BW (1978) Characterization of an apparently nononcogenic Marek's disease virus. J Natl Cancer Inst 60:1075–1082

Schat KA, Chen C-LH, Calnek BW, Char D (1991) Transformation of T-lymphocyte subsets by Marek's disease herpesvirus. J Virol 65:1408–1413

Sharma JM (2008) Marek's disease. In: Dufour-Zavala L, Swayne DE, Glisson JR, Pearson JE, Reed WM, Jackwood MW, Woolcock PR (eds) A laboratory manual for the isolation and identification of avian pathogens, 5th edn. American Association of Avian Pathologists, Jacksonville, pp 99–105

Shehata AA, Luschow D, Hafez HM (2021) History and current status of Marek's disease in turkeys. Ger J Vet Res 1:1–6

Solomon JJ, Witter RL, Nazerian K, Burmester BR (1968) Studies on the etiology of Marek's disease. I. Propagation of the agent in cell culture. Proc Soc Exp Biol Med 127:173–177

Spatz SJ, Schat KA (2011) Comparative genomic sequence analysis of the Marek's disease vaccine strain SB-1. Virus Genes 42:331–338. https://doi.org/10.1007/s11262-011-0573-0

Tulman ER, Afonso CL, Lu Z, Zsak L, Rock DL, Kutish GF (2000) The genome of a very virulent Marek's disease virus. J Virol 74:7980–7988

Venugopal K, Howes K, Flannery DMJ, Payne LN (2000) Subgroup J avian leukosis virus infection in turkeys: induction of rapid onset tumours by acutely transforming virus strain 966. Avian Pathol 29:319–325. https://doi.org/10.1080/03079450050118449

Voelckel K, Bertram E, Gimeno IM, Neumann U, Kaleta EF (1999) Evidence for Marek's disease in turkeys in Germany: detection of MDV-1 using the polymerase chain reaction. Acta Virol 43:143–147

Voute RJ, Wagenaar-Schaafsma AE (1974) A condition bearing a resemblance of Marek's disease in table turkeys in The Netherlands. Tijdschrift Diergeneeskunde 99:166–169

Witter RL (1972) Turkey herpesvirus: lack of oncogenicity for turkeys. Avian Dis 16:666–670

Witter RL (1983) Characteristics of Marek's disease viruses isolated from vaccinated commercial chicken flocks: association of viral pathotype with lymphoma frequency. Avian Dis 27:113–132

Witter RL (1997) Increased virulence of Marek's disease virus field isolates. Avian Dis 41:149–163

Witter RL, Solomon JJ (1971) Epidemiology of a herpesvirus of turkeys: possible sources and spread of infection in Turkey flocks. Infect Immun 4:356–361

Witter RL, Nazerian K, Purchase HG, Burgoyne GH (1970) Isolation from turkeys of a cell-associated herpesvirus antigenically related to Marek's disease virus. Am J Vet Res 31:525–538

Witter RL, Solomon JJ, Sharma JM (1974) Response of turkeys to infection with virulent Marek's disease viruses of Turkey and chicken origins. Am J Vet Res 35:1325–1332

Witter RL, Silva RF, Lee LF (1987) New serotype 2 and attenuated serotype 1 Marek's disease vaccine viruses: selected biological and molecular characteristics. Avian Dis 31:829–840

Witter RL, Calnek BW, Buscaglia C, Gimeno IM, Schat KA (2005) Classification of Marek's disease viruses according to pathotype: philosophy and methodology. Avian Pathol 34:75–90

Witter RL, Gimeno IM, Pandiri AR, Fadly AM (2010) Tumor diagnosis manual: the differential diagnosis of lymphoid and myeloid tumors in the chicken. pp. 1–84): American Association of Avian Pathologists.

Zelnik V, Harlin O, Fehler F, Kaspers B, Gobel TW, Nair VK et al (2004) An enzyme-linked immunosorbent assay (ELISA) for detection of Marek's disease virus-specific antibodies and its application in an experimental vaccine trial. J Vet Med B Infect Dis Vet Public Health 51:61–67

Reticuloendotheliosis and Lymphoproliferative Disease

13

Retrovirus-Induced Tumors of Turkeys

John Dunn

Abstract

Retrovirus-induced tumors, specifically reticuloendotheliosis (RE) and lymphoproliferative disease (LPD), are the most common virus-induced tumors in turkeys. And the diseases are not RE and LPD in turkeys, with emphasis on clinical signs and diagnosis.

Frequently diagnosed, the viruses appear widespread and thus important to rule out. This review aims to describe the history and current situation of.

Keywords

Reticuloendotheliosis · Lymphoproliferative disease · Turkeys · Tumors · Virus-induced · Lymphoma

13.1 Introduction

Reticuloendotheliosis (RE) and lymphoproliferative disease (LPD), both caused by retroviruses, are the most common virus-induced tumors in turkeys. While the clinical disease is not frequently diagnosed, both virus types appear to be widespread in turkeys. In wild turkeys in the United States, the reticuloendotheliosis virus has been frequently found alone or part of mixed infections with lymphoproliferative disease virus or fowl poxvirus (Allison et al. 2014; Alger et al. 2017; Willis et al.

J. Dunn (✉)
Southeast Poultry Research Laboratory, US National Poultry Research Center, Agricultural Research Service, United States Department of Agriculture, Athens, GA, USA
e-mail: John.dunn@usda.gov

H. M. Hafez, A. A. Shehata (eds.), *Turkey Diseases and Disorders Volume 2*,
https://doi.org/10.1007/978-3-031-63322-5_13

2022). Marek's disease, caused by a herpesvirus, can also affect turkeys and is reviewed separately in the previous chapter.

Reticuloendotheliosis is the most common virus-induced tumor of turkeys and can cause significant economic losses due to tumor mortality and loss of production. In addition to turkeys, RE can naturally affect chickens, ducks, geese, pheasants, Japanese quail, peafowl, and prairie chickens. RE encompasses multiple strains of related viruses that can cause chronic lymphoid neoplasia, immunosuppression, runting disease syndrome, or acute reticulum cell neoplasia. The ability of REV to integrate into the host genome as well as large DNA viruses has led to recent issues with vaccine contamination (Jackson et al. 1977; Fadly et al. 1996) and is also suggested to be the mechanism of initial introduction and spread within avian species (Niewiadomska and Gifford 2013).

Lymphoproliferative disease was first described in the United Kingdom, followed soon after by reports of outbreaks in several other European countries in Israel in the 1970s. While primarily a disease of turkeys, both chickens and turkeys were able to be experimentally infected (Ianconescu et al. 1983). The disease was last reviewed in the tenth edition of the textbook *Diseases of Poultry* (Biggs 1997) but was removed from subsequent editions due to sporadic incidence. In 2009, LPDV was found in the United States in a wild turkey, and a follow-up survey suggested a widespread distribution among asymptomatic wild turkeys (Allison et al. 2014). Subsequent studies have confirmed widespread distribution in North American wild turkeys (Thomas et al. 2015; Kreh and Palamar 2022; MacDonald et al. 2022; Thiemann et al. 2022; Shea et al. 2022; Cox et al. 2022).

13.2 Reticuloendotheliosis

13.2.1 Etiology

Reticuloendotheliosis virus (REV) is a member of the family *Retroviridae*, subfamily *Orthoretrovirinae,* genus *Gammaretrovirus.* REV is a single-stranded, positive-sense RNA virus with genome of non-defective strains approximately 9.0 kb (Barbosa et al. 2007), whereas replication-defective strains are approximately 5.7 kb due to deletions in the gag-pol and env regions (Coffin 1982). The first strain of REV, known as the defective strain T (REV-T), was isolated from turkey visceral lymphomas in 1957 (Robinson and Twiehaus 1974), with the designation as Reticuloendotheliosis based on the rapid proliferation of mononuclear cells of the reticuloendothelial system (Theilen et al. 1966). While REV-T led to the naming of the disease, the REV group also includes chick syncytial virus (Cook 1969), duck infectious anemia virus (Ludford et al. 1972), and spleen necrosis virus (Trager 1959). REV strains are classified as either defective or non-defective based on their ability to replicate alone or if they require a non-defective RE helper virus (Hoelzer et al. 1979). REV-T stocks contain a non-defective helper designated REV-A that lacks acute oncogenic properties but replicates in chicken fibroblasts. REV-T contains an oncogene, *v-rel*, which is responsible for the acute oncogenicity of REV-T (Bose 1992).

13.2.2 Transmission

Horizontal and vertical transmission are both important for REV, as well as transmission via contaminated vaccines. In turkeys, both horizontal (Paul et al. 1977) and vertical transmission (McDougall et al. 1980) have been demonstrated through experimental infection. In other species, REV has been detected in feces, cloacal swabs, body fluids, and litter (Zavala and Nair 2020). Insect transmission may also be an important consideration as REV has been recovered from mosquitoes and several other species after contact with infected chickens (Motha et al. 1984). The antibody and virus status of the turkey hens appears important on the level of vertical transmission to progeny (Witter and Salter 1989). Infected tom turkeys can also vertically transmit the virus through infected semen (McDougall et al. 1980, 1981). While not commonly administered in turkeys, REV-contaminated fowl pox and Marek's disease vaccines not only have also led to unintentional REV transmission (Kawamura et al. 1976; Fadly et al. 1996; Liu et al. 2009; Wei et al. 2012) but has have been detected in Newcastle disease and infectious bronchitis vaccines (Wei et al. 2012).

13.2.3 Clinical Signs and Gross Lesions

Lymphomas in turkeys (reticular cell tumor or T-cell lymphoma) most often affect the liver, intestine, or spleen. An outbreak of RE in turkey breeders featured diarrhea between 8 and 12 weeks of age, with more than 20% mortality (McDougall et al. 1978). The affected turkeys had liver enlargement as the prominent lesion. Another outbreak in a commercial turkey flock included clinical signs of lethargy, weakness, anorexia, diarrhea, and reduced egg production, with an almost 30% mortality rate prior to termination of the flock (Okoye et al. 1993). Necropsy revealed neoplastic nodules or gray foci in the liver, intestines, and spleen. Other flocks have reported lymphomas with much lower mortality rates (Witter and Salter 1989). Disseminated T-cell lymphomas were diagnosed in a turkey breeding flock that was experiencing increased mortality and decreased egg production for more than 20 weeks (Crespo et al. 2002). A comprehensive review of the pathogenesis of RE syndromes can be found elsewhere (Zavala and Nair 2020).

13.2.4 Diagnosis

The diagnosis of RE is challenging, but because REV is not ubiquitous in turkeys, there is greater value in the detection of REV infection and REV antibodies. REV-induced lymphomas are important to differentiate from tumors caused by the lymphoproliferative disease virus, Marek's disease virus, or avian leukosis virus. Typical microscopic lesions include aggregates of large lymphoblastoid cells with pale vesicular nuclei, prominent nucleoli, moderate basophilic cytoplasm, and mitotic figures that are numerous through the lesion (McDougall 1993; Crespo et al. 2002).

Virus isolation can be carried out in cultures of chicken embryo fibroblasts, although chicken kidney cells, duck embryo fibroblasts, turkey embryo fibroblasts, or quail embryo fibroblasts can also be used (Zavala et al. 2016). Cellular inoculum is preferred from tumor or tissue suspensions, blood, plasma, splenocytes, or white blood cells. Anti-REV positive sera (Ianconescu 1977) or monoclonal antibodies (Cui et al. 1986) can be used for the identification of REV in culture, as well as an ELISA immunoassay using the same monoclonal antibodies (Cui et al. 1988), immunoperoxidase (Calvert and Nazerian 1994), or complement fixation (Smith et al. 1977). More recently, the detection of viral RNA or DNA by PCR has become common in virus-infected cell cultures as well as paraffin-embedded tissues, blood, and tumor tissues (Aly et al. 1993; Cao et al. 2013). PCR has been particularly beneficial in the differential diagnosis of tumor tissues and in evaluating vaccines for REV contamination.

Serology is also of value in the diagnosis of RE, especially for the purposes of ruling out REV. Commercial ELISA antibody kits are available and commonly used to determine specific pathogen-free (SPF) flock status or confirm breeder flock status for exporting progeny.

In differentiating REV-induced tumors from other viral-induced tumors, PCR from the tumor tissue is of high value since none of the tumor viruses are ubiquitous in turkeys, with caution that some turkeys may be infected with multiple viruses. Determining the lymphoma cell type would be valuable in excluding Marek's disease (T-cell lymphoma), although REV can induce both B- and T-cell lymphomas in chickens, so it shouldn't be assumed that REV can only cause T-cell lymphomas in naturally infected turkeys. A better strategy for diagnosing the tumor by immunohistochemistry would be labeling for virus antigens. Figure 13.1 illustrates labeling of a lymphoblastoid cell line designated as RECC-RP13 that was established from a liver tumor of a chicken inoculated with chick syncytial virus (Nazerian et al. 1982). The monoclonal antibodies 11A25 and 11B118 provide strong cytoplasmic labeling specific to REV glycoprotein 62 (Ahmed et al. 2018).

13.2.5 Treatment and Control

There is no known treatment for RE in turkeys. There are no vaccines or other specific control methods for RE in turkeys. The most common preventative measures currently in use include screening of poultry biologicals and strict biosecurity and screening in SPF flocks (Witter 2006). Insect control may also be important based on possible horizontal transmission through infected mosquitoes and other insects, as mentioned above. Fortunately, REVs degrade quickly outside the host at ambient temperatures. If the disease were to become endemic in domestic turkeys,

Fig. 13.1 IHC of RECC-RP13 cell line illustrating strong cytoplasmic labeling with REV monoclonal antibody (11A25)

minimizing vertical transmission through test and cull procedures paired with small-group rearing may be beneficial, as has been accomplished with avian leukosis virus in chickens (Nair et al. 2020).

13.3 Lymphoproliferative Disease

13.3.1 Etiology

Lymphoproliferative disease virus (LPDV) appears to be a type C retrovirus of the family *Retroviridae*, subfamily *Orthoretrovirinae,* genus *Alpharetrovirus* (Chajut et al. 1992; Allison et al. 2014). While there were a variety of early reports describing conditions that may have been caused by LPDV, Biggs et al. were the first to experimentally reproduce the lesions from a natural outbreak in the United Kingdom and narrow down the cause to either a herpesvirus or C-type particle (Biggs et al. 1974). Type C refers to the morphological structure, with C-type particles having a central electron-dense core, as is typical for most oncoviruses and endogenous viruses, forming at the surface of the cell at the site of budding. The LPDV genome is 7143-bp long, and compared to other oncoviruses, LPDV is mostly closely related to the avian leukosis/sarcoma viruses (Gak et al. 1991; Sarid et al. 1994; Allison et al. 2014). Like other oncoviruses, the genome contains three genes, *gag*, *pol*, and *env*, which code for structural proteins. There is a fourth open reading frame (ORF), which codes the protease, and four additional ORFs of unknown function. Unlike other oncogenic retroviruses, however, LPDV does not contain a putative viral oncogene (Gak et al. 1989). Similar to avian leukosis virus (ALV), LPDV integrates into the host genome, but unlike ALV, it is unclear whether oncogenesis for LPDV is triggered by insertion into a specific region (Allison et al. 2014). Virus isolation has been unsuccessful from attempts using embryo fibroblasts of chickens, turkeys, ducks, and quail, as well as kidney cells of chicks and turkeys (McDougall et al. 1978).

13.3.2 Transmission

Although the exact mechanism of transmission is uncertain, early studies on experimental transmission of LPDV have demonstrated that turkeys inoculated with spleen homogenates from a natural outbreak were able to transmit the disease through direct contact (McDougall et al. 1978). In one report, LPD outbreaks were reported in subsequent flocks and only ceased once the producer switched to a different strain of turkey (Biggs et al. 1978).

13.3.3 Clinical Signs and Gross Lesions

LPDV induces generalized clinical signs, if any, including ruffled feathers, anorexia, and reluctance to move (Biggs 1997). In many cases, there are no clinical signs detected prior to death. Death is the usual sequela to outward clinical signs, with the highest incidence between 10 and 18 weeks and mortality reaching as high as 25% in affected flocks (Biggs et al. 1978). In wild turkey surveillance, age is a strong predictor of LPDV infection, with juveniles less likely to test positive than adults (Alger et al. 2017). Gross splenomegaly is the most common lesion, which may be accompanied by hepatomegaly and enlargement of gonad or thymus. Miliary foci may be present on the liver, pancreas, thymus, kidneys, gonads, intestinal wall, lungs, or myocardium. The brachial and sciatic plexuses are enlarged in some birds.

13.3.4 Diagnosis

The diagnosis of LPD is challenging, given the inability to isolate and grow the virus in cell cultures or embryos and the inability to detect antibodies in infected chickens. Clinical signs and gross lesions are important, as well as characteristic microscopic lesions that consist of lymphoproliferation of pleomorphic cells. These lesions primarily consist of lymphocytes, lymphoblasts, reticulum cells, and plasma cells. PCR is a valuable diagnostic tool using primers specific to the *gag* gene (Sarid et al. 1994; Allison et al. 2014). From surveillance of wild turkey populations, bone marrow was found most effective for detecting LPDV proviral DNA in paired tissue testing compared to spleen and liver (Thomas et al. 2015), whereas whole blood provides high levels of detection from live birds (Alger et al. 2015).

13.3.5 Treatment and Control

There is no known treatment for LPD in turkeys. There are no vaccines or other specific control methods for LPD in turkeys. Susceptibility appears variable depending on turkey strain, which suggests that selecting for host resistance could be utilized as a control method (McDougall et al. 1978). This finding is consistent with the report noted above that switching to a different strain of turkey put an end to one outbreak that was recurring in subsequent flocks.

References

Ahmed H, Mays J, Kiupel M, Dunn JR (2018) Development of reliable techniques for the differential diagnosis of avian tumour viruses by immunohistochemistry and polymerase chain reaction from formalin-fixed paraffin-embedded tissue sections. Avian Pathol 47(4):364–374. https://doi.org/10.1080/03079457.2018.1451620

Alger K, Bunting E, Schuler K, Jagne J, Whipps CM (2015) Diagnosing lymphoproliferative disease virus in live wild turkeys (*Meleagris gallopavo*) using whole blood. J Zoo Wildl Med 46(4):806–814. https://doi.org/10.1638/2015-0037.1

Alger K, Bunting E, Schuler K, Whipps CM (2017) Risk factors for and spatial distribution of lymphoproliferative disease virus (LPDV) in wild turkeys (*Meleagris gallopavo*) in New York State, USA. J Wildl Dis 53(3):499–508. https://doi.org/10.7589/2016-06-137

Allison AB, Kevin Keel M, Philips JE, Cartoceti AN, Munk BA, Nemeth NM et al (2014) Avian oncogenesis induced by lymphoproliferative disease virus: a neglected or emerging retroviral pathogen? Virology 450:2–12. https://doi.org/10.1016/j.virol.2013.11.037

Aly MM, Smith EJ, Fadly AM (1993) Detection of reticuloendotheliosis virus infection using the polymerase chain reaction. Avian Pathol 22(3):543–554. https://doi.org/10.1080/03079459308418942

Barbosa T, Zavala G, Cheng S, Villegas P (2007) Full genome sequence and some biological properties of reticuloendotheliosis virus strain APC-566 isolated from endangered Attwater's prairie chickens. Virus Res 124(1–2):68–77. https://doi.org/10.1016/j.virusres.2006.10.002

Biggs PM (1997) Lymphoproliferative disease of turkeys. In: Calnek BW, Barnes HJ, Beard CW, McDougald LR, Saif YM (eds) Diseases of poultry, 10th edn. Iowa State University Press, Ames, pp 485–489

Biggs PM, Milne BS, Frazier JA, McDougall JS, Stuart JC (1974) Lymphoproliferative disease in turkeys. World's Poultry Congress, pp 55–55

Biggs PM, McDougall JS, Frazier JA, Milne BS (1978) Lymphoproliferative disease of turkeys 1. Clinical aspects. Avian Pathol 7(1):131–139. https://doi.org/10.1080/03079457808418265

Bose HR (1992) The Rel family: models for transcriptional regulation and oncogenic transformation. Biochim Biophys Acta (BBA) Rev Cancer 1114(1):1–17. https://doi.org/10.1016/0304-419X(92)90002-G

Calvert JG, Nazerian K (1994) An immunoperoxidase plaque assay for reticuloendotheliosis virus and its application to a sensitive serum neutralization assay. Avian Dis 38(1):165–171

Cao W, Mays J, Dunn J, Fulton R, Silva R, Fadly A (2013) Use of polymerase chain reaction in detection of Marek's disease and reticuloendotheliosis viruses in formalin-fixed. Paraffin Embedded Tumor Tissues Avian Dis 57(4):785–789. https://doi.org/10.1637/10542-032713-ResNote.1

Chajut A, Sarid R, Yaniv A, Smythers GW, Tronick SR, Gazit A (1992) The lymphoproliferative disease virus of turkeys represents a distinct class of avian type-C retrovirus. Gene 122(2):349–354. https://doi.org/10.1016/0378-1119(92)90225-e

Coffin JM (1982) Structure of the retroviral genome. RNA tumor viruses. In: Weiss R, Teich N, Varmus H, Coffin JM (eds) Molecular biology of tumor viruses. Cold Spring Harbor Laboratory, Cold Spring Harbor, pp 261–368

Cook MK (1969) Cultivation of a filterable agent associated with Marek's disease. Journal of the National Cancer Institute 43(1):203–212

Cox F, Hardin J, Dittmar R, Edwards D (2022) Molecular surveillance for lymphoproliferative disease virus and reticuloendotheliosis virus in Rio Grande wild turkeys (*Meleagris gallopavo* intermedia) in Texas, USA. J Wildl Dis 58(4):909. https://doi.org/10.7589/JWD-D-22-00023

Crespo R, Woolcock PR, Fadly AM, Hall C, Shivaprasad HL (2002) Characterization of T-cell lymphomas associated with an outbreak of reticuloendotheliosis in turkeys. Avian Pathol 31(4):355–361. https://doi.org/10.1080/03079450220141624

Cui ZZ, Lee LF, Silva RF, Witter RL (1986) Monoclonal antibodies against avian reticuloendotheliosis virus: identification of strain-specific and strain-common epitopes. J Immunol 136(11):4237–4242

Cui ZZ, Lee LF, Smith EJ, Witter RL, Chang TS (1988) Monoclonal-antibody-mediated enzyme-linked immunosorbent assay for detection of reticuloendotheliosis viruses. Avian Dis 32(1):32–40

Fadly AM, Witter RL, Smith EJ, Silva RF, Reed WM, Hoerr FJ et al (1996) An outbreak of lymphomas in commercial broiler breeder chickens vaccinated with a fowlpox vaccine contaminated with reticuloendotheliosis virus. Avian Pathol 25(1):35–47. https://doi.org/10.1080/03079459608419118

Gak E, Yaniv A, Chajut A, Ianconescu M, Tronick SR, Gazit A (1989) Molecular cloning of an oncogenic replication-competent virus that causes lymphoproliferative disease in turkeys. J Virol 63(6):2877–2880. https://doi.org/10.1128/jvi.63.6.2877-2880.1989

Gak E, Yaniv A, Sherman L, Ianconescu M, Tronick SR, Gazit A (1991) Lymphoproliferative disease virus of turkeys: sequence analysis and transcriptional activity of the long terminal repeat. Gene 99(2):157–162. https://doi.org/10.1016/0378-1119(91)90122-r

Hoelzer JD, Franklin RB, Bose HR (1979) Transformation by reticuloendotheliosis virus: development of a focus assay and isolation of a nontransforming virus. Virology 93(1):20–30. https://doi.org/10.1016/0042-6822(79)90272-1

Ianconescu M (1977) Reticuloendotheliosis antigen for the agar gel precipitation test. Avian Pathol 6(3):259–267. https://doi.org/10.1080/03079457708418234

Ianconescu M, Yaniv A, Gazit A, Perk K, Zimber A (1983) Susceptibility of domestic birds to lymphoproliferative disease virus (LPDV) of turkeys. Avian Pathol 12(3):291–302. https://doi.org/10.1080/03079458308436172

Jackson CAW, Dunn SE, Smith DI, Gilchrist PT, MacQueen PA (1977) Proventriculitis, "Nakanuke" and reticuloendotheliosis in chickens following vaccination with herpesvirus of turkeys (HVT). Aust Vet J 53:457–458

Kawamura H, Wakabayashi T, Yamaguchi S, Taniguchi T, Takayanagi N (1976) Inoculation experiment of Marek's disease vaccine contaminated with a reticuloendotheliosis virus. Natl Inst Anim Health Q (Tokyo) 16(4):135–140

Kreh CD, Palamar MB (2022) Prevalence of lymphoproliferative disease virus in wild turkeys (*Meleagris gallopavo*) in North Carolina. Wildl Soc Bull 46(2):1263. https://doi.org/10.1002/wsb.1263

Liu Q, Zhao J, Pu J, Zhang G, Liu J (2009) Full genome sequences of two reticuloendotheliosis viruses contaminating commercial vaccines. Avian Dis 53:341–346

Ludford CG, Purchase HG, Cox HW (1972) Duck infectious anemia virus associated with plasmodium lophurae. Exp Parasitol 31(1):29–38. https://doi.org/10.1016/0014-4894(72)90044-6

MacDonald AM, Johnson JB, Casalena MJ, Nemeth NM, Kunkel M, Blake M, Brown JD (2022) Active and passive disease surveillance in wild turkeys (*Meleagris gallopavo*) from 2008 to 2018 in Pennsylvania, USA. Wildl Soc Bull 46(2):1289. https://doi.org/10.1002/wsb.1289

McDougall JS (1993) Tumor viruses of turkeys. In: McFerran JB, McNulty MS (eds) Virus infections of vertebrates, 4. Virus infections of birds. Elsevier Science Publishers B.V., Amsterdam, pp 455–463

McDougall JS, Biggs PM, Shilleto RW, Milne BS (1978) Lymphoproliferative disease of turkeys II. Experimental transmission and aetiology. Avian Pathol 7(1):141–155. https://doi.org/10.1080/03079457808418266

McDougall JS, Shilleto RW, Biggs PM (1980) Experimental infection and vertical transmission of reticuloendotheliosis virus in the turkey. Avian Pathol 9(3):445–454. https://doi.org/10.1080/03079458008418428

McDougall JS, Shilleto RW, Biggs PM (1981) Further studies on vertical transmission of reticuloendotheliosis virus in turkeys. Avian Pathol 10(2):163–169. https://doi.org/10.1080/03079458108418470

Motha MX, Egerton JR, Sweeney AW (1984) Some evidence of mechanical transmission of reticuloendotheliosis virus by mosquitoes. Avian Dis 28(4):858–867

Nair V, Gimeno IM, Dunn JR, Zavala G, Williams SM, Reece RL (2020) Neoplastic diseases. In D.E. Swayne, M. Boulianne, C.M. Logue, L.R. McDougald, V. Nair, D.L. Suarez, S.D. Wit, T. Grimes, D. Johnson, M. Kromm, T.Y. Prajitno, I. Rubinoff & G. Zavala (Eds.), Diseases of poultry. Wiley, pp. 548–715

Nazerian K, Elmubarak A, Sharma JM (1982) Establishment of B-lymphoblastoid cell lines from Marek's disease virus-induced tumors in turkeys. Int J Cancer 29(1):63–68. https://doi.org/10.1002/ijc.2910290111

Niewiadomska AM, Gifford RJ (2013) The extraordinary evolutionary history of the reticuloendotheliosis viruses. PLoS Biol 11(8):e1001642. https://doi.org/10.1371/journal.pbio.1001642

Okoye JOA, Ezema W, Agoha JN (1993) Naturally occurring clinical reticuloendotheliosis in turkeys and chickens. Avian Pathol 22(2):237–244. https://doi.org/10.1080/03079459308418917

Paul PS, Sautter JH, Pomeroy BS (1977) Susceptibility of turkeys to Georgia strain of Marek's disease virus of chicken origin. Am J Vet Res 38(10):1653–1656

Robinson FR, Twiehaus MJ (1974) Historical note: isolation of the avian reticuloendothelial virus (strain T). Avian Dis 18(2):278. https://doi.org/10.2307/1589142

Sarid R, Chajut A, Gak E, Kim Y, Hixson CV, Oroszlan S, Tronick SR et al (1994) Genome organization of a biologically active molecular clone of the lymphoproliferative disease virus of turkeys. Virology 204(2):680–691. https://doi.org/10.1006/viro.1994.1584

Shea SA, Gonnerman M, Blomberg E, Sullivan K, Milligan P, Kamath PL (2022) Pathogen survey and predictors of lymphoproliferative disease virus infection in wild turkeys (*Meleagris gallopavo*). J Wildl Dis 58(3):537. https://doi.org/10.7589/JWD-D-21-00152

Smith EJ, Solomon JJ, Witter RL (1977) Complement-fixation test for reticuloendotheliosis viruses: limits of sensitivity in infected avian cells. Avian Dis 21(4):612. https://doi.org/10.2307/1589420

Theilen GH, Zeigel RF, Twiehaus MJ (1966) Biological studies with RE virus (strain T) that induces reticuloendotheliosis in turkeys, chickens, and Japanese quail. J Natl Cancer Inst 37(6):731–743

Thiemann R, Dalton MF, Rose H, Baughman B, Butler A, Adcock K et al (2022) An investigation of the cause of wild turkey mortality in Mississippi. Avian Dis 66(2):237. https://doi.org/10.1637/aviandiseases-D-22-00007

Thomas JM, Allison AB, Holmes EC, Phillips JE, Bunting EM, Yabsley MJ et al (2015) Molecular surveillance for lymphoproliferative disease virus in wild turkeys (*Meleagris gallopavo*) from the eastern United States. PLoS One 10(4):e0122644. https://doi.org/10.1371/journal.pone.0122644

Trager W (1959) A new virus of ducks interfering with development of malaria parasite (*Plasmodium lophurae*). Exp Biol Med 101(3):578–582. https://doi.org/10.3181/00379727-101-25023

Wei K, Sun Z, Zhu S, Guo W, Sheng P, Wang Z et al (2012) Probable congenital transmission of reticuloendotheliosis virus caused by vaccination with contaminated vaccines. PLoS One 7(8):e43422. https://doi.org/10.1371/journal.pone.0043422

Willis B, Trautman C, Cox F, Lujan T, Hardin J, Dittmar R et al (2022) Genome sequence of Fowlpox virus-integrated reticuloendotheliosis virus from a Rio Grande wild turkey

(*Meleagris gallopavo intermedia*). Microbiol Resour Announc 11(6):e00174–e00122. https://doi.org/10.1128/mra.00174-22

Witter RL (2006) Prevention and control of reticuloendotheliosis virus infection: rationale and strategies. pp 81–89

Witter RL, Salter DW (1989) Vertical transmission of reticuloendotheliosis virus in breeder turkeys. Avian Dis 33(2):226–235

Zavala G. Fadly AM, Hunt H (2016) "Oncornaviruses leukosis/sarcoma and reticuloendotheliosis." A laboratory manual for the isolation, identification and characterization of avian pathogens. 6th ed. Jacksonville (FL): American Association of Avian Pathologists 269–283

Zavala G, Nair V (2020) Reticuloendotheliosis. In: Swayne DE, Boulianne M, Logue CM, McDougald LR, Nair V, Suarez DL (eds) Diseases of poultry, 14th edn. Wiley Blackwell, Hoboken, pp 625–637

Arbovirus Infection

14

Awad A. Shehata and Hafez M. Hafez

Abstract

The term "arbovirus viruses" describes viruses transmitted to vertebrates by arthropod vectors, including mosquitos, ticks, and flies. Several arboviruses have been described; however, this chapter will discuss arboviruses infecting domestic and wild birds. These viruses belong to three families, namely, ***Togaviridae*** (Highlands J Virus, Eastern equine encephalitis, Western equine encephalitis), ***Flaviviridae*** (Turkey meningoencephalitis, Tembusu virus, West Nile virus, and Usutu virus), and ***Bunyaviridae*** (Turlock-like Bunyavirus). Highlands J (HJ), Eastern equine encephalitis (EEE), Western equine encephalitis (WEE), Israel turkey meningoencephalitis (ITM), and Western (WN) Nile viruses infect turkeys. Eastern equine encephalitis and Western equine are zoonotic and cause flu-like illness with sudden onset fever, headache, and fatigue. Some patients may develop neuroinvasive disease, with typical signs of meningitis, encephalitis, paralysis, convulsions, and coma in humans. The horse is the primary host of these two viruses. Arboviruses can be isolated in newly born mice. However, newborn chicks are highly susceptible, but it is not recommended for arbovirus isolation to avoid potential aerosol transmission. Virus identification can be done using PCR and IFA.

A. A. Shehata (✉)
TUM School of Natural Sciences, Bavarian NMR Center (BNMRZ), Structural Membrane Biochemistry, Technical University of Munich, Garching, Germany
e-mail: awad.shehata@tum.de

H. M. Hafez
Institute of Poultry Diseases, Faculty of Veterinary Medicine, Free University of Berlin, Berlin, Germany
e-mail: hafez.mohamed@fu-berlin.de

H. M. Hafez, A. A. Shehata (eds.), *Turkey Diseases and Disorders Volume 2*,
https://doi.org/10.1007/978-3-031-63322-5_14

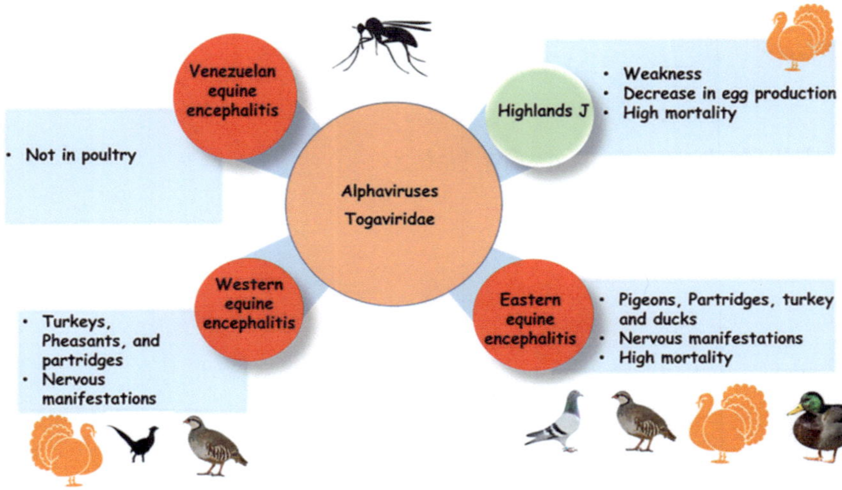

Fig. 14.1 Alphaviruses, family *Togaviridae*, as arboviruses infecting domestic poultry and wild birds. All these viruses are transmitted via mosquitos. The green color indicates viruses that cause disease in birds, while the red color indicates arboviruses that cause disease in birds and humans

Table 14.1 Summary diagnostic methods of arboviruses

Virus isolation	Laboratory animal	• Newly born mice • Inoculation of suspected samples intracranially • Observation of mice for 10 day • Brains of dead mice is used for virus identification
	Cell culture	• Chicken and duck embryo fibroblasts, Vero, baby hamster kidney (BHK-21), and rabbit kidney (RK-13) • Incubation of cells for 7 days and examination for cytopathic effects
	Embryonated eggs	• Embryonated chicken eggs can be used for isolation • Via yolk sac route • Embryonic death is common at 2–4 days after inoculation • Infected embryos with Israel Turkey meningoencephalitis are cherry red color in cases o
Virus identification	EEE	• IFA, antigen capture-ELISA, complement fixation test, plaque-reduction test (RRN test), PCR
	WEE	• IFA, antigen capture-ELISA, plaque-reduction test (RRN test), PCR
	HJ	• IFA, PCR
	WN	• IFA, antigen capture-ELISA, PCR
	IT	• IFA, haemagglutination inhibition test, neutralization test, PCR

Additionally, a hemagglutination inhibition test can be used for the diagnosis of ITM (ITM agglutinates goose red blood cells). Differential diagnosis of arboviruses from closely related agents, such as diseases causing nervous manifestations (i.e., Newcastle disease, avian encephalomyelitis, and *Clostridium botulinum*) and diseases causing a drop in egg production (i.e., avian influenza and turkey rhinotracheitis) should be considered.

Keywords
Arboviruses · *Togaviridae* · *Flaviviridae* · *Bunyaviridae*

14.1 Alphaviruses

14.1.1 Eastern Equine Encephalitis

Eastern equine encephalitis virus (EEEV) was first isolated in 1933 from the brain of a horse with encephalitis (TenBroeck and Merrill 1933). EEEV is a zoonotic arbovirus transmitted by mosquitoes (*Culiseta melanura*), affecting mainly pheasants and emus, but it can also affect pigeons, partridges, quail, ducks, chukar, partridges, and turkeys (Spalatin et al. 1961; Ficken et al. 1993; Johnson et al. 2003). Humans and horses are accidental hosts. Both quails and chickens are experimentally sensitive to EEEV.

14.1.1.1 Etiology

The disease is caused by the EEEV, genus Alphavirus of the family *Togaviridae. EEEV is enveloped,* ss RNA of 9.7–11.8 kilobases (kb). Togaviridae comprises two genera, *Alphavirus* and *Rubivirus*, but only the Alphavirus genus contains arboviruses.

14.1.1.2 Epidemiology

The disease occurs primarily in the eastern parts of North America, throughout Central America and the Caribbean, and in eastern parts of South America. EEEV could not be isolated from Europe. Outbreaks of EEE in avian species have been reported primarily in pheasants (Johnson et al. 2003).

14.1.1.3 Transmission

The virus is transmitted through the bite of infected mosquitos. Cannibalism, pecking, and artificial insemination may play a role in the virus spread and transmission. Mosquito bites can infect humans; however, care should be taken to avoid contact or droplet exposure when handling suspect-infected birds (Guy 2018).

14.1.1.4 Clinical Signs and Lesions

In turkeys, EEE is characterized by drowsiness, incoordination, progressive weakness, paralysis of legs and wings, and low mortality. A significant drop in egg production without mortality may also occur. Histopathologically, calcification of blood vessel walls in the cerebral cortex, the cerebellar folia, and the basal part of the medulla have been reported. Neurological lesions in intracerebrally inoculated birds included lymphocytic perivascular infiltration, neuronal degeneration, and endothelial cell swelling (Guy 2018).

14.1.1.5 Diagnosis

The diagnosis is based on virus isolation from blood, brain, spleen, liver, and heart on Vero cells, BHK-21 cells, and duck and chicken embryo fibroblasts. The virus can also be isolated in mice (IC) or one-old chicks (SC or IM), causing death within 2–5 days post-inoculation. EEEV can also be isolated in 6–8-day-old embryonated chicken eggs via the yolk sac route (Guy 2018). RT-PCR, immunohistochemistry, virus neutralization, and ELISA tests can be used for virus identification. Serological investigations can be done using ELISA and hemagglutination inhibition (HI) using goose or 1-day-old chicken erythrocytes. It is recommended to remove the nonspecific inhibitors of hemagglutination of serum using kaolin before performing the HI tests. EEEV should be differentiated from neurologic diseases in poultry and game birds, such as the Highlands J (HJ) virus, NDV, AEV, and botulism. Drop in egg production in turkeys can be caused by EEE, WEE, HJ, NDV, AI, AE, PM-3, TCO, and TRT viruses.

14.1.1.6 Vaccination

An inactivated EEE vaccine, developed for use in horses, has been used to protect pheasants. However, further studies are needed to evaluate its efficacy in birds.

14.1.2 Western Equine Encephalitis (WEEV)

The WEEV was first identified as the cause of encephalitis and high mortality in turkeys in Wisconsin (Woodring 1957). It is transmitted by *Culiseta tarsalis* and circulated mainly in western parts of North America, Central America, and South America. The WEEV is genetically close to EEEV; however, unlike EEEV, WEEV is rarely associated with disease in avian species. WEEV is associated with a drop in egg production in turkeys in California (Cooper and Medina 1999). Affected birds produce small, white-shelled, and shell-less eggs. However, neither signs nor clinical signs were observed (Guy 2018). Both WEE and EEE infections have been reported in ratites, particularly emus, causing hemorrhagic enteritis.

14.1.3 Highlands J Virus Infection (HJV)

HJ virus was first isolated in 1960 from blue jays in Florida (Henderson et al. 1962). Since then, it has been detected in chukar partridges and turkeys. The virus is closely related to WEEV. In turkey hens, HJV is associated with an acute drop in egg production (Wages et al. 1993) and mortality in young turkeys (Ficken et al. 1993). The clinical and pathologic characteristics of HJV infection in turkeys are closely similar to those of EEEV. The HJV can be differentiated from WEEV serologically using polyclonal and monoclonal antibodies (Karabatsos et al. 1988).

14.2 Flaviviruses

14.2.1 Israel Turkey Meningoencephalitis

14.2.1.1 Etiology

Turkey meningoencephalitis, belonging to the family *Flaviviridae*, was first described in 1958 in Israel (Komarov and Kalmar 1960). The virus has been identified in Israel, South Africa, and Spain (Komarov and Kalmar 1960; Barnard et al. 1980; Aguero 2011).

Turkeys are the only species to be naturally infected with turkey meningoencephalitis. The disease is characterized by drowsiness, incoordination, torticollis, tremors, progressive paresis, and paralysis. The mortality rate is 15–30% and may reach 80%. It also causes a drop in egg production without any negative effect on the egg quality, fertility, and hatchability. The virus is transmitted by mosquitoes (*Aedes* and *Culex* spp.), with seasonal incidence from August to September. Pheasants and partridges are also susceptible to natural infection. Chickens, ducks, geese, and pigeons are refractory to infection. Experimentally, young poults, Japanese quail, and suckling mice are highly susceptible to the turkey meningoencephalitis virus inoculated by IC or IM routes (Guy 2018).

14.2.1.2 Clinical Signs

Turkeys older than 10–12 weeks old exhibit clinical signs. The most common symptoms are paresis, incoordination, reluctance, or inability to walk, resting on the breasts, with legs stretched front, wings spread laterally, and greenish diarrhea. Morbidity and mortality rates vary from 15 to 30%, but they could exceed 80%. Although egg quality, fertility, and hatchability are unaltered, turkey breeder hens show a significant decline in egg production. Following infection recovery, egg production resumes as usual (Guy 2018).

14.2.1.3 Postmortem Lesions

The primary postmortem lesions are splenomegaly, or atrophy of the spleen, catarrhal enteritis, and myocarditis. Ovarian regression, ruptured ovarian follicles, and peritonitis are observed in breeder hens (Banet-Noach et al. 2003).

14.2.1.4 Histopathology

The primary microscopic lesions are non-purulent meningoencephalitis characterized by sub-meningeal and perivascular lymphocytic infiltration and focal myocardial necrosis (Guy 2018).

14.2.1.5 Diagnosis

The diagnosis of turkey meningoencephalitis is based on virus isolation and identification and detection of viral RNA using RT-PCR (Davidson et al. 2000). Brain, spleen, liver, and ovary samples are recommended for virus isolation and serum for serological testing. Virus isolation can be done in 6–8-day-old embryonated chicken eggs via the yolk sac route or on monolayers of chicken embryo fibroblasts (CEF). Before embryo mortality is seen, one or more passages in embryonated chicken eggs may be necessary. Embryos die 3–6 days post-inoculation and show a distinct cherry-red discoloration. The virus can also be isolated in suckling mice by inoculating samples intracerebrally or intramuscularly (Ianconescu 1989). CEF cells are less sensitive than embryonated chicken eggs or suckling mice for isolation of turkey meningoencephalitis.

14.2.1.6 Vaccination

Turkey encephalitis can be controlled by vaccination. However, there is a risk of vaccination failure. Perelman and others (Perelman et al. 2012) recommended some instructions to avoid vaccination failure: (1) Use only high-quality TME vaccines, (2) pre-cool (7 °C) the diluent, (3) use the vaccine within 60 min, (4) vaccinate during the day when it is possible to follow and monitor the vaccinating team quality of work, and (5) adapt the size of the needle to the size of the bird to be injected.

14.2.2 West Nile Virus (WNV)

West Nile virus (WNV) was first isolated in 1937 from a febrile woman in Uganda (Smithburn et al. 1940). The disease was detected for the first time in 1999 in avian species, horses, and human beings in North America (Steele et al. 2000). The virus is endemic in several countries in Europe, Asia, Africa, North America, and Central America. Outbreaks are common in late summer and autumn. The virus has circulated in Germany since 2018. In 2020, four birds tested positive for the WNV genome at Zoopark Erfurt (Thuringia) (Bergmann et al. 2023).

14.2.2.1 Etiology

The virus belongs to the family *Flaviviridae* and has two distinct lineages based on genetic analysis. Lineage I contained WN viruses isolated in Europe, Africa, and North America; Lineage II contained viruses isolated in Africa, Madagascar, and, most recently in Central Europe (Bakonyi et al. 2006).

14.2.2.2 Reservoir

WNV has a wide host range, including birds, reptiles, amphibians, mammals, and mosquitos. Geese are the primary natural host. Humans and mammals, including equids, are considered the dead-end hosts for WNV due to their low-level viremia (McLean et al. 2002; Troupin and Colpitts 2016). It was reported in geese (*Anser anser domesticus*) for the first time in 1997 in Israel (Lanciotti et al. 1999).

14.2.2.3 Clinical Signs and Lesions

The primary clinical signs are neurological signs, including incoordination, apathy, torticollis, opisthotonos, recumbency, and leg and wing paralysis (Swayne et al. 2001; Glávits et al. 2005). Postmortem lesions include pallor of the myocardium and occasionally the kidneys, splenomegaly, and hepatomegaly. Histopathologically, lymphocytic perivascular infiltration and neuronal degeneration of the brain tissues are common. In turkeys, no clinical signs or mortality have been reported. Three-week–old turkeys inoculated S.C. showed no clinical signs (Swayne et al. 2000).

14.2.2.4 Diagnosis

A summary of the diagnostic methods of arboviruses is shown in Table 14.1. The recommended samples for WNV isolation and detection are the brain, spleen, and kidneys. The diagnosis is based on virus isolation and identification, viral RNA detection using RT-PCR, and serological Assay. Virus isolation can occur in 7-day-old embryonated chicken eggs via the yolk sac route. The virus causes embryonic mortalities within 2–6 days post-inoculation. Samples can also be inoculated intra-cerebrally into suckling mice, Mice develop ataxia within 4–7 days, and CPE can be seen within 48–72 h. The virus isolation can also be carried out in Vero cell cultures and/or mosquito cell cultures. The WNV can be identified using IFA and RT-PCR (Braverman et al. 1981; Lanciotti et al. 2000). Serological diagnosis is based on HI, VN, or ELISA tests (Figs. 14.1 and 14.2).

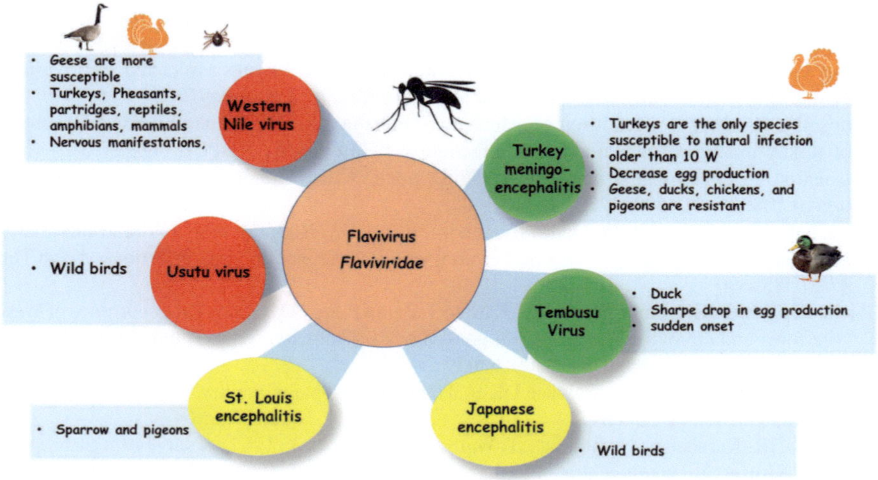

Fig. 14.2 Alphaviruses, family *Togaviridae*, as arboviruses infecting domestic poultry and wild birds. All these viruses are transmitted via mosquitos. However, the Western Nile virus can be transmitted through mosquitoes and ticks. The green color indicates viruses that cause disease in birds. The red color indicates zoonotic viruses. Yellow color indicates viruses that infect birds and cause disease in humans

14.2.2.5 Vaccination

Inactivated WNV vaccines such as EQUIP WNV® (Zoetis Belgium SA, Louvain-la-Neuve, Belgium) are licensed only for use in horses. However, commercial vaccines or therapy for birds are unavailable. Recombinant canarypox vaccine Proteq West Nile® (Boehringer Ingelheim Vetmedica GmbH, Ingelheim am Rhein, Germany) was also proven to be used in horses. In the absence of licensed vaccines for birds, some bird holders and veterinarians have attempted to vaccinate endangered and threatened bird species against WNV using equine vaccines, but the safety and efficacy of vaccination protocols are unknown. Recently, it was suggested that inactivated vaccines are safe for avian species (Bergmann et al. 2023).

References

Aguero M (2011) Bagaza virus in partridges and pheasants, Spain, 2010. Emerg Infect Dis 17:1498. https://doi.org/10.3201/eid1708.110077

Bakonyi T, Ivanics É, Erdélyi K, Ursu K, Ferenczi E, Weissenböck H et al (2006) Lineage 1 and 2 strains of encephalitic West Nile virus, Central Europe. Emerg Infect Dis 12(4):618–623. https://doi.org/10.3201/eid1204.051379

Banet-Noach C, Simanov L, Malkinson M (2003) Direct (non-vector) transmission of West Nile virus in geese. Avian Pathol 32(5):489–494. https://doi.org/10.1080/0307945031000154080

Barnard BJ, Buys SB, Du Preez JH, Greyling SP, Venter HJ (1980) Turkey meningoencephalitis in South Africa. Onderstepoort J Vet Res 47(2):89–94

Bergmann F, Fischer D, Fischer L, Maisch H, Risch T, Dreyer S et al (2023) Vaccination of zoo birds against West Nile virus—a field study. Vaccine 11(3):652. https://doi.org/10.3390/vaccines11030652

Braverman Y, Rubina M, Frish K (1981) Pathogens of veterinary importance isolated from mosquitoes and biting midges in Israel. Int J Trop Insect Sci 2(03):157–161. https://doi.org/10.1017/S1742758400000953

Cooper GL, Medina HA (1999) Egg production drops in breeder turkeys associated with Western equine encephalitis virus infection. Avian Dis 43(1):136. https://doi.org/10.2307/1592773

Davidson I, Grinberg R, Malkinson M, Mechani S, Pokamunski S, Weisman Y (2000) Diagnosis of turkey meningoencephalitis virus infection in field cases by RT-PCR compared to virus isolation in embryonated eggs and suckling mice. Avian Pathol 29(1):35–39. https://doi.org/10.1080/03079450094252

Ficken MD, Wages DP, Guy JS, Quinn JA, Emory WH (1993) High mortality of domestic turkeys associated with highlands J virus and eastern equine encephalitis virus infections. Avian Dis 37(2):585–590

Glávits R, Ferenczi E, Ivanics E, Bakonyi T, Mató T, Zarka P, Palya V (2005) Co-occurrence of West Nile fever and circovirus infection in a goose flock in Hungary. Avian Pathol 34(5):408–414. https://doi.org/10.1080/03079450500268039

Guy JS (2018) Arboviruses. In: Swayne DE et al (eds) In "disease of poultry", 14th edn. Blackwell Publishing, Iowa, pp 507–547

Henderson JR, Karabatsos N, Bourke AT, Wallis RC, Taylor RM (1962) A survey for arthropod-borne viruses in south-central Florida. Am J Trop Med Hyg 11:800–810. https://doi.org/10.4269/ajtmh.1962.11.800

Ianconescu M (1989) Turkey meningo-encephalitis. In: Purchase HG, Arp LH, Domermuth CH, Pearson JE (eds) A laboratory manual for the isolation and identification of avian pathogens, 3rd edn. American Association of Avian Pathologists, Kennett Square, pp 163–164

Johnson AJ, Langevin S, Wolff KL, Komar N (2003) Detection of anti-West Nile virus immunoglobulin M in chicken serum by an enzyme-linked immunosorbent assay. J Clin Microbiol 41(5):2002–2007. https://doi.org/10.1128/JCM.41.5.2002-2007.2003

Karabatsos N, Lewis AL, Calisher CH, Hunt AR, Roehrig JT (1988) Identification of highlands J virus from a Florida horse. Am J Trop Med Hyg 39(6):603–606. https://doi.org/10.4269/ajtmh.1988.39.603

Komarov A, Kalmar E (1960) A hitherto undescribed disease-Turkey meningoencephalitis. Vet Rec 72:257–261

Lanciotti RS, Roehrig JT, Deubel V, Smith J, Parker M, Steele K et al (1999) Origin of the West Nile virus responsible for an outbreak of encephalitis in the northeastern United States. Science 286(5448):2333–2337. https://doi.org/10.1126/science.286.5448.2333

Lanciotti RS, Kerst AJ, Nasci RS, Godsey MS, Mitchell CJ, Savage HM et al (2000) Rapid detection of west nile virus from human clinical specimens, field-collected mosquitoes, and avian samples by a TaqMan reverse transcriptase-PCR assay. J Clin Microbiol 38(11):4066–4071. https://doi.org/10.1128/JCM.38.11.4066-4071.2000

McLean RG, Ubico SR, Bourne D, Komar N (2002) West Nile virus in livestock and wildlife. Curr Top Microbiol Immunol 267:271–308. https://doi.org/10.1007/978-3-642-59403-8_14

Perelman S, Kin E, Heifez S, Benet-Noah C, Even-Chen T, Ashash U et al (2012) Investigation study of Turkey meningoencephalitis (TME) vaccine failure: causes and solutions in the field. Isr J Vet Med 7(3):154–161

Smithburn KC, Hughes TP, Burke AW, Paul JH (1940) A neurotropic virus isolated from the blood of a native in Uganda. Am J Trop Med 20:471–493

Spalatin J, Karstad L, Anderson JR, Lauerman L, Hanson RP (1961) Natural and experimental infections in Wisconsin turkeys with the virus of eastern encephalitis. Zoonoses Res 1:29–48

Steele KE, Linn MJ, Schoepp RJ, Komar N, Geisbert TW, Manduca RM et al (2000) Pathology of fatal West Nile virus infections in native and exotic birds during the 1999 outbreak in New York City, New York. Vet Pathol 37(3):208–224. https://doi.org/10.1354/vp.37-3-208

Swayne DE, Beck JR, Zaki S (2000) Pathogenicity of West Nile virus for turkeys. Avian Dis 44(4):932–937

Swayne DE, Beck JR, Smith CS, Shieh WJ, Zaki SR (2001) Fatal encephalitis and myocarditis in young domestic geese (*Anser anser domesticus*) caused by West Nile virus. Emerg Infect Dis 7(4):751–753. https://doi.org/10.3201/eid0704.010429

TenBroeck C, Merrill MH (1933) A serological difference between eastern and western equine encephalomyelitis virus. Proc Soc Exp Med 31:217–220

Troupin A, Colpitts TM (2016) Overview of West Nile virus transmission and epidemiology. Methods Mol Biol 1435:15–18. https://doi.org/10.1007/978-1-4939-3670-0_2

Wages DP, Ficken MD, Guy JS, Cummings TS, Jennings SR (1993) Egg-production drop in turkeys associated with alphaviruses: eastern equine encephalitis virus and highlands J virus. Avian Dis 37(4):1163–1166

Woodring FR (1957) Naturally occurring infection with equine encephalomyelitis virus in turkeys. J Am Vet Med Assoc 130:511–512

Part II
Parasitic Diseases

Parasitic Infections in Turkeys

15

Endoparasites Affecting Turkeys

Xochitl Hernandez-Velasco, Guillermo Tellez-Isaias, Daniel Hernandez-Patlan, Bruno Solis-Cruz, Víctor M. Petrone-García, Inkar Castellanos-Huerta, Jesús A. Maguey-González, Juan D. Latorre, Saeed El-Ashram, Wolfgang Eisenreich, Hafez M. Hafez, and Awad A. Shehata

X. Hernandez-Velasco (✉)
Departamento de Medicina y Zootecnia de Aves, Facultad de Medicina Veterinaria y Zootecnia, UNAM, Mexico City, Mexico
e-mail: xochitlh@fmvz.unam.mx

G. Tellez-Isaias · I. Castellanos-Huerta · J. A. Maguey-González · J. D. Latorre
Department of Poultry Science, University of Arkansas, Fayetteville, AR, USA
e-mail: gtellez@uark.edu; icastell@uark.edu; jm201@uark.edu; jl115@uark.edu

D. Hernandez-Patlan · B. Solis-Cruz
Laboratory 5: LEDEFAR, Multidisciplinary Research Unit, National Autonomous University of Mexico-Superior Studies Faculty at Cuautitlan (UNAM-FESC), Cuautitlan Izcalli, Mexico State, Mexico

Nanotechnology Engineering Division, Polytechnic University of the Valley of Mexico, Tultitlan, Mexico State, Mexico
e-mail: danielpatlan@comunidad.unam.mx; bruno_sc@comunidad.unam.mx

V. M. Petrone-García
Departamento de Ciencias Pecuarias, FESC, UNAM, Cuautitlán, Estado de Mexico, Mexico

S. El-Ashram
College of Life Science and Engineering, Foshan University, Foshan, Guangdong, China

Faculty of Science, Kafrelsheikh University, Kafr El-Sheikh, Egypt

W. Eisenreich · A. A. Shehata
Bavarian NMR Center, Structural Membrane Biochemistry Department of Chemistry, Technische Universität München, Garching, Germany
e-mail: wolfgang.eisenreich@mytum.de; awad.shehata@tum.de

H. M. Hafez
Institute of Poultry Diseases, Faculty of Veterinary Medicine, Free University of Berlin, Berlin, Germany
e-mail: hafez.mohamed@fu-berlin.de

H. M. Hafez, A. A. Shehata (eds.), *Turkey Diseases and Disorders Volume 2*,
https://doi.org/10.1007/978-3-031-63322-5_15

Abstract

Parasites are one of the most significant threats to the turkey industry, as they can cause severe economic losses. The prevalence of parasitosis has decreased over time, mainly in countries and regions where intensive production is carried out under conditions of greater biosecurity and sanitary control, in addition to the use of more efficient diagnostic techniques and deworming programs; however, the increase in the number of birds raised in extensive or free-range systems has favored an increase in cases of parasitism and makes their control more difficult. Controlling and eradicating parasites is not easy; proof of this is that with 220 million people affected annually by malaria, efforts have been made to stop its transmission through prophylaxis, pesticide-based eradication of mosquito vectors, and vaccine development. Nonetheless, drug resistance, pesticide resistance in mosquitoes, and vaccine failures due to this parasite's complex life cycles and naturally occurring genetic polymorphism have all proven to be significant issues. Poultry is infected by numerous kinds of parasites, some of which live on the surface of the body and others internally, so parasites of poultry may be divided into two general groups: Internal parasites include protozoa and worms, and the second group is made up of arthropods including lice, mites, fleas, and ticks. This chapter summarizes the most important endoparasites, including *Sporozoea*, *Zoomastigophorea*, nematodes, and cestodes that affect commercial turkeys.

Keywords

Endoparasites · Ectoparasites · Sporozoea · Zoomastigophorea · Nematodes · Cestodes

15.1 Endoparasites

The main endoparasites described in turkeys and their localization are presented in Table 15.1.

15.1.1 Sporozoea Affecting Turkeys

Parasitic protozoa are a diverse group of unicellular eukaryotic microorganisms with one or more nuclei and cytoplasm, cytoskeleton, organelles, and other structures with diverse functions. The locomotion and reproduction mechanisms of protozoan parasites are diverse; some have specialized locomotion structures and different forms of reproduction, even between the phases of the reproductive cycle of the same species. Protozoa can be found in the lumen of the intestinal tract, intracellularly in the blood, or within the cells of many tissues of their host. Some

Table 15.1 Intermediate host and site of infection of the main endoparasites that affect turkeys

Phylum/class	Genus	Species	Intermediate host	Site of infection
Phylum: Apicomplexa Class: Sporozoea	*Eimeria*	*Eimeria* spp.	No	Intestine
	Cryptosporidium	*C. meleagridis*	No	Respiratory tract and intestine
		C. baileyi	No	
	Haemoproteus	*H. meleagridis*	Louse flies	Blood
	Leucocytozoon	*L. somondii*	Black flies	Blood
Phylum: Sarcomastigophora Class: *Zoomastigophorea*	*Histomonas*	*H. meleagridis*	No	Caecum and liver
		H. wenrich	No	Caecum
	Hexamita	*H. meleagridis*	No	Intestine
	Trichomonas	*T. gallinarum*	No	Caecum
	Chilomastix	*C. gallinarum*	No	Caecum
Nematodes	*Ascaridia*	*Ascaridia* spp.	–	Intestine
	Heterakis	*Heterakis gallinarum*	–	Intestine
	Subulura		–	Intestine
	Capillaria spp.		–	Intestine
	Tetrameres americana		–	Intestine
	Syngamus trachea		–	*Trachea*
Tapeworms (Cestodes)	*Davainea*	*D. meleagridis*	Unknown	Duodenum
	Raillietina	*R. echinobothrida*	Ants	Duodenum
		R. cesticillus	Beetle species	Duodenum
		R. magninumida	Beetle	Duodenum
	Amoebotaenia	*A. cuneata*	Earthworm	Duodenum
	Choanotaenia	*C. infundibulum*	Beetle, house fly	Duodenum
	Hymenolepis	*H. carioca*	Beetle species	Duodenum
		H. cantaniana	Beetle species	Duodenum
	Metroliasthes	*M. lucida*	Locusts	Duodenum

species of parasitic protozoa are of high economic importance due to their prevalence and the damage they can cause.

15.1.2 Coccidiosis (Eimeriosis)

15.1.2.1 Summary

Coccidiosis in turkeys is an intestinal disease characterized by massive destruction of the intestinal mucosa, leading to great economic losses. The disease is caused by host-specific *Eimeria* spp.; five species are known to cause coccidiosis in turkeys, namely, *E. adenoeides*, *E. gallopavonis*, *E. meleagrimitis*, *E. meleagridis*, and *E. dispersa. E. adenoeides* primarily induce lesions in the ceca, while *E. meleagrimitis* induces lesions in the duodenum and jejunum of turkeys. Furthermore, two species, namely, *E. innocua* and *E. subrotunda*, are considered nonpathogenic. Coccidia parasites invade and replicate within the intestinal epithelial cells, leading to tissue destruction, inflammation, and hemorrhage. The severity of the disease is related to the number of parasites, the host's immune status, and the parasite's virulence. Oocysts are extremely resistant to common disinfectants. Only disinfectants containing very small lipophilic molecules that can pass through the pores in the oocyst wall are effective. The control of coccidiosis in turkeys is based primarily on hygienic measures and coccidiostats as feed additives, usually for the first 8 weeks of life. Ionophores such as narasin and salinomycin are effective against sporozoites; however, they are highly toxic for turkeys and ineffective in treating clinical coccidiosis. Live attenuated and subunit vaccines are also available for turkeys.

15.1.2.2 Etiology

Coccidiosis is a protozoan disease caused by *Eimeria*, affecting a wide range of animal species. The disease is characterized by severe enteritis and diarrhea, leading to reduced growth rates, increased mortality rates, and significant economic losses in the turkey industry (Chapman 2008). The most important ones are *E. adenoeides, E. meleagrimitis, E. gallopavonis, E. meleagridis*, and *E. dispersa* (Hafez 2008; El-Sherry et al. 2019) (Table 15.2 and Fig. 15.1).

Eimeria adenoeides. *E. adenoeides* is one of the most pathogenic species and can cause significant damage to the turkey's intestinal lining. It is commonly found in the ileum, ceca (Fig. 15.2a, b), and rectum and can cause weight loss, decreased feed intake, and increased mortality in young turkeys.

Eimeria meleagrimitis. *E. meleagrimitis* is another important species that can cause severe damage to the intestinal tract. It is commonly found in the duodenal but in the small intestine in heavy infections and can cause diarrhea, dehydration, and reduced weight gain. In severe cases, it can also lead to death (Fig. 15.2c, d).

Eimeria gallopavonis. *E. gallopavonis* is commonly found in the lower intestinal tract, posterior ileum, ceca, and rectum and can cause moderate to severe

Table 15.2 Location and pathogenicity of *Eimeria* spp. infecting turkeys

Host	*Eimeria*	Location	Pathogenicity[a]
Turkeys	*E. adenoeides*	Caecum	+++
	E. dispersa	Duodenum, jejunum	+
	E. gallopavonis	Rectum	++
	E. innocua	Duodenum, jejunum	–
	E. meleagridis	Caecum	+
	E. meleagrimitis	Duodenum, jejunum	+++
	E. subrotunda	Duodenum, jejunum	–

[a]– nonpathogenic, + low pathogenic, ++ moderately pathogenic, +++ highly pathogenic (Hafez 2008)

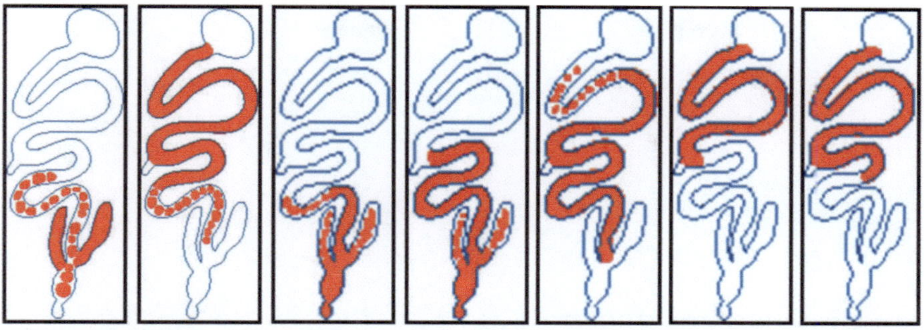

Fig. 15.1 Site of infection of *Eimeria* species affecting turkeys

damage. It can lead to bloody diarrhea, reduced feed intake, weight loss, inflammation, and ulceration of the ileum.

Eimeria meleagridis. *E. meleagridis* is commonly found in the upper small intestine and ceca and can cause mild to moderate damage. It can lead to reduced weight gain and feed intake but is generally less pathogenic than other species. Transient edema, inflammation, and caseous exudates in the ceca may occur (Fig. 15.2e).

Eimeria dispersa. *E. dispersa* is commonly found in the duodenum and subsequently spreads to the lower intestine and occasionally in the cecal necks and can cause mild to moderate damage (Fig. 15.2f). It can lead to diarrhea, reduced feed intake, and weight loss (Chapman 2008).

15.1.2.3 Epidemiology and Life Cycle

Coccidiosis in turkeys is a worldwide disease, with high prevalence rates reported in intensive production systems. The disease is more common in young birds than in older birds. The severity of the disease is related to the *Eimeria* species, age of birds, stocking density, and rearing form (indoor or free range) (Vrba and Pakandl 2014). Coccidiosis is transmitted by the ingestion of sporulated oocysts shed in feces. *Eimeria* spp. sporulated outside the host within 1–2 days. It contains four sporocysts, each of which contains two sporozoites. Coccidia oocysts are highly resistant to environmental stressors and can survive for extended periods in the environment.

The cycle of *Eimeria* species in turkeys begins when the bird ingests the sporulated oocysts of the parasite. These oocysts can be found in the environment, including in litter and soil, and can survive for long periods under favorable conditions. Once the sporocysts reach the turkey's digestive system, the oocysts release sporozoites, the parasite's infective stage. The sporozoites then invade the cells lining the intestinal tract, where they multiply rapidly. As the sporozoites continue to replicate, they undergo several stages of development, ultimately forming mature oocysts that contain multiple infectious units known as sporocysts. These mature oocysts

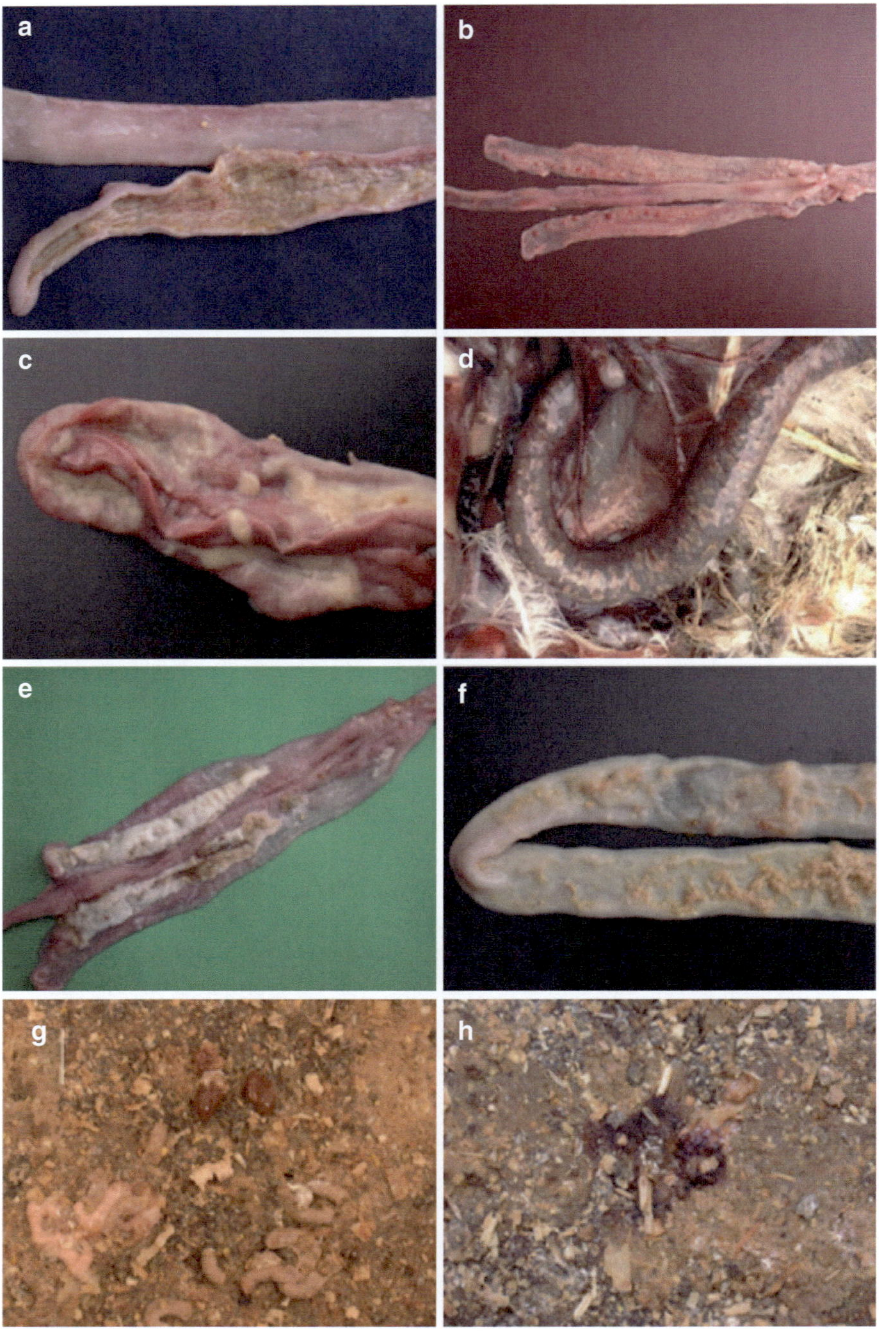

Fig. 15.2 Intestinal macroscopic lesions, feces, and litter of poults with *Eimeria* spp. (**a**, and **b**) Ceca infected with *E. adenoides,* showing petechial hemorrhages and white mucous content. (**c**, **d**) Congested duodenal mucosa with a large amount of mucus. (**e**) Caseous exudates in the ceca. (**f**) Flaccidity and thinning of the intestinal wall, with cream-colored mucus. (**g**, **h**) Watery feces of poults with mucus and blood. Note that the litter is packed due to excess moisture in the feces (Figs. A to C and E to H, X. Hernandez-Velasco; Fig. D Hafez M Hafez)

are then shed in the turkey's feces, infecting other birds and contaminating the environment (Edgar and Flanagan 1979) (Fig. 15.3).

The cycle of *Eimeria* species in turkeys can be particularly problematic since some species can cause severe damage to the intestinal tract. This damage can lead to reduced feed intake and feed conversion rate, weight loss, and increased susceptibility to secondary infections. In severe cases, coccidiosis can even result in high mortalities.

15.1.2.4 Clinical Signs

The clinical signs of coccidiosis in turkeys vary depending on the severity of the infection. Early signs include decreased feed intake, lethargy, and watery droppings (Fig. 15.2g). As the disease progresses, diarrhea becomes more severe, and the droppings become bloody (Fig. 15.2h). The birds become dehydrated and develop a

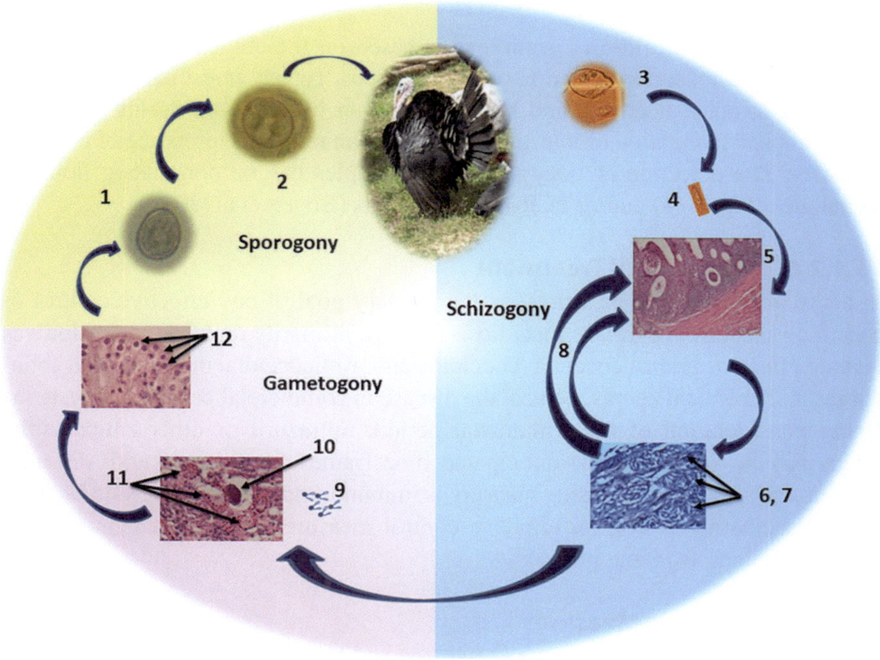

Fig. 15.3 The life cycle of Eimeria spp. Exogenous phase (**sporogony**): Non-sporulated oocysts (1), under normal temperature conditions (15–30 °C) and moisture, are sporulated in 1–2 days. Sporulated (infective) oocysts) (2). **Excystation** after ingestion of sporulated oocyst: sporozoites are released by mechanical and enzymatic digestion (3). **Cell invasion**. Free sporozoites (4) (initial infective units) infect intestinal cells and change to trophozoites (5). **Schizogony**. The schizont will multiply by asexual multiple division (endogenous and asexual phase) (6) with Merozoites (7) inside and will have at least two generations of infection (8). **Gametogony**. The male gametocytes multiply by sexual multiple division and release microgametes into the lumen of the gut: free microgametes (9), male microgametes (10), fertilize female macrogamete (11) to produce a zygote (12). **Transmission**. The mature oocysts are excreted with the feces (1). Sporulated oocysts can potentially produce 2–3 millions of new oocysts within 5–7 days

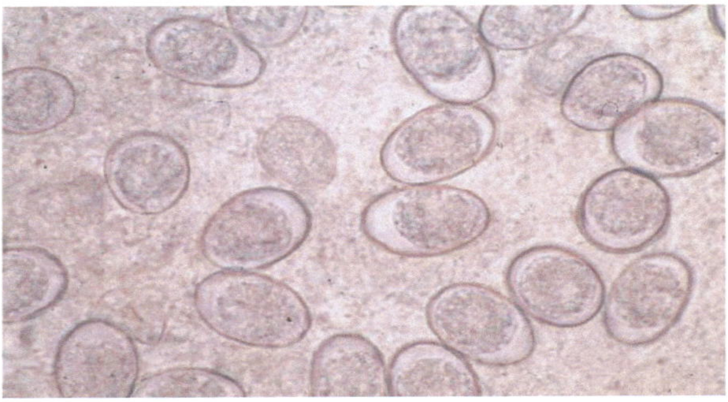

Fig. 15.4 Non-sporulating oocysts of *Eimeria* spp. from a fresh feces sample (H.M. Hafez)

hunched posture. Mortality rates can be as high as 30%, with the highest rates seen in young birds.

15.1.2.5 Diagnosis

The diagnosis of coccidiosis in turkeys is based on the clinical signs, gross and microscopic lesions, and detecting oocysts in feces. Typical intestinal lesions, such as villous atrophy, necrosis, and hemorrhage, can be observed during necropsy. Microscopic examination of intestinal scrapings can reveal the presence of coccidia parasites. The detection of oocysts in fecal samples can be done using flotation techniques (Fig. 15.4) and/or PCR-based methods (Boyett et al. 2022).

15.1.2.6 Control and Treatment

In addition to hygienic measures, as the primary goal of any effective control of coccidiosis, control of turkey coccidiosis is based primarily on using feed additives that are effective against coccidia (coccidiostats). Anticoccidial drugs, such as ionophores and chemical agents, control the disease in commercial production systems. Under consideration of the withdrawal period, toltrazuril or other anticoccidial drugs may be applied in the anticipated time frame of infections with different *Eimeria* spp. Another promising measure is immunoprophylaxis using suitable vaccines. Vaccination is also an effective control measure, with live attenuated and subunit vaccines available in turkeys.

15.1.2.7 Hygienic Measures

Good management practices, such as hygiene and biosecurity measures, are essential to prevent the spread of the disease (Chapman 2008). Hygienic measures include cleaning and disinfection of the farm with an effective disinfectant before each restocking to reduce the infection pressure by reducing the number of infectious oocysts. Special attention must be paid to the dryness of the litter. In particular, the damp island occurring in the drinking trough area must always be eliminated. Furthermore, repeated working through the litter leads to diluting the oocyst concentration in the house and thus reduces the infection pressure. The effective disinfectants against *Eimeria* spp. are those containing very small lipophilic molecules that could pass through the pores in the oocyst wall and thereby kill the sporozoites such as H_2S, chloroform, KOH, ammonia gas, and methyl bromide. Temperatures above 56 °C or below −15 °C kill coccidia oocysts within a short time. In the litter, the survival time is limited to a few days due to the development of ammonia.

15.1.2.8 Anticoccidial Drugs

Several coccidiostats are available in turkeys, including chemical drugs and ionophores. Chemicals (synthetic compounds) include halofuginone, robenidine, diclazuril, decoquinate, nicarbazin, toltrazuril, clopidol, nequinate, ethopabate, amprolium, sulfadimethoxine, and sulfaquinoxaline. Each drug has a unique mode of action, for example, (1) inhibition of parasite mitochondrial respiration (decoquinate, clopidol), (2) inhibition of the folic acid pathway (sulfonamides), (3) competitive inhibition of thiamine uptake (amprolium), (4) nucleoside analog (Diclazuril and toltrazuril), and (5) oxidative phosphorylation inhibitor (Robenidine). Unknown mode of action (e.g., halofuginone, nicarbazin) (Hafez 2008).

Chemical drugs are usually administered in feed or water. The drawbacks of these drugs are: (1) Surviving *Eimeria* spp. became quickly resistant. (2) They do not allow the host to develop immunity against coccidiosis. Toltrazuril belongs to triazines and acts on intracellular life cycle stages undergoing schizogony and gametogony. It has a high efficacy without negatively affecting the formation of immunity. Moreover, it can be used in all poultry species.

Ionophores (polyether antibiotics), such as monensin, salinomycin, lasalocid, narasin, maduramicin, and semduramicin, are produced by the fermentation of *Streptomyces* spp. or *Actinomadura* spp. On the other hand, ionophores such as lasalocid, monensin, and maduramicin alter the gut environment to make it less hospitable to coccidian (Garton 2014). These drugs disrupt ion gradients across the cell membrane of the parasite. Ionophores can be classified into (1) monovalent ionophores (monensin, narasin, salinomycin), (2) monovalent glycosidic ionophores (maduramicin, semduramicin), and (3) divalent ionophore (lasalocid). Ionophores develop resistance very slowly. However, ionophore intoxication has been reported in turkeys due to (1) high doses of ionophores (e.g., monensin >300 ppm); (2) pharmacological incompatibility when applied with tiamulin, chloramphenicol, and sulphaquinoxaline (Hafez 2008). The main clinical causes of ionophores intoxication in turkeys are anorexia with depression, weakness, complete paralysis, posterior paralysis with legs extended, dyspnea, dehydration, and mortality (variable but may exceed 70%).

Coccidiostats such as monensin, salinomycin, and halofuginone are not effective in treating clinical coccidiosis. The use of coccidiostats in turkeys is a controversial issue, with some experts advocating for their use as a necessary tool for preventing and controlling coccidiosis, while others argue that they can contribute to the development of drug-resistant coccidia strains and have negative effects on the turkey's gut microbiome. Several programs are known for using coccidiostats:

1. **Single (straight) program**: A single drug is used in the feed of a single flock. Using this program can lead to a reduction in drug efficacy and the development of resistance.

2. **Shuttle or dual program**. In this program, one anticoccidial drug is used in the starter ration, usually synthetic; then another drug in the growers; then a third drug in the finisher diet; and finally, a fourth type during withdrawal. The strongest drug may be used in the starter feed. Using this program improves the efficacy of anticoccidial drugs and reduces the development of drug resistance.
3. **Rotation program**. This program is widely used to reduce the incidence of resistance development. The anticoccidial drugs are changed every interval of months (every 6 months), in successive flocks. Rotations are only possible if drugs with different modes of action follow each other.

Despite the controversy, many turkey producers continue to use coccidiostats as a part of their management practices. When used correctly, coccidiostats can effectively prevent and control coccidiosis, reducing the economic impact of the disease. However, this should be used in combination with good management practices, such as maintaining good hygiene, litter management, and providing adequate nutrition and water, which is essential. Coccidiostats are a valuable tool in preventing and controlling coccidiosis in turkeys. However, their use should be carefully monitored and integrated into a comprehensive disease management plan to ensure their effectiveness and minimize the risk of negative side effects. It is essential to work with a veterinarian to determine the most appropriate coccidiostat regimen for your flock, considering factors such as the turkey's age, strain, and health status (Chapman et al. 2004).

15.1.2.9 Alternative Anticoccidials in Turkeys

Several trials have been done to develop alternative natural anticoccidial drugs using prebiotics and probiotics, plant extracts, fungal extracts, and essential oils. Most natural products do not directly target *Eimeria* spp.; however, they restore the intestinal microbiota and strengthen the immune response (Shehata et al. 2022). Several phytogenic substances, such as *Artemisia*, clove, tea tree, and thyme (Kadykalo et al. 2018), effectively reduce coccidiosis *Cinnamomum verum* (Qaid et al. 2021).

15.1.2.10 Anticoccidial Vaccines for Turkeys

Coccidia vaccines have been developed for turkeys and have been widely used by turkey producers. *Eimeria* vaccines are live attenuated and/or non-attenuated oocysts administered orally to turkeys. The vaccines stimulate mainly the bird's cellular immune system and, in lesser regard, the humoral immune system to produce antibodies that protect against the disease. The vaccines are available as monovalent or multivalent preparations containing different *Eimeria* strains (Chapman 2018). The use of *Eimeria* vaccines in turkeys has several advantages: (1) reducing the incidence and severity of coccidiosis; (2) improving bird health and productivity; (3) reducing the need for antimicrobials that can contribute to developing antimicrobial resistance; (4) reducing the environmental contamination with oocysts, which can infect other birds and animals; and (5) vaccination against coccidiosis effectively that prevents necrotic enteritis in a mixed infection (Bangoura et al. 2014).

Two vaccines with different attenuated but reproducible turkey coccidia species are commercially produced in Canada. These are the Immuncox T1 vaccine, which includes *E. adenoides* and *E. meleagrimitis*, and the Immuncox T2 vaccine, which includes *E. gallopavonis*. The vaccine is administered between the third and fifth day old via drinking water.

However, *Eimeria* vaccines have some limitations: (1) They require careful handling and storage to maintain their viability. (2) They require adequate timing and dosage right application to achieve optimal protection, and failure to do so may result in inadequate immunity. (3) They may not provide complete protection against all *Eimeria* strains, and additional control measures may be required (Chapman et al. 2005).

15.1.3 Cryptosporidiosis

15.1.3.1 Summary

Cryptosporidiosis is a parasitic infection caused by the protozoan *Cryptosporidium* spp., which affects a wide range of animal species, including humans. They are ubiquitous protozoal organisms that persist in the environment for various durations depending on temperature and humidity. Turkeys are one of the livestock species that are susceptible to Cryptosporidiosis, and the infection can result in significant economic losses to the turkey industry. *C. baileyi* is the most prevalent species in domestic poultry, causing respiratory and intestinal infections. While *C. meleagridis* infection causes mild to severe diarrhea, both *C. parvum* and *C. galli* were detected in chickens or turkeys without clinical disease. *C. meleagridis* is a zoonotic parasite, particularly in immune-compromised patients and children. Moreover, zooanthroponosis (from human-to-animals) of *C. parvum* has also been reported. Currently, no specific drugs are available for the treatment of *Cryptosporidiosis* in turkeys. Some oocysts are self-infecting without leaving the host, which favors the chronicity of the disease or lethal infections in immune-deficient birds. Good hygiene practices include regular cleaning and disinfection of the environment and equipment, as well as avoiding overcrowding and stress in the birds (Abbassi and Répérant 2015).

15.1.3.2 Etiology

Several species are known to cause cryptosporidiosis in birds, *C. baileyi* and *C. meleagridis* being the most common in turkeys and *C. parvum* and *C. galli* rarely occurring (Sulaiman et al. 2003; Berrilli et al. 2012), Table 15.3. Oocysts are ellipsoids, 5.6–6.3 × 4.5–4.8 μ (Taylor et al. 2016). The life cycle of *Cryptosporidium* is very similar to other coccidia.

15.1.3.3 Epidemiology

Cryptosporidium is a ubiquitous parasite that is prevalent worldwide. The oocysts can survive for extended periods in the environment and are also resistant to the majority of common disinfectants. The life cycle of *Cryptosporidium* involves

Table 15.3 The main Cryptosporidium spp. in poultry

Species	Host	Site of infection
C. baileyi	Chickens, turkeys, ducks	Bursa of Fabricius, cloaca, respiratory system
C. galli	Chickens, finches	Proventriculus
C. meleagridis[a]	Turkeys	Small intestine
Cryptosporidium spp.	Pigeon	Small intestine

[a]*C. parvum* is not commonly seen in poultry. There is evidence that *C. meleagridis* is synonymous with *C. parvum* (McDougald 2020)

asexual and sexual phases and culminates in oocyst production. In the host, **the oocyst forms four sporozoites without sporocysts**. The infection is usually transmitted by the respiratory and fecal-oral route through thick-walled sporulated oocysts, while thin-walled oocysts are responsible for endogenous self-infections (Helmy and Hafez 2022). The endogenous cycle is short (4–7 days), the endogenous stages are small (4–7 mcm), and the parasites are just beneath the epithelial cell membranes.

Cryptosporidiosis is more severe in turkeys than in chickens and is frequently fatal in quail. In turkeys, the infection is mostly seen in young birds, especially those under 6 weeks old (Gharagozlou et al. 2006; McDougald 2020). The incidence of Cryptosporidiosis in turkeys varies widely depending on the region and farming practices, ranging from 5 to 50% (Goodwin et al. 1988). The most common species in turkeys are *C. baileyi* and *C. meleagridis* (McDougald 2020). In 2017, a nested PCR detected *Cryptosporidium* in 9.3% (8/86) of turkeys. Sequence analysis revealed that *C. parvum* was the most frequently identified in 5.1% (13/256) of all poultry species, including 8.1% (7/86) of turkeys, 3.2% (5/158) of broilers, and 8.3% (1/12) of layers. *Cryptosporidium baileyi* was detected in only 1.3% (2/256) of the broilers (Helmy et al. 2017).

15.1.3.4 Clinical Signs and Pathology

The clinical signs of Cryptosporidiosis in turkeys include diarrhea, dehydration, anorexia, and lethargy. **The most common clinical signs include diarrhea and dehydration**. The diarrhea is usually watery and may be accompanied by mucus or blood. The severity of the disease can range from mild to severe, with mortality rates of up to 10% reported in some cases. The pathology of the infection includes villous atrophy, inflammation, and crypt hypertropic in the small intestine (de Graaf et al. 1999), in the cloacal bursa (Fig. 15.5) and/or respiratory disease with swelling of infraorbital sinuses and serous conjunctivitis. Microscopic lesions of the infected tissues included deciliation of the epithelium and inflammation (McDougald 2020) (Fig. 15.6). Cryptosporidium have been found in the sinuses, trachea, bronchi, cloaca, and bursa.

15.1.3.5 Diagnosis

The diagnosis of Cryptosporidiosis in turkeys is based on detecting the parasite in fecal samples or mucosal scrapings using the microscopic examination. Phase

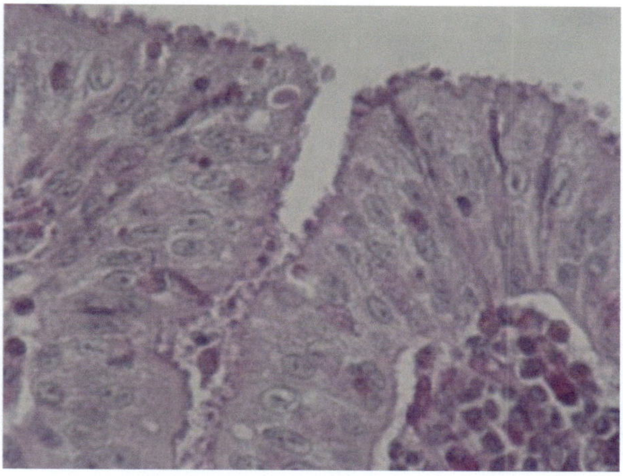

Fig. 15.5 Histological section of a bursa of Fabricius from a turkey poult. Presence of *Cryptosporidium* sp. on the surface of the epithelium, epithelial hyperplasia, and subepithelial mononuclear lymphocytic infiltrate (X. Hernandez-Velasco)

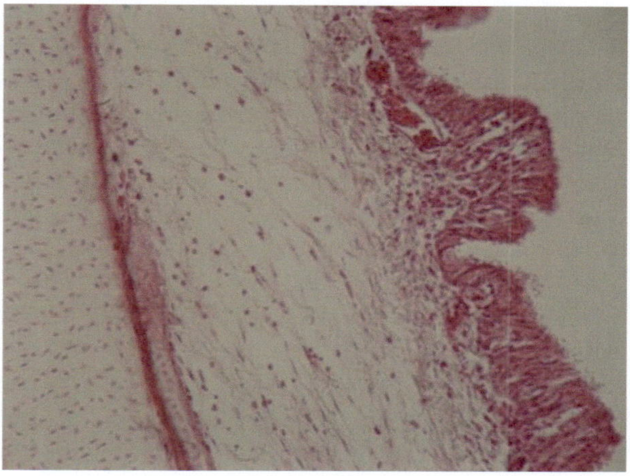

Fig. 15.6 Histological lesions of the trachea with *Cryptosporidium* spp. Loss of the brush border in the epithelium is observed due to the presence of the parasite, in addition to lymphocytic infiltrate and hyperplasia of the epithelium (X. Hernandez-Velasco)

contrast microscopy and several staining procedures [e.g., Giemsa stain, acid-fast smear (Fig. 15.7), methylene blue] are helpful in increasing optical contrast, immunofluorescence, or molecular methods such as PCR (Bowman 2019). However, detecting *Cryptosporidium* in fecal samples does not necessarily indicate the clinical disease, as some turkeys can harbor the parasite without showing any clinical signs (Wages and Ficken 1989).

15.1.3.6 Control and Treatment

Currently, no specific drugs are available to treat Cryptosporidiosis in turkeys. Managing the disease mainly involves supportive care such as rehydration, electrolyte therapy, and nutritional support. Hygiene practices are recommended, such as regular cleaning and disinfecting of the environment and equipment and avoiding overcrowding and stress in the birds (Bermudez et al. 1988; Helmy and Hafez 2022).

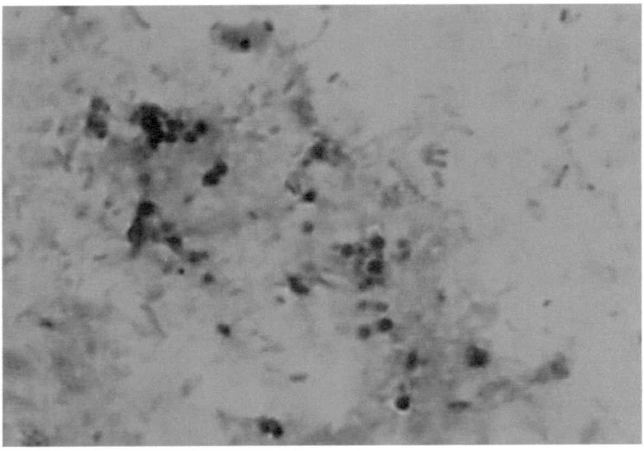

Fig. 15.7 Detection of *Cryptosporidium* sp. oocysts in smear droppings (acid-fast stained) (X. Hernandez-Velasco)

15.1.4 Haemoproteus

Haemoproteus parasites, belonging to the family *Haemoproteidae*, are a sister group to malaria agents of the genus *Plasmodium* (*Plasmodiidae*). However, *Haemoproteus* parasitize only birds and reptiles. Haemoproteus and Plasmodium do not digest hemoglobin completely, leading to the accumulation of residual pigment in red blood cells. This characteristic feature distinguishes Haemoproteus from *Leucocytozoidae* and *Garniidae*, which do not develop residual pigment in red blood cells. *Haemoproteus* spp. are transmitted by Culicoides biting midges (*Ceratopogonidae*), and a few species are vectored by louse flies (*Hippoboscidae*) (Valkiūnas and Iezhova 2022). Experimentally, *H. meleagridis* causes lameness, diarrhea, anorexia, and depression in turkeys. Anemia, depression, and lameness can be observed. Histologic lesions were associated mainly with megaloschizont development in the musculature. The diagnosis is based on stained blood smears and observation of large, pigmented gametocytes in mature RBCs. Merozoites are not seen in the peripheral blood. Schizogony can be detected in the endothelial cells of the lung, liver, and spleen. PCR tests for *Haemoproteus* can also be used (Valkiūnas et al. 2022). No drug is approved for commercial use. Antimalarial drugs reduce the parasitemia but do not eliminate the parasite. Chloroquine, primaquine, quinacrine, and buparvaquone have been used in pigeons. Combinations of chloroquine, primaquine, and mefloquine have been used to treat owls. Treatment is not recommended in asymptomatic birds. Hygienic measures to control the vectors help to control the disease.

15.1.5 Leucocytozoon

15.1.5.1 Summary

Leucocytozoon is an obligate intracellular protozoan parasite of wild and domestic *birds.* There are approximately 86 species of *Leucocytozoon; many of them are host-specific, and all are transmitted by simuliids. Leucocytozoon smithi causes* the disease known as leucocytozoonosis or turkey malaria. *The pathogenicity of this parasite is usually low but can be particularly pathogenic to young turkeys. Diagnosis is based on finding the gamonts in Giemsa-stained blood smears or meronts in tissue sections.* Raising turkeys in poultry houses and improvements in biosecurity have reduced the incidence of leucocytozoonosis.

15.1.5.2 Epidemiology

It is distributed mainly in Europe and North America. *L. smithi* is transmitted by blood-sucking black fly (*Simuliidae meridionale* and *S. slossonae*) (Peirce 2005; Pramual et al. 2020).

15.1.5.3 Life Cycle

The sporozoites infect a new host when an infected female blackfly feeds on it. The sporozoites reach the blood and invade cells of various tissues and organs, where they form endogenous stages (schizonts and later meronts), causing necrosis (Peirce 2005). In the liver, meronts form merozoites, and these enter blood cells and form gamonts. Gametocytes in the blood of the host are acquired by a blackfly. Schizogony occurs in tissues cells, but not in circulating blood cells. Sporogony is confined to blackflies. In the gut of the blackfly, oocysts are formed. The sporozoites then migrate to the fly's salivary glands and infect the avian host as the fly feeds (Steele and Noblet 1993).

15.1.5.4 Signs

Clinical signs include dehydration, anorexia, depression, hemoglobinuria, and hemolytic anemia. Sick birds decrease feed intake, lose weight, and have difficulty moving or incoordination. Severe infections cause emaciation, convulsions, and death. Birds that survive 3 days after showing signs of the disease generally recover. Hens infected had higher mortality, decrease in egg production, and hatch of fertile eggs (Atkinson and Van Riper III 1991).

15.1.5.5 Pathogeny

The pathogenicity of *Leucocytozoon* infections is usually low. High levels of infection can occur without clinical signs of disease. *L. smithi* can be severely pathogenic, especially in poults. The degree of tissue damage determines the survival or death of the bird. In severe cases, it can cause hemolytic anemia and death. Surviving turkeys may remain carriers for years, and sometimes, the infection continues causing lethargy and persistent coughing. Affected birds have enlarged spleen and liver, enteritis, pale heart muscle, hemosiderosis, pneumonia, and congested lungs (Taylor et al. 2016; Pramual et al. 2020).

15.1.5.6 Diagnostic

The parasites (elongated gametocytes) can be seen in Giemsa-stained blood smears. Gametocytes develop into erythroblasts, erythrocytes, monocytes, and leukocytes and cause great distortion of the host cell and its nucleus. On postmortem, the spleen and liver are enlarged, and there may be enteritis. Schizonts can be seen in histological sections of liver or other infected tissues such as kidney, muscle, and brain (Steele and Noblet 1993).

15.1.5.7 Treatment and Control

There are no reported effective treatments. Disease prevention depends on blackfly control. Turkeys should not be raised in areas with high populations of blackflies.

15.1.6 Histomoniasis

15.1.6.1 Summary

Histomoniasis, also known as blackhead disease or typhlohepatitis, is a parasitic disease caused by the flagellate protozoan parasite *Histomonas meleagridis*. The disease is distributed worldwide. The parasite exists in two forms in the host: amoeboid (lacks flagella and is present in tissue) and flagellated form (has flagella and is present in the lumen of caeca). The parasite is transmitted by the earthworms that carry the *Heterakis gallinarum* eggs containing histomonads. Histomoniasis affects several bird species, but it is mostly seen in turkeys, which can lead to severe economic losses due to the high mortality rate of 80–100%. The disease diagnosis is based on clinical signs and postmortem lesions of unilateral or bilateral cecal lesions associated with circular necrotic areas in the liver. Flagellated *H. meleagridis* can be demonstrated in wet smears from the ceca. Although nitroimidazoles (dimetridazole, ipronidazole, and ronidazole) were effective against histomoniasis, they are no longer permitted (banned) in several countries. Available products such as anticoccidial drugs roxarsone, antibiotics, and benzimidazole derivatives are ineffective in treating histomoniasis. The prophylaxis depends on implementing biosecurity measures and deworming of *Heterakis gallinarum.*

15.1.6.2 Etiology

Histomoniasis is caused by the flagellated protozoan *Histomonas meleagridis*, a unicellular parasite belonging to the Phylum *Parabasalia*, Class *Tritrichomonas*, Order *Tritrichomonadida*, and Family *Dientamoebidae.* The parasite has two morphological forms: amoeboid and flagellated forms. The amoeboid form, in tissues, is roundish or oval (pleomorphic), 8–19 μm in size, and lacks flagella; this flagellum is lost upon mucosal invasion with the development of pseudopods. The flagellar form, in the cecal lumen, is similar to the amoeboid form but has one, rarely two, flagella. Reproduction of both forms occurs via binary fission. The flagella are sensitive to the external environment and survive in excreted feces for a few hours. However, they remain infective in *Heterakis gallinarum* eggs for several years.

15.1.6.3 Epidemiology

Histomoniasis of turkeys was first described by Smith (1895). Although the disease occurs in chickens and numerous poultry species such as chickens, pheasants, and peafowls, turkeys are the most susceptible host (Landim de Barros et al. 2022).

Therefore, the separate rearing of poultry species is critical because chickens frequently serve as asymptomatic carriers and reservoirs of *H. meleagridis*-infected heterakid eggs.

Turkeys can be infected at any age, but the disease is most common between third and 12th weeks of age. The course of the disease is mainly influenced by the age and the intestinal flora of the host. In addition, the disease is usually accompanied by secondary bacterial infections such as *E. coli* and *Clostridium perfringens*. *H. meleagridis* is distributed worldwide.

The susceptibility of three different turkey lines, namely, wild Canadian turkey (WCT), British United turkey (BUT-Big6), and Kelly–Bronze turkeys (KBT) after experimental infection with *H. meleagridis,* was investigated. The mortality rate was 95% in WCT, 78% in BUT-Big6, and 75% in KBT. The obtained results demonstrate that all tested turkey lines are susceptible to infection; however, the mortality rate for the wild Canadian turkey is statistically significantly higher compared to the other tested two lines (Hafez et al. 2010).

15.1.6.4 Transmission

Histomonas meleagridis can be directly transmitted through the cloacal drinking transfer of materials from the vent region into the caeca through waves of reverse peristalsis. Horizontal transmission of *H. meleagridis* can occur through contact of infected and uninfected turkeys, even in the absence of *Heterakis gallinarum*. *Histomonas meleagridis* can be transmitted by the earthworms that carry the *H. gallinarum* eggs containing Histomonads. *Heterakis gallinarum nematode is* commonly found in infected turkeys' caeca, and the eggs are resistant to environmental conditions. Once the nematode ingests the protozoan, it becomes infected and releases the parasite into the host's cecal contents. The protozoan then invades the cecal wall, causing ulcers and inflammation, which can lead to the formation of nodules in the liver, spleen, and other organs (Hess et al. 2008) (Fig. 15.8).

The incubation period for natural infection is 7–12 days. The disease is characterized by high morbidity and mortality rates (70–90%).

15.1.6.5 Clinical Signs and Postmortem Lesions

The clinical signs of histomoniasis in turkeys can vary depending on the severity of the infection. Early signs may include depression, decreased appetite, and diarrhea, which can progress to more severe symptoms such as weight loss, lethargy, and anemia. In severe cases, turkeys may develop a characteristic "dropped wing" posture, which is caused by inflammation and damage to the liver (Beer et al. 2022b).

The main postmortem lesions are observed in the caecum and liver. The lumen of the caecum is dilated, and the thickened cecal mucosa is covered with fibrinous to diphtheroid inflammation that can pulplike to fill the lumen. Hepatic lesions are highly variable and appear 6–8 days after infection. The necrotic target-like lesions with raised edges and a depressed center on the liver are pathognomonic lesions for *H. meleagridis*. In some cases, the liver will appear green or tan. Further lesions might also be seen in other organs, such as the kidneys, bursa of Fabricius, spleen, and pancreas (Figs. 15.9 and 15.10).

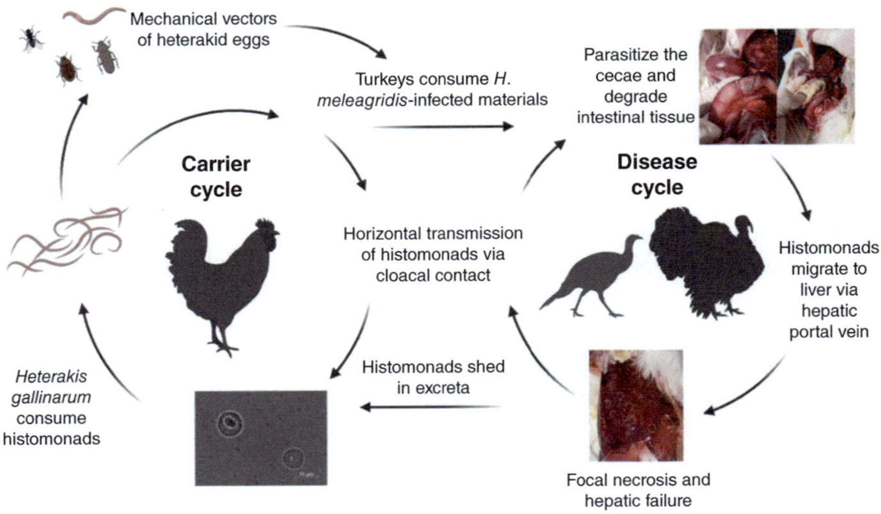

Fig. 15.8 Complex transmission of *Histomonas meleagridis*. Histomonads can be consumed by *Heterakis gallinarum* and subsequently incorporated into the nematode ova. Once inside the intestine, the histomonads migrate to the ceca, replicating and degrading the cecal lining. Carrier birds such as chickens can harbor the cecal worms and shed infected heterakid eggs into the environment. Earthworms, flies, and other invertebrates can be intermediate mechanical vectors of infected heterakid ova. Turkeys may ingest infected materials such as excreta or invertebrates contaminated with the protozoa. Direct transmission can occur rapidly from bird to bird due to cloacal drinking and reverse peristalsis movement of materials into the vent region. [Adapted from (Beer et al. 2022b)]

15.1.6.6 Diagnosis

Diagnosis of histomoniasis in turkeys can be challenging, as clinical signs can be similar to other poultry diseases. A definitive diagnosis can be made through isolation (Hauck et al. 2010) and histopathological examination, in situ hybridization or immunohistochemistry of affected tissues, such as the cecal wall, liver, or spleen, where characteristic lesions caused by the protozoan can be observed. Suspected tissue sections can be stained with hematoxylin and eosin, or periodic acid–Schiff to investigate the *Histomonas meleagridis* trophozoites (El-Wahab et al. 2021). *Histomonas meleagridis* can be isolated on a medium containing Medium 199 rice flour and horse serum, balanced with Hank or Earle salts, under microaerophilic conditions.

Additionally, molecular diagnostic techniques such as polymerase chain reaction (PCR) can be used to detect the presence of the parasite (Grabensteiner and Hess 2006; Hafez et al. 2005). Molecular typing based on sequencing various genes has been described (Bilic et al. 2014; Hauck et al. 2010). At the most, four types were distinguished, and it is unclear if the detected types relate to differences in virulence of the host spectrum.

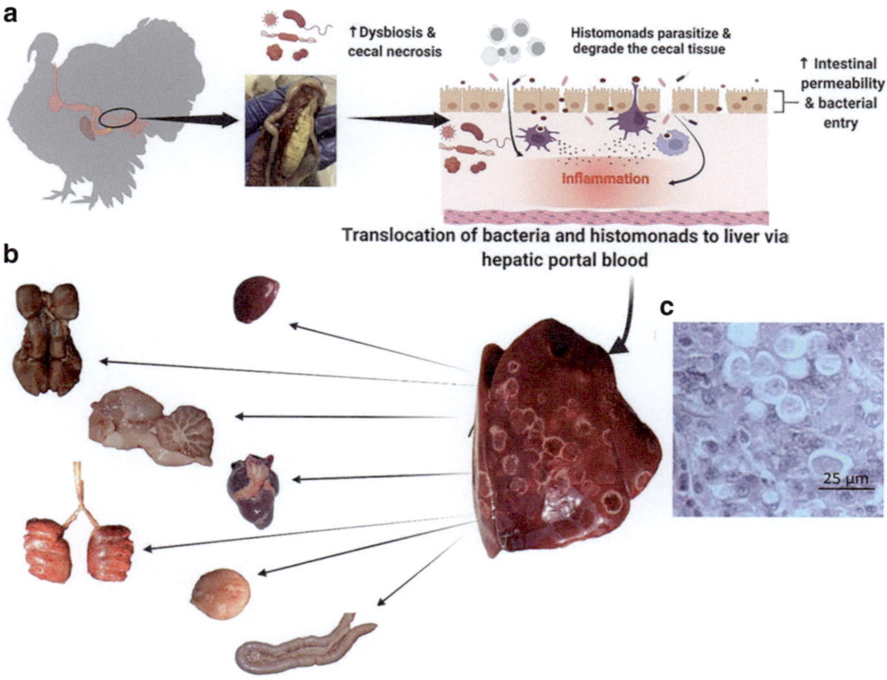

Fig. 15.9 Pathogenesis of histomoniasis. The parasite induces a severe inflammatory reaction in the ceca. The inflammatory reaction is followed by necrosis, with dysbiosis causing increased permeability in the ceca (**b**). This allows bacterial and parasitic translocation to the liver via the hepatic portal blood; the resulting pathognomonic lesions are exhibited as target-like liver lesions and caseous cecal cores (**a**). *Histomonas meleagridis* in the liver of a turkey, periodic acid–Schiff (PAS), 40 × (**c**). From the liver, bacteria, and histomonads migrate to other parenchymal organs (spleen, heart, kidneys, pancreas, lungs, brain, bursa of Fabricius), causing chronic systemic inflammation and multiple organ failure. [Adapted from Beer et al. 2022b)]

15.1.6.7 Control

Control of histomoniasis in turkeys requires a multifaceted approach, including biosecurity measures to prevent the introduction and further spread of the infection and contamination of equipment and treatment of infected birds.

Paromomycin, an aminoglycoside antibiotic with activity against protozoa, is currently registered for food-producing animal species. In a pilot study, the efficacy of different doses of paromomycin in the feed against *Histomonas meleagridis* in experimentally challenged turkey poults was evaluated. Groups consisting of 30 birds each were given feed with 100, 200, and 400 ppm paromomycin, respectively, starting on day 1 via feed to day 42. On day 21, all birds were infected intra-cloacally with *H. meleagridis*. One group of 30 birds was left untreated. Additionally, ten birds were kept as a noninfected and non-treated control group. Before the challenge, there was no significant difference between untreated and treated groups concerning feed consumption and feed conversion rate. After the challenge, mortality

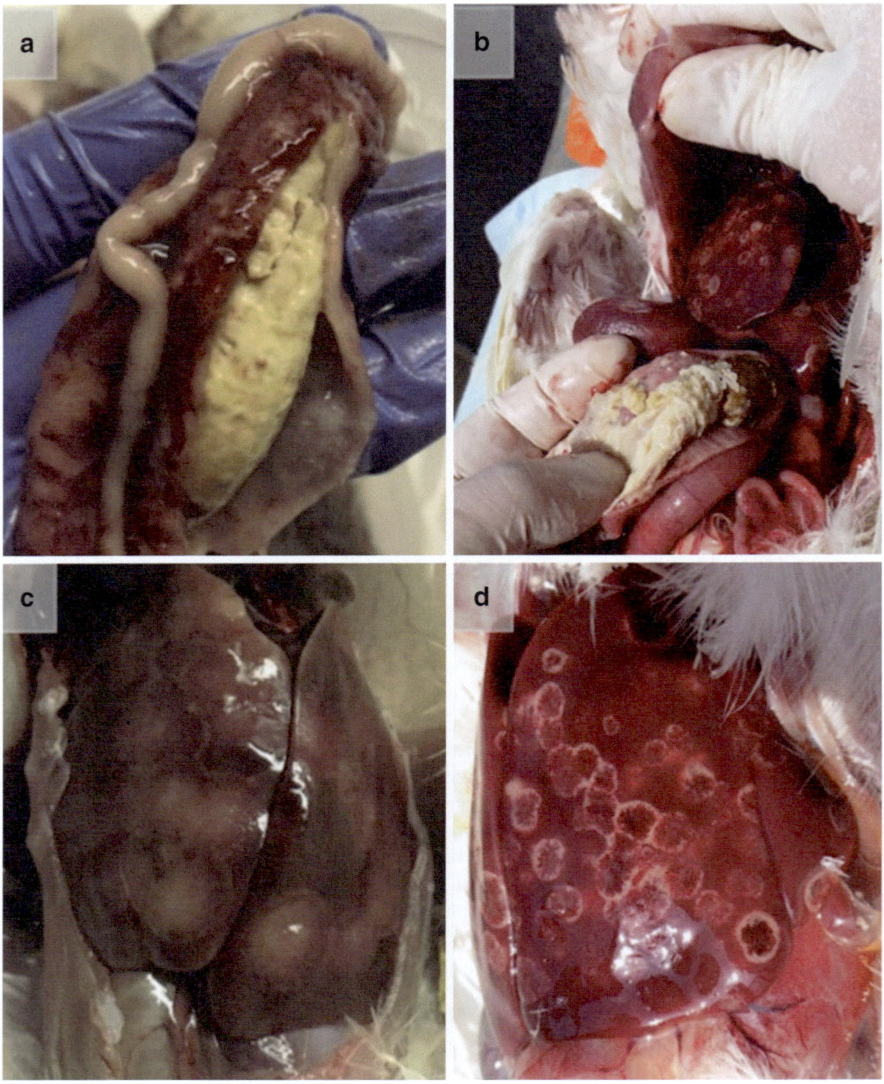

Fig. 15.10 Classic lesions resulting from *Histomonas meleagridis* infection. (**a**, **b**) Caseous cheese-like cecal core; (**c**, **d**) focal necrosis resulting in target-like liver lesions [Adapted from Beer et al. 2022b)]

was 80% in the infected non-treated birds. In the treated groups, the mortality rate was 73.3%, 43.3%, and 20%, respectively. No histomonal DNA was found in the caeca and livers of the surviving birds. In addition, higher doses of paromomycin (200 and 400 ppm) led to reduced *Clostridium perfringens* count in the droppings (Hafez et al. 2010).

In conclusion, a prophylactic application of paromomycin as a feed additive was effective against the challenge with *H. meleagridis* under experimental conditions (Hafez et al. 2010). The most effective control measure is the use of a live attenuated vaccine, which can provide immunity to the protozoan and reduce the severity of the disease. Also, management practices such as removing wet litter and using effective disinfectants can help reduce the risk of infection (Liebhart et al. 2017; Beer et al. 2022a).

15.1.7 Trichomonas

15.1.7.1 Summary

Trichomoniasis is a parasitic disease that affects a wide range of animals, including birds. Among birds, it primarily affects pigeons, and turkeys are particularly susceptible to this disease. Trichomoniasis in turkeys is caused by a protozoan parasite known as *Trichomonas gallinae*. This disease is often characterized by the presence of yellowish, cheesy masses in the oral cavity and upper digestive tract of infected birds (Villeneuve and Brugère-Picoux, 2015). The diagnosis depends on demonstrating the large number of *Trichomonas gallinae* in wet smears prepared from oropharyngeal swabs. For prophylaxis, asymptomatic carriers and sick birds must be removed from the flock to control the disease.

15.1.7.2 Etiology

Trichomoniasis in turkeys is caused by the protozoan parasite *Trichomonas gallinae, Parabasalia: Trichomonadida.* The body is elongated or pyriform with four anterior flagella, 5–19 × 2–9 µm. *Trichomonas* reproduces by binary fission and does not form cysts (Taylor et al. 2016). The parasite can persist for up to 1 h in water sources such as gutters and drinkers (Purple et al. 2015). In unfavorable conditions, the parasite can form a pseudocyst; however, the moist environment is essential to maintain its infectivity (Amin et al. 2014).

15.1.7.3 Epidemiology

The parasite is worldwide distributed and an obligate, anaerobic, flagellated organism that resides in the upper digestive tract of birds, particularly in the oral cavity, esophagus, crop, and proventriculus. *Columbiformes, Accipitriformes, Strigiformes, Psittaciformes, Falconiformes,* and *Passeriformes* are susceptible. Pigeon is the main reservoir for *T. gallinae* and acts as the main carrier for other birds, such as chickens and turkeys.

15.1.7.4 Transmission

The parasite is transmitted from bird to bird by ingesting contaminated food, water, or saliva. Wet litters and overcrowding promote transmission. It can also be transmitted through contact with infected oral secretions or recently contaminated water.

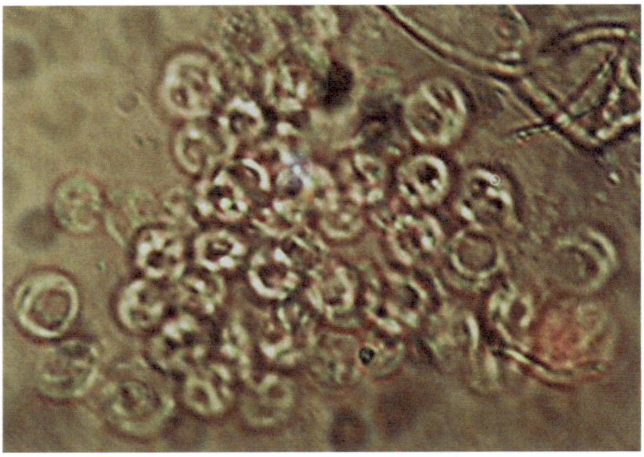

Fig. 15.11 *Trichomonas gallinae* (X. Hernandez-Velasco)

15.1.7.5 Pathogenesis

Trichomonas is a normal inhabitant (commensal) of the upper gastrointestinal tract. In the turkey lesions, most commonly occur in the crop, esophagus, and pharynx and are uncommon in the mouth (Taylor et al. 2016). However, depending on strain virulence and host susceptibility, it might cause mild to severe lesions (Marx et al. 2017). After ingestion, the parasites enter the upper digestive tract of turkeys and start to reproduce rapidly. The parasites attach themselves to the surface of the host's epithelial cells and cause inflammation and necrosis of the affected tissues. This results in the formation of yellowish, cheesy masses in the digestive tract of infected birds. Sometimes, the parasite spreads to the liver and other visceral organs.

15.1.7.6 Clinical Signs and Lesions

The clinical signs of trichomoniasis in turkeys can vary depending on the severity of the infection. The most common signs include lethargy, anorexia, weight loss, regurgitation, and difficulty swallowing. Affected birds may also exhibit a yellowish discoloration of the feathers around the beak, and a greenish fluid may be found in the mouth and a foul-smelling odor from the oral cavity.

15.1.7.7 Diagnosis

The diagnosis of trichomoniasis in turkeys is based on a combination of clinical signs, gross pathology, and laboratory tests. The presence of yellowish, cheesy masses in the oral cavity and upper digestive tract is a strong indicator of trichomoniasis. The diagnosis can be confirmed by microscopic examination of fresh wet-mount preparations or stained smears of the affected tissues (Fig. 15.11). Trichomonas has twisting and wriggling movement. PCR-based methods can also be used for the detection of *T. gallinae* DNA in the affected tissues (Sigrist et al. 2022).

15.1.7.8 Control and Treatment

Prevention of trichomoniasis in turkeys involves implementing good biosecurity measures such as proper sanitation, disinfection of equipment, and control of bird movement. Infected birds should be isolated and treated promptly to prevent the spread of the disease to other birds. The use of probiotics and immunostimulants may also be beneficial in preventing trichomoniasis in turkeys.

The antiprotozoal drugs such as enheptin, dimetridazole, metronidazole, carnidazole, ronidazole, and ornidazole are effective against trichomoniasis. These drugs are administered orally or in drinking water for 5–7 days. Some authors recommended that mixing the drugs with food is a better choice, since the crop empties slower, and the flavor of drinking water can be altered, and birds refuse to drink. Several strains develop resistance to nitroimidazoles, highlighting the need to develop new natural products. Natural alternatives such as chitosan and several phytogenic substances (extracts or essential oils) are also effective against *T. gallinae* (Gómez-Muñoz et al. 2022).

15.1.8 Cochlosomiasis

Cochlosoma is a protozoan genus belonging to the Trichomonadida Order. They have a prominent adhesive disc, axostyle, costa, and six flagella, one of which is attached to a membrane that runs laterally along the body and undulates. A large **sucking organ** is present on the anterior ventral surface covering one third to one half of the body length. Members of this genus have an oval body (6–12 × 4–7 μ). The movement is erratic and jerky with horizontal rotation, no spiraling action (Clipsham 1995). *Cochlosoma* species inhabit the intestinal tracts of birds and mammals. They have been linked to vomiting and enteritis in young turkeys and ducks. The genus currently contains five species, with *C. anatis*, a parasite of ducks and turkeys, being the most notable (Cooper et al. 1995).

Cochlosoma species infection has also been linked to retching and catarrhal enteritis in domestic and wild turkey poults and ducklings. *Cochlosoma* adheres to the host via suction, leaving tiny *lesions* and swelling on the epithelial surface (Lindsay et al. 1999).

Although *Cochlosoma* species have been identified as the primary culprit in numerous instances, they are frequently seen in association with Coccidia, *Salmonella*, and *Hexamita*. *Cochlosoma* outbreaks can be avoided by minimizing overcrowding and environmental pollution. Both metronidazole and ronidazole are potent anti-infective agents (Evans et al. 2006).

15.1.9 Hexamitiasis (Spironucleosis)

Hexamitiasis is an acute protozoal disease of turkeys, pheasants, peafowl, and quail caused by *Spironucleus meleagridis* (formerly *Hexamita meleagridis*) characterized by catarrhal enteritis. Pigeons can also be affected by *Spironucleus columbae*. The disease incidence of this parasite is low in commercial turkeys. *Spironucleus meleagridis* is spindle-shaped, approximately 6–10 μm × 2–4 μm, with four anterior, two anterolateral, and two posterior flagella. Principally, a disease of poults under 10 weeks. The transmission occurs via the fecal-oral route, or by ingestion of contaminated water. The main clinical sign is chronic watery diarrhea, and white or tan foamy droppings are common. Birds may die in comas or convulsions. Postmortem examination reveals watery distension of the small intestine with catarrhal inflammation (Clipsham 1995; Villeneuve and Brugère-Picoux, 2015). The diagnosis depends on microscopical examination to demonstrate the presence of the parasite in the intestinal mucous (scrapings of the duodenal and jejunal mucosa). The small size, large number of flagella, and linear locomotion are important differences from other flagellated enteric protozoa. There is no effective treatment or vaccine for hexamitiasis, although antibiotics may be used to control secondary infections. Hygienic measures and an all-in-all-out principle are recommended to avoid parasite transmission from adult carriers to young poults. Also, it is recommended to raise single-aged and single-species flocks (Clipsham 1995).

15.2 Nematode (Roundworms) Parasites Affecting Turkeys

Poultry worms are divided into three main groups: (1) roundworms that includes large roundworms (e.g., *Ascaridia*), cecal worms (e.g., *Heterakis*), capillary worms (e.g., *Capillaria*), and Gapeworms (e.g., *Syngamus*); (2) tapeworms (e.g., *Raillietina* and *Choanotaenia*); and (3) flukes, rare in occurrence and therefore, none of economic importance. Roundworms (nematodes) are elongated, cylindrical, unsegmented worms. The body is covered with a chitinous cuticle plain striated or ornamented.

Nematodes have a simple and complete alimentary tract, and in contrast to tapeworms, are male or female in sex. Males usually smaller than females and usually have two testes, copulatory organs (spicules), and a membrane structure (bursa) at the posterior extremity of the body. Ovary and uterus form a continuous structure. Some of the nematode parasites have shown to have a marked pathological effect on the hosts *Heterakis* acts as a vector for *H. meleagridis,* the protozoa causing blackhead in turkey and chickens.

15.2.1 *Ascaridia* spp.

15.2.1.1 Summary

Ascaridia spp. are large-sized parasitic nematodes commonly found in the intestines of chickens and turkeys. *Ascaridia* infection reduces efficiency of feed utilization as primary damage, but in severe infections, it can cause death. Control of *Ascaridia* infections can be conducted through implementation of management practices such as regular cleaning, use of disinfectants, and chemotherapy, and proper disposal of the used litter.

15.2.1.2 Epidemiology

Ascaridia galli is one of the most common parasitic roundworms of poultry. ***A. dissimilis*** and *A. galli* are the most common species of this round worm in **turkeys** (McDougald 2020) (Fig. 15.12).

15.2.1.3 Pathogenesis

In general, it does not cause much damage, but under favorable conditions, these worms can cause significant health problems for turkeys, leading to reduced growth rates, poor feed conversion, and even death in severe cases. At the necropsy of these birds, it is possible to find numerous amounts of nematodes obstructing the intestine (Norton et al. 1992; McDougald 2020) (Fig. 15.13).

15.2.1.4 Life Cycle

Ascaridia spp. are soil-transmitted parasites that infect birds through ingestion of infective eggs. There are four larval stages of Ascaridia in the life cycle. The first and second larval stages of *A. galli* and *A. dissimilis* develop within the egg outside the host. Once inside the bird's intestinal tract, the eggs hatch, and the larvae penetrate the intestinal wall and migrate to the liver (via the hepatic portal circulation), heart, and lungs before returning to the intestine, where they mature into adult worms. The second larval stage of *Ascaridia dissimilis* becomes adult inside the host within 1 month. Adult worms mate and produce eggs, which are shed in the feces, completing the life cycle (Hemsley 1971).

15.2.1.5 Clinical Signs of Infection

In young turkeys, *Ascaridia* spp. infections can cause a range of clinical signs, including reduced feed intake, poor growth rates, diarrhea, and increased mortality rates. In severe cases, *Ascaridia* spp. infections can cause intestinal obstruction and perforation, leading to death (Collins and Andersen 2022). *Ascaridia dissimilis* is associated with white spots in the liver.

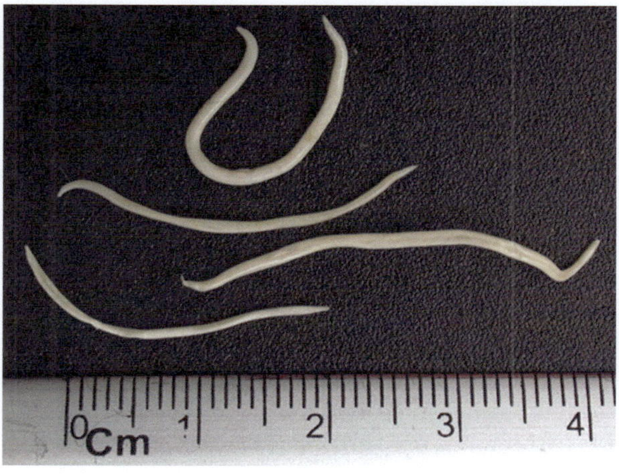

Fig. 15.12 *Ascaridia galli* (X. Hernandez-Velasco)

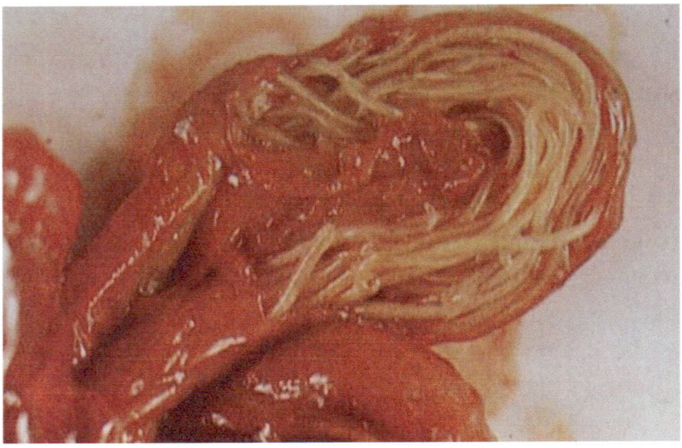

Fig. 15.13 Numerous specimens of *Ascaridia* species occupying the entire lumen of a turkey intestine (H.M. Hafez)

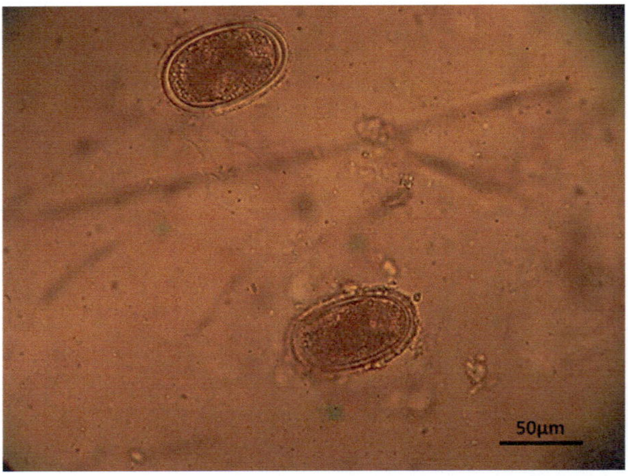

Fig. 15.14 *Ascaridia galli* eggs (X. Hernandez-Velasco)

15.2.1.6 Diagnosis

Diagnosis of *Ascaridia* spp. infections in turkeys can be challenging, as clinical signs can be nonspecific and may be confused with other poultry diseases. A fecal examination is the most common diagnostic method, and the *Ascaridia* egg is very similar to that of *Heterakis*. The egg of *Ascaridia* is slightly larger and more barrel-shaped than *Heterakis* eggs (Zajac 2021) (Fig. 15.14). The detection of *Ascaridia* spp. eggs in feces are not always reliable, as eggs may be shed intermittently or in low numbers. Serological tests and postmortem examinations can also be used for diagnosis (Băcescu et al. 2011).

15.2.1.7 Control Measures and Treatment

Preventive measures are critical in controlling *Ascaridia* spp. infections in turkey flocks. These include regular cleaning and disinfection of the environment, minimizing contact with contaminated soil and water sources, and implementing biosecurity measures to prevent the introduction of infected birds to the flock (Hemsley 1971). Ascaridia eggs are killed at 45 °C for 12 h.

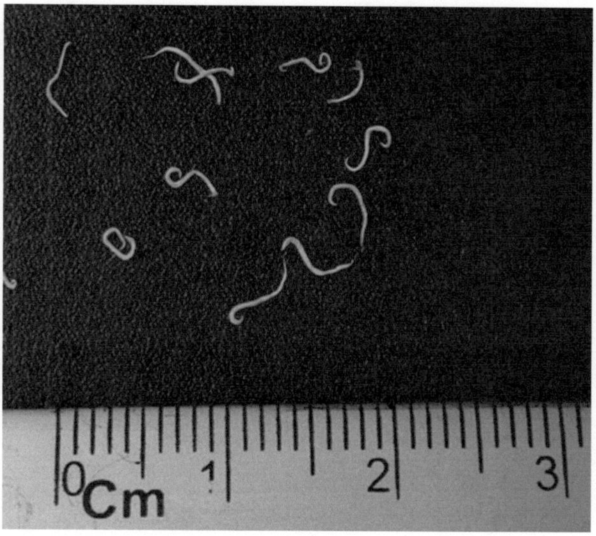

Fig. 15.15 *Heterakis gallinarum* (X. Hernandez-Velasco)

Regular deworming programs should also be implemented, and the efficacy of these programs should be monitored through fecal examinations. Several anthelmintic drugs are effective against *Ascaridia* spp. infections in turkeys, including **benzimidazoles**, **levamisole**, and **ivermectin**. The use of plant extracts with anthelmintic potential has been investigated in this and other genera of parasitic nematodes (Hassanain et al. 2009). Fenbendazole and levamisole are effective against the adult worms as well as larval stages. Piperazine is effective against adult worms. Phenothiazine is not effective against Ascaridia.

15.2.2 *Heterakis gallinarum*

15.2.2.1 Summary

Heterakis gallinarum is a small and thin parasitic nematode (Fig. 15.15) that commonly infects turkeys, chickens, pheasant, and quail causing a variety of health problems and economic losses. *Heterakis gallinarum* is the most important nematode in turkeys because it serves as a transport host for *Histomonas meleagridis*, the causative agent of Histomoniasis (blackhead disease). *H. gallinarum* has a direct life cycle, meaning it can complete its life cycle within a single host, although earthworms can ingest *H. gallinarum* eggs and carry them to the intestine, and birds may become infected by eating such earthworms. The adult worms parasitize the cecum and large intestine of turkeys, where they lay eggs that pass out in the feces. The eggs then develop into infective larvae, which can survive in the soil for several months. When turkeys ingest contaminated soil or feed, the larvae are released in the intestine and migrate to the cecum, where they develop into adult worms. Control is based on hygiene.

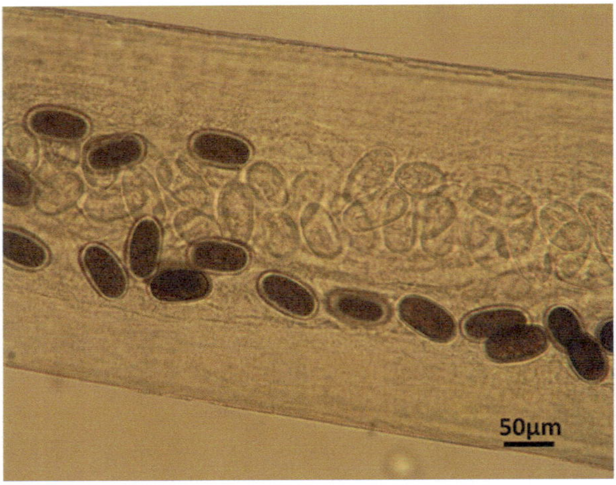

Fig. 15.16 *Heterakis gallinarum* eggs (X. Hernandez-Velasco)

15.2.2.2 Epidemiology

H. gallinarum is prevalent in turkeys worldwide, particularly in intensive production systems, where birds are housed in large numbers. The infection is more common in young turkeys, with the highest prevalence occurring between 6 and 12 weeks of age. The parasite can persist in the environment for long periods, making it difficult to control (Cupo and Beckstead 2019) *Heterakis gallinarum.*

15.2.2.3 Clinical Signs

H. gallinarum infection can cause a range of clinical signs in turkeys, including diarrhea, decreased appetite, weight loss, and reduced growth rates. The severity of the clinical signs depends on the intensity of the infection and the age of the birds. *H. gallinarum* can cause inflammation and thickening of the cecal wall and the formation of nodules in the mucosa and submucosa in severe infections (Brener et al. 2006; McDougald 2020).

15.2.2.4 Diagnosis

The diagnosis of *H. gallinarum* infection in turkeys is based on the identification of eggs, which are slightly smaller than *Ascaridia* eggs and with flattened walls (Zajac 2021) (Fig. 15.16) or adult worms in the feces and/or in intestinal contents of infected birds (Fig. 15.17). The eggs are oval-shaped and have a thick shell with a characteristic pitted surface. They can be identified using standard fecal flotation techniques.

15.2.2.5 Prevention and Treatment

Prevention of *H. gallinarum* infection in turkeys involves good management practices, including proper hygiene and sanitation, regular cleaning and disinfection of the environment, and avoidance of overcrowding. The use of anthelmintic drugs should be limited to cases where the infection is diagnosed, and the drugs should be rotated to reduce the risk of resistance.

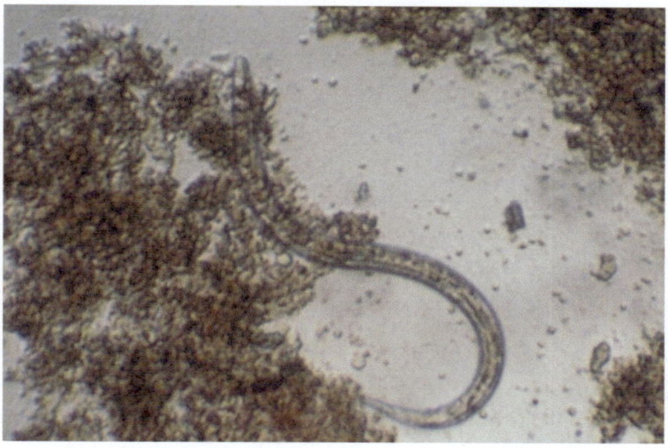

Fig. 15.17 *Heterakis gallinarum* seen under the light microscope in a feces sample (X. Hernandez-Velasco)

Several anthelmintic drugs are effective against *H. gallinarum* infection in turkeys, including benzimidazoles, tetrahydropyrimidines, and imidazothiazoles. These drugs can be administered orally or in the feed. However, resistance to anthelmintic drugs is a growing concern, particularly in intensive production systems, where the parasites are exposed to sublethal doses of drugs (Seddiek et al. 2011).

15.2.3 Subulura

15.2.3.1 Summary

Subulura brumpti is a small-sized nematode parasite that affects the gut (usually ceca) of turkeys and chickens. *Subulura brumpti* has indirect life cycles, and intermediate hosts are certain beetle (e.g., *Alphitobius* spp.) and cockroach species. These worms are less frequent than *Heterakis* or *Ascaridia*, usually not or only mildly pathogenic, and most infections do not cause clinical signs. The diagnosis is based on the identification of the eggs, since the nematodes are very similar to *Heterakis*, in addition to the fact that both species are found in the lumen of the ceca.

15.2.3.2 Etiology

Subulura brumpti belongs to the Family *Subuluridae* and is an obligate parasite that requires a host to complete its life cycle. The adult nematodes are slender and measure about 8–18 mm in length. They have a cylindrical body with a tapered anterior end and a blunt posterior end. The eggs of *S. brumpti* are almost spherical and measure about 66–76 × 82–86 μm, fully embryonated when laid. The larvae of *S. brumpti* are free-living and can survive in the environment for up to 6 months. The life cycle of *S. brumpti* involves an indirect cycle, with the turkey being the definitive host and the earthworm, cockroaches, and beetles serving as the intermediate host (Taylor et al. 2016; McDougald 2020).

15.2.3.3 Epidemiology

Subulura brumpti is found in turkeys and quails worldwide, in general with low prevalence, depending on the geographical location and management practices. The infection is more common in turkeys raised on the ground rather than those raised on wire floors. The transmission of *S. brumpti* occurs via ingestion of intermediate hosts containing infective larvae (Farhang 2012).

15.2.3.4 Pathogenesis

The infection caused by *S. brumpti* can lead to chronic respiratory disease, which is characterized by coughing, sneezing, nasal discharge, and respiratory distress. The nematodes can cause damage to the respiratory system by obstructing the airways, leading to reduced airflow and oxygen uptake. The chronic respiratory disease caused by *S. brumpti* can lead to decreased productivity and mortality, especially in young turkeys (Youssefi et al. 2018).

15.2.3.5 Diagnosis

The diagnosis of subuluriasis can be made by microscopic examination of the ceca contents. The eggs of *S. brumpti* are readily identifiable under the microscope and can be distinguished from other nematode eggs based on their size and shape. Serological tests have also been developed for the detection of *S. brumpti* antibodies in turkeys (Belete et al. 2016).

15.2.3.6 Control

The control of subuluriasis in turkeys involves a combination of management practices and anthelmintic (benzimidazoles, levamisole, and ivermectin) treatment. Good management practices, such as maintaining clean and dry litter, preventing overcrowding, and minimizing exposure to intermediate hosts, can reduce the risk of infection. Anthelmintic treatment is recommended for infected turkeys and can be administered via feed or water. The choice of anthelmintic should be based on the sensitivity of the nematode to the drug and the withdrawal period (Udoh et al. 2014).

15.2.4 *Capillaria* spp.

15.2.4.1 Summary

Capillaria spp. are parasitic nematodes that infect various species of birds, including turkeys. These worms commonly parasite the crop and the intestine and cause weakness and emaciation in infected birds. They are hairlike worms, very fine, and measure half inch to several inches long. In many cases, they can go unnoticed.

15.2.4.2 Epidemiology

Capillaria spp. are ubiquitous parasites found in the soil. They can survive in the environment for several months and are resistant to many disinfectants. Infections

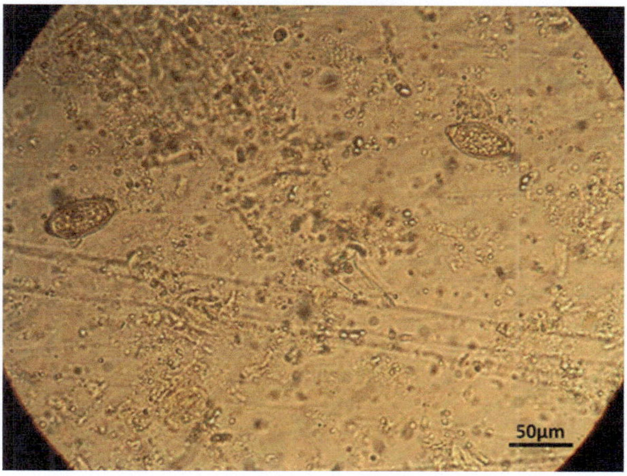

Fig. 15.18 Eggs of *Capillaria* sp. in a feces sample. The eggs have prominent opercula at both ends (X. Hernandez-Velasco)

with these parasites are more common in free-ranging birds. Turkeys become infected by ingesting earthworms with the infective stage. *Capillaria* spp. can infect turkeys of all ages, but young turkeys are more susceptible than adults.

15.2.4.3 Pathogenesis

After ingestion, the eggs hatch in the intestinal tract. *Capillaria* buries its anterior ends in the mucosa of the esophagus and the crop of infected birds; due to this, even a few parasites can produce catarrhal inflammation and thickening of the wall of these organs. Heavy infections can cause diphtheritic inflammation and marked thickening of the esophageal wall and crop, anemia, and weight loss.

15.2.4.4 Clinical Signs

The clinical signs of *Capillaria* spp. infections in turkeys can vary depending on the severity of the infection. Mild infections may not show any clinical signs, while heavy infections can lead to significant morbidity and mortality. The most common clinical signs of *Capillaria* spp. infections in turkeys include diarrhea, weight loss, anemia, and decreased feed intake. Other less common clinical signs may include lethargy, ruffled feathers, and poor growth (Hurst et al. 1979).

15.2.4.5 Diagnosis

The diagnosis of *Capillaria* spp. infections in turkeys is based on a combination of clinical signs, fecal examination, and careful postmortem examination of the esophagus and crop for the presence of the worms. Fecal examination is the most common method of diagnosis and involves the detection of *Capillaria* spp. eggs in the feces (Fig. 15.18). The eggs are barrel-shaped with bipolar plugs, 20–28 × 40–65 μ. *Syngamus* eggs also have two polar caps but are much larger than *Capillaria* eggs. Postmortem examination can also be used to confirm the diagnosis and assess the severity of the infection (Villeneuve and Brugère-Picoux, 2015; Udoh et al. 2014).

15.2.4.6 Treatment

There are several anthelmintic drugs that can be used to treat *Capillaria* spp. infections in turkeys. The most used drugs include benzimidazoles, fenbendazole, levamisole, and ivermectin. Treatment should be initiated as soon as possible to prevent further damage to the intestinal mucosa and organs.

15.2.4.7 Prevention

Prevention of *Capillaria* spp. infections in turkeys are primarily based on good husbandry practices, including proper sanitation and hygiene. Regular cleaning and disinfection of housing, equipment, and feeders can reduce the risk of contamination. Additionally, turkeys should be kept on clean and dry litter to reduce exposure to infective eggs (McDougald 2020).

15.2.5 *Tetrameres americana*

15.2.5.1 Summary

Tetrameres americana (globular roundworm) are small nematodes (no more than 5 mm in length), with marked sexual dimorphism. The male is slender and whitish, while the female is round and bright red. *Tetrameres* lodge in the upper digestive tract, primarily in the proventriculus glands of chickens, turkeys, ducks, pigeons, and quails.

15.2.5.2 Epidemiology

T. americana is distributed mainly in Africa and North America and infects most domestic birds. Intermediate hosts are earthworms, beetles, grasshoppers (*Melanoplus femurrubrum and M. differentialis*), or cockroaches (*Blattella germanica*), so *Tetrameres* infection is not common in housed birds (McDougald 2020).

15.2.5.3 Etiology

Females are hematophagous and globular in shape; they can be seen with the naked eye as dark-colored circular spots on the mucosa of the proventriculus, even without opening it (Fig. 15.19). Males inhabit the mucosal surface, are thin, and are difficult to find due to their diminutive size (Fig. 15.20).

15.2.5.4 Pathogenesis

The bird becomes infected following ingestion of the intermediate host and the parasite locate in the glands of the proventriculus. The females are embedded in the mucosal glands and can cause glandular atrophy, anemia, and erosion; however, in general, they are low pathogenic parasites (Taylor et al. 2016).

15.2.5.5 Clinical Signs and Lesions

The signs it causes in birds are generally mild and nonspecific. The lesions they cause are thickening and congestion of the proventriculus wall, erosions, and anemia.

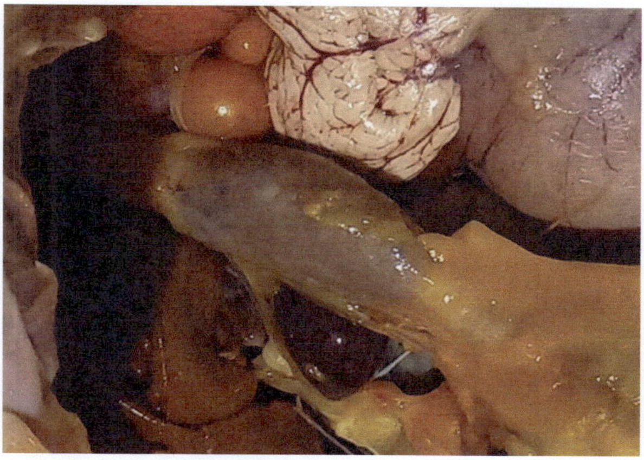

Fig. 15.19 *Tetrameres americana* viewed from the serosal surface of the proventriculus (X. Hernandez-Velasco)

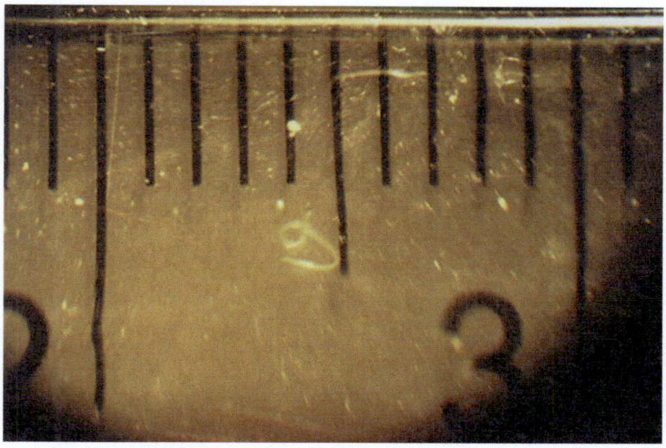

Fig. 15.20 Male *Tetrameres americana* (X. Hernandez-Velasco)

15.2.5.6 Diagnosis

The diagnosis is made during necropsy. Eggs measure 42–50 × 24 μm and are embryonated when passed (Villeneuve and Brugère-Picoux, 2015) (Fig. 15.21).

15.2.5.7 Treatment and Control

As in most helminth infections, control depends on regular anthelmintics treatment such as fenbendazole or levamisole, along with control of intermediate hosts and sanitation measures.

15.2.6 *Syngamus trachea*

15.2.6.1 Summary

Syngamus trachea is a roundworm inhabits the trachea of chickens, turkeys, and other domestic and wild birds. *S. trachea* is often called the red-worm or forked-worm because of its red color and because the smaller male worm attaches itself, in

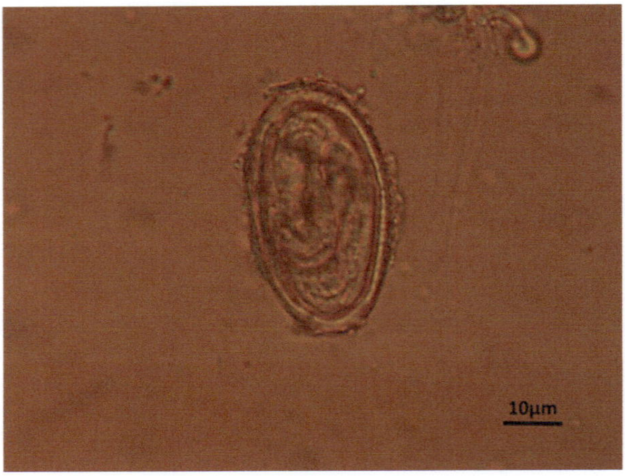

Fig. 15.21 *Tetrameres americana* embryonated egg (X. Hernandez-Velasco)

permanent copulation, to the female and together look like the letter Y. Both male and female worms attach to the lining of the bird's trachea and obstruct the passage of air through it, causing the bird to breathe with its mouth open. Therefore, *S. trachea* is also commonly known as gapeworm. Infection begins when a bird ingests embryonated eggs or the intermediate hosts containing gapeworms larvae. After being eaten by the bird, the gapeworms larvae hatch in its digestive system and migrate from it to the trachea. Gapeworms attach to tracheal lining and reproduce, and the female lays eggs. Subsequently, the eggs are coughed up, swallowed, and excreted along with the feces.

15.2.6.2 Etiology

S. trachea is a medium-sized, red roundworm. The female is larger than the male, measuring from a quarter to an inch in length. The male gapeworm can reach a length of a quarter of an inch. The life cycle of *S. trachea* may be direct by ingestion of embryonated eggs or infective larvae, or indirect involving an intermediate host such as earthworms (*Eisenia fetida* and *Allolobophora caliginosus*), slugs, and snails (McDougald 2020).

15.2.6.3 Epidemiology

This disease is more frequent in birds raised in open spaces where birds have access to the intermediate hosts. Gapeworms affect young birds most severely. The mortality rate among poults and chicks due to an infestation of *S. trachea* is usually high. The soil can remain infected for years.

15.2.6.4 Pathogenesis

The gapeworms clog the trachea of the infested birds, and due to an insufficient supply of air, the infected poults open their beaks to breathe. A convulsive shaking of the head may be observed, with a wheezing cough due to labored breathing. Continuous effort to obtain air prevents infested birds from feeding. Death can

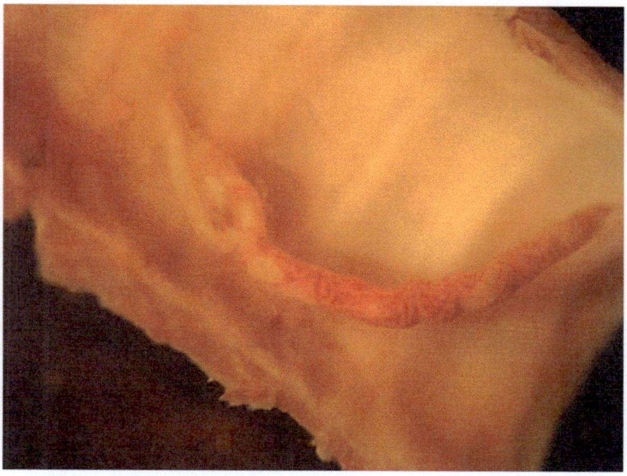

Fig. 15.22 *Syngamus trachea* male and female on the mucosa or the trachea (X. Hernandez-Velasco)

result from suffocation when enough amounts accumulate in the trachea to impede the passage of air or from general weakness. At necropsy, the mucosa of the trachea appears inflamed, and sometimes, there are nodules.

15.2.6.5 Diagnosis

Diagnosis of *S. trachea* infection is based on signs that are highly suggestive of the disease and on dead birds by identification of parasitic nematodes in the trachea (Fig. 15.22). A fecal examination is one of the most widely used methods to detect or confirm the disease. *Syngamus* eggs are large, segmented, and with slightly bulging poles (Zajac 2021) (Fig. 15.23).

15.2.6.6 Control

The prevention of gapeworms can be done with good sanitation, using clean soil and litter, and avoiding the presence of intermediate hosts. Young growing turkeys must be separated from old turkeys and in areas not recently used by other birds. Some drugs effective for eliminating gapeworms are fenbendazole and tiabendazole.

15.3 Tapeworm (Cestodes) Parasites Affecting Turkeys

Tapeworms are flat, white, and segmented parasites. Most tapeworms are large enough to be easily seen; some may be several inches long. However, few species are very small, so they could be often overlooked in postmortem examinations. The cestodes do not have both mouth and a digestive tract. The head region (scolex) may be armed with a special structure (cup) and hooks by which the tapeworm is attached to the intestinal wall. The tapeworm grows at the neck region and becomes numerous segments (proglottids); the segments farthest from the head are the oldest ones. Each segment contains both male and female sexual organs. When the oldest segments mature, they contain a lot of eggs (gravid segments), are released, and

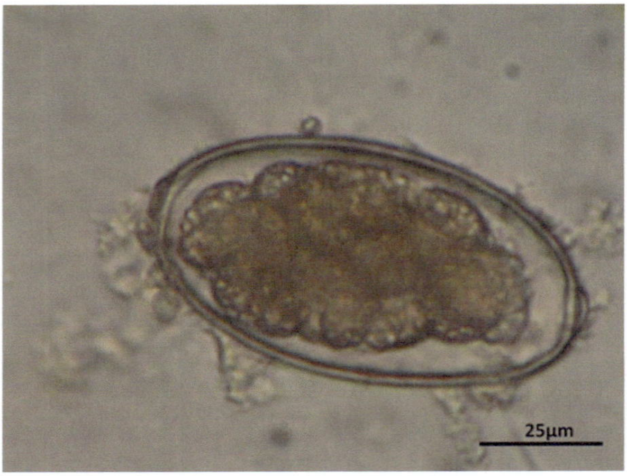

Fig. 15.23 The *Syngamus trachea* egg is segmented and has two polar caps (X. Hernandez-Velasco)

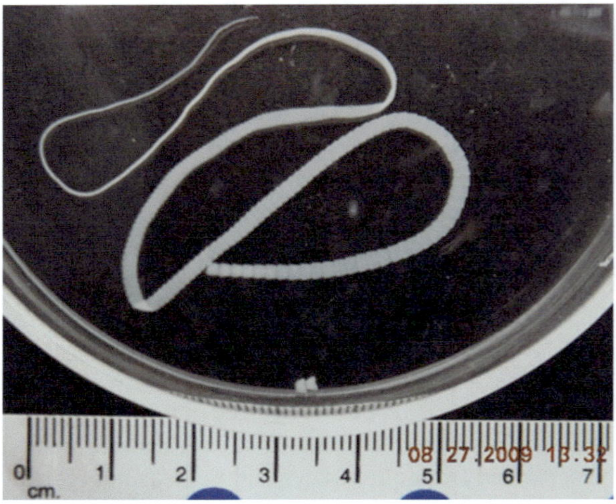

Fig. 15.24 *Raillietina cesticillus* (X. Hernandez-Velasco)

excreted in the droppings. These eggs must be eaten by an intermediate host (the type of intermediate host depends on the tapeworm species). Poultry acquires the infection by eating the infected intermediate host. Various genera can infect poultry. However, tapeworms most commonly encountered are *Raillietina cesticillus* and *Choanotaenia infundibulum.*

15.3.1 *Raillietina cesticillus*

Raillietina cesticillus is a cestode parasite that affects turkeys, chickens, and other domesticated birds (Fig. 15.24). This parasitic infection can cause significant economic losses in the poultry industry due to reduced weight gain, decreased egg production, and mortality (El-Dakhly et al. 2016).

15.3.1.1 Epidemiology

Raillietina cesticillus infection is widespread in domesticated birds, including turkeys. The prevalence of this infection varies depending on the geographical location, management practices, and age of the birds. In Turkey, the prevalence of *Raillietina cesticillus* infection in turkeys has been reported to be between 6.7 and 44.8% (Kurt and Acici 2008).

15.3.1.2 Life Cycle

The life cycle of *Raillietina cesticillus* involves an intermediate host, which is usually an arthropod such as beetles, ants, or grasshoppers. The eggs of the parasite are shed in the feces of infected birds and are ingested by the intermediate host. The eggs hatch in the gut of the intermediate host, and the larvae migrate to the muscles, where they form cysts. When the infected intermediate host is ingested by a bird, the cysts in the muscles are digested, and the larvae are released into the gut. The larvae then attach to the gut wall and develop into adult worms, which can grow up to 10 cm in length (Reid and Nugara 1961).

15.3.1.3 Clinical Signs

Raillietina cesticillus infection in turkeys is often asymptomatic, but in severe cases, it can cause weight loss, decreased egg production in breeder flocks, diarrhea, and mortality. The severity of clinical signs depends on the number of worms present in the gut of the bird. Large numbers of worms can cause blockages in the gut, which can lead to deaths (Nadakal et al. 1973).

15.3.1.4 Diagnosis

The diagnosis of *R. cesticillus* infection in turkeys is based too on the detection of eggs in the feces of infected birds. The parasite's eggs are oval and have a thick shell (Fig. 15.25). They can be identified using a microscope. In addition, postmortem examination of infected birds can also reveal the presence of adult worms in the gut

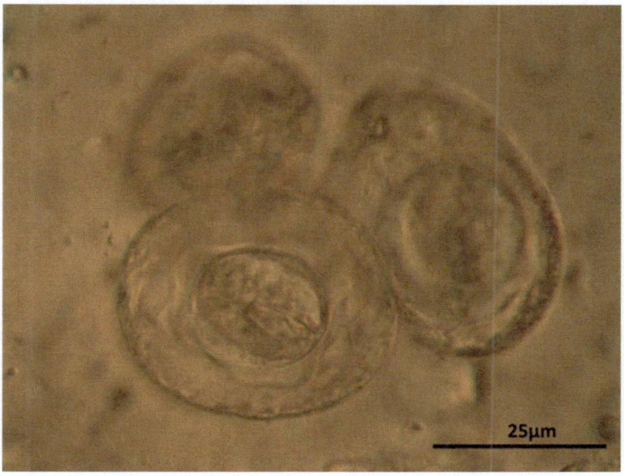

Fig. 15.25 Eggs of *Raillietina cesticillus* showing distinctive funnel-shaped structures between membranes (X. Hernandez-Velasco)

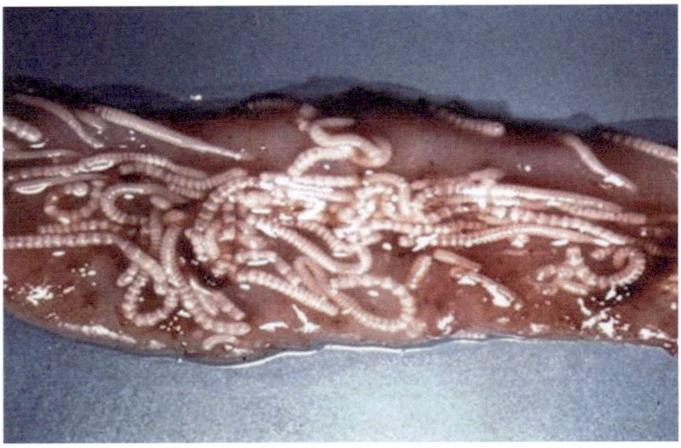

Fig. 15.26 Turkey intestine with large numbers of *Raillietina* spp. (H.M. Hafez)

(Reid and Nugara 1961) (Fig. 15.26). When evaluating *R. cesticillus* under a microscope, the broad and large scolex and the rostellum wide are observed (Fig. 15.27).

15.3.1.5 Control Measures

The control of *Raillietina cesticillus* infection in turkeys involves a combination of management practices and drug therapy. Good management practices such as proper sanitation, hygiene, and the use of insecticides can reduce the risk of infection. The use of anthelmintic drugs such as fenbendazole, praziquantel, niclosamide, and albendazole can effectively treat *Raillietina cesticillus* infestation in turkeys (McDougald 2020).

Raillietina echinobothrida (nodular tapeworm) (Fig. 15.28) can also affect turkeys and induce hyperplastic enteritis and caseous nodules where the scolex adheres to the wall of the small intestine (Villeneuve and Brugère-Picoux 2015). Its life cycle, epidemiology, and control are similar to that of *R. cesticillus*. *Raillietina echinobothrida* can be distinguished from *R. cesticillus* under the microscope because it has circular suckers on the scolex and two rows of hooks on the rostellum (Fig. 15.29). Both species can also be differentiated by identifying the eggs in the feces of infected animals (McDougald 2020) (Fig. 15.30).

15.3.2 *Choanotaenia infundibulum*

Choanotaenia infundibulum is a tapeworm that infects chickens, turkeys, and wild birds. The parasite is primarily found in the upper small intestine and can cause intestinal damage, malnutrition, and reduced growth rates.

15.3.2.1 Life Cycle

The life cycle of *Choanotaenia infundibulum* begins when adult tapeworms release gravid proglottids containing eggs into the lumen of the turkey's small intestine. The eggs are then passed out of the host's body via feces. These eggs can be ingested by an intermediate host, which is usually houseflies (*Musca domestica*), beetles

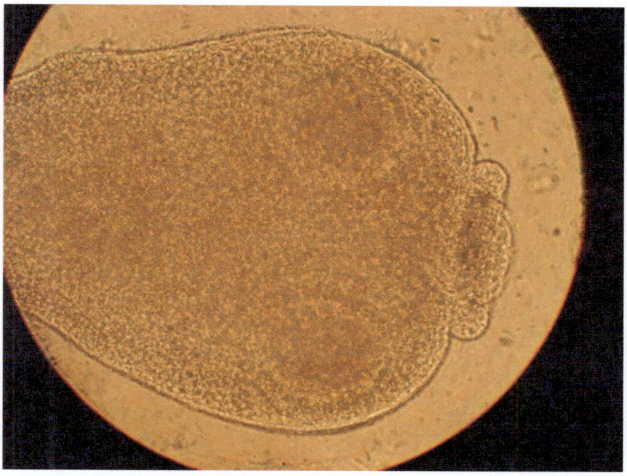

Fig. 15.27 *Raillietina cesticillus* scolex. The unarmed suckers are not prominent, and the rostellum is wide and armed with several hooks (X. Hernandez-Velasco)

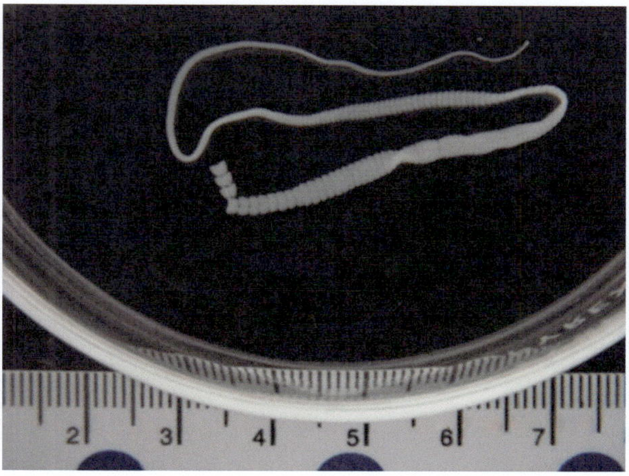

Fig. 15.28 *Raillietina echinobothrida* (X. Hernandez-Velasco)

(*Aphodius, Calathus, Geotrupes,* and *Tribolium*), earthworms, and grasshoppers (Taylor et al. 2016). Inside the intermediate host, the eggs hatch and release oncospheres, which penetrate the gut wall and develop into cysticercoids. When the turkey ingests the infected intermediate host, the cysticercoids are released in the small intestine, where they develop into adult tapeworms, completing the life cycle (McDougald 2020).

15.3.2.2 Epidemiology

This parasite is cosmopolitan.

15.3.2.3 Pathogenesis

Choanotaenia infundibulum infections can cause significant damage to the small intestine of turkeys. The tapeworms attach themselves to the intestinal wall, causing inflammation and damage to the gut lining. This can lead to malabsorption of

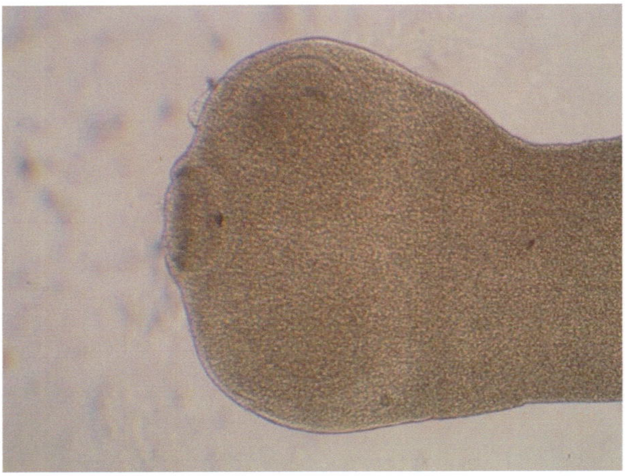

Fig. 15.29 *Raillietina echinobothrida* scolex (X. Hernandez-Velasco)

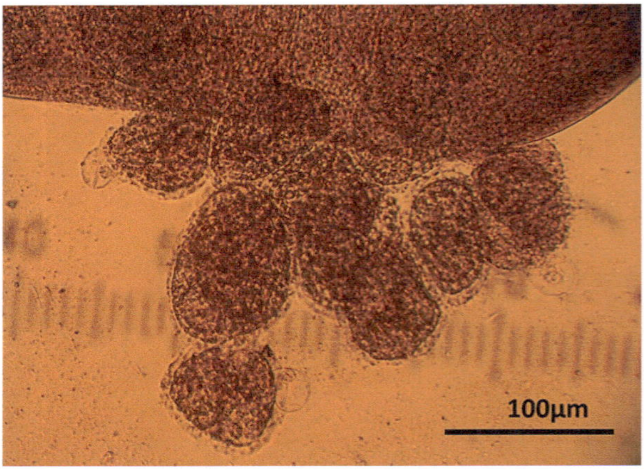

Fig. 15.30 Proglottid of *Raillietina echinobothrida*, capsules containing eggs and some free eggs are observed (X. Hernandez-Velasco)

nutrients, leading to malnutrition and reduced growth rates in infected turkeys. Heavy infestations can also lead to intestinal obstruction, causing morbidity and mortality in severe cases.

15.3.2.4 Diagnosis

The diagnosis of *Choanotaenia infundibulum* in turkeys can be challenging. The most common method of diagnosis is the detection of tapeworm eggs in fecal samples using microscopy. The rostellum is ringed with slender hooks (Fig. 15.31). Eggs measure about 45 × 55 μ and possess two long distinctive filaments (Taylor et al. 2016) (Fig. 15.32).

15.3.2.5 Control

Control of *Choanotaenia infundibulum* in turkeys can be achieved through good management practices. Measures such as proper hygiene, control of intermediate hosts, and the use of anthelmintic drugs can effectively reduce infection rates.

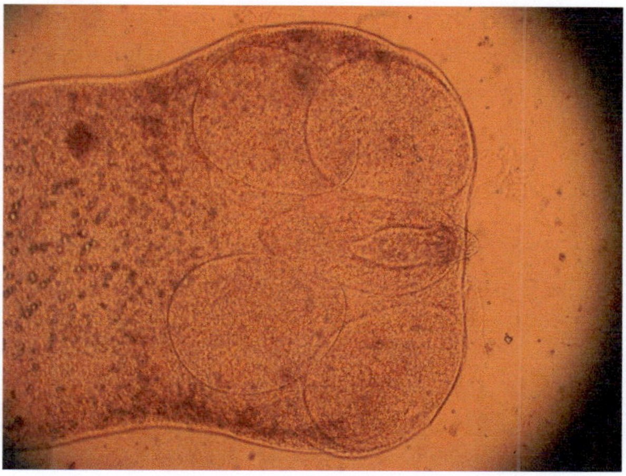

Fig. 15.31 *Choanotaenia infundibulum* scolex (X. Hernandez-Velasco)

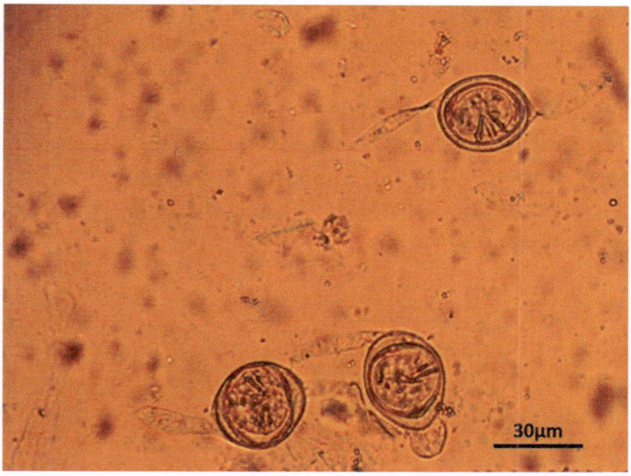

Fig. 15.32 *Choanotaenia infundibulum* eggs with outer membranes prolonged in two elongated filaments (X. Hernandez-Velasco)

Regular deworming with effective anthelmintics is an essential control measure. However, the development of anthelmintic resistance in *Choanotaenia infundibulum* has been reported, and the judicious use of anthelmintics is necessary to avoid resistance development (Dixon and Hansen 1965).

15.3.3 *Metroliasthes lucida*

Metroliasthes lucida is a parasitic cestode that infects turkeys, guinea fowl, and chickens. The parasite has been reported in commercial and wild turkeys (Rebolloso et al. 2006; Udoh et al. 2014).

15.3.3.1 Life Cycle

Metroliasthes lucida uses grasshoppers as intermediate hosts. The infective third-stage larvae being ingested by the turkey through contaminated food or water. The larvae penetrate the small intestinal wall, where they develop into adults. The adult worms lay eggs that pass out in the feces of the infected turkey, which serves as a source of infection for other birds (McDougald 2020).

15.3.3.2 Etiology

Methroliasthes lucida is a long parasite (20 cm) without rostellum and presents four suckers (Rebolloso et al. 2006).

15.3.3.3 Epidemiology

Methroliasthes lucida is prevalent in turkey flocks worldwide, with varying levels of infection rates depending on the management practices of the farm. Factors that increase the risk of infection include overcrowding, poor sanitation, and inadequate biosecurity measures. The parasite is more common in wild and older birds, with infection rates increasing with age (Rebolloso et al. 2006; Karaye et al. 2018).

15.3.3.4 Clinical Signs

The clinical signs of *Metroliasthes lucida* infection in turkeys include decreased feed intake, weight loss, diarrhea, and increased mortality. Moderate infections can lead to anemia, and massive infestation can cause intestinal occlusion and death (Rebolloso et al. 2006).

15.3.3.5 Diagnosis

Fecal examination is the most common diagnostic method, with the detection of adult worms. In gravid proglottids, the uterus consists of two sacs side by side (para-uterine organ) (Fig. 15.33). The eggs have three membranes, 75 × 50 mm (McDougald 2020).

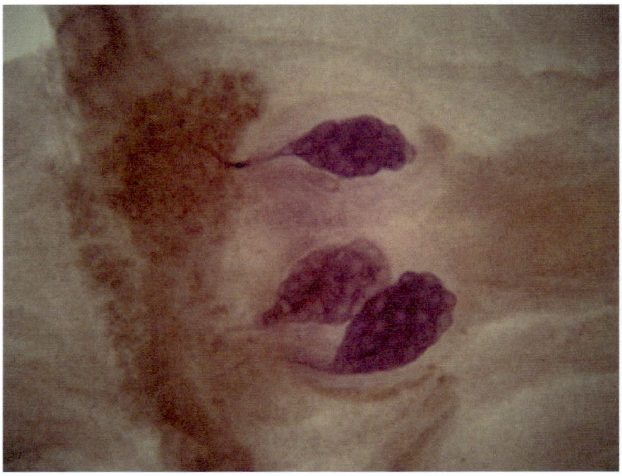

Fig. 15.33 Parauterine organ of *Metroliasthes lucida,* stained with Meyer's hemalum

15.3.3.6 Treatment and Control Measures

Control measures for *Metroliasthes lucida* infection in turkeys include improved management practices, such as reducing overcrowding, improving sanitation, and implementing effective biosecurity measures. The use of anthelmintics is also an effective control measure, with several drugs being available for the treatment of turkey nematode infections. However, the widespread use of anthelmintics can lead to the development of drug resistance, which can reduce their effectiveness (McDougald 2020).

15.4 Conclusion

Parasitic infestations and/or infections are a significant threat to the turkey industry, causing significant losses in productivity and revenue. Coccidiosis, blackhead disease, worms, and external parasites are some of the most common parasites affecting turkeys. Prevention and treatment of parasitic infections are critical in maintaining the health and productivity of turkey flocks. Proper sanitation, biosecurity, and parasite control programs are essential in preventing parasite infestations, while the use of anthelmintic drugs and other parasiticides can help treat infected turkeys (Table 15.4).

Table 15.4 Principal parasites affecting turkeys

Kind of parasite and disease	Etiology	Control	Treatment
Protozoan			
Coccidiosis	*Eimeria adenoeides, E. meleagrimitis, E. gallopavonis, E. meleagridis, E. dispersa*	Prevent the buildup of oocytes in the environment by removing damp litter and preventing wet spots Use of anticoccidial drugs (sulfadimethoxine plus ormetoprim, amprolium, and ethopabate, amprolium, sulfaquinoxaline and ethopabate, halofuginone, monensin, lasalocid, diclazuril, maduramicin, monensin)	Amprolium or sulfonamide
Histomoniasis (blackhead)	*Histomonas meleagridis*	Use good sanitation practices in the brooding area. Rotate range areas frequently. Use of vaccines	Benzimidazole anthelmintics reduce exposure to both worms and histomonads. Paromomycin sulfate

(continued)

Table 15.4 (continued)

Kind of parasite and disease	Etiology	Control	Treatment
Trichomonas	*Trichomonas gallinae*	Eliminate possible sources of infection. Prevent or minimize contact with pigeons and doves. Sick birds should be removed from a flock	Nitroimidazole compounds (dimetridazole and metronidazole)
Cryptosporidiosis	*C. baileyi and C. meleagridis*	Good hygiene practices, such as regular cleaning and disinfection of the environment and equipment, as well as avoiding overcrowding and stress in the birds	There is no known method of prevention or treatment other than sanitation
Cochlosomiasis	*Cochlosoma*	Can be avoided by minimizing overcrowding and environmental pollution	Metronidazole and ronidazole
Nematodes			
Ascaridiosis	*Ascaridia* spp.	Regular cleaning, use of disinfectants, and proper disposal of the used litter Roundworm eggs can remain viable for up to a year in the soil	Benzimidazoles, levamisole, and ivermectin
Subuluriasis	*Subulura*	Good management practices, such as maintaining clean and dry litter, preventing overcrowding, and minimizing exposure to earthworms, can reduce the risk of infection	Anthelmintic treatment. Benzimidazoles, levamisole, and ivermectin
Heterakidosis	*Heterakis gallinarum*	Proper hygiene and sanitation	Benzimidazoles, tetrahydropyrimidines and imidazothiazoles
Capillariosis	*Capillaria*	Good husbandry practices, including proper sanitation and hygiene. Eliminate the intermediated host and its breeding habitat	Fenbendazole, mebendazole
Tetrameriasis	*Tetrameres americana*	Earthworms, beetles, grasshoppers, and cockroaches	Fenbendazole or levamisole, together with control of intermediate hosts and sanitation measures

Table 15.4 (continued)

Kind of parasite and disease	Etiology	Control	Treatment
Syngamosis	*Syngamus trachea*	Proper hygiene and sanitation and avoiding the presence of intermediate hosts	Fenbendazole, Thiabendazole
Methroliasthes lucida infection	*Methroliasthes lucida*	Improved management practices include reducing overcrowding, improving sanitation, and implementing effective biosecurity measures	Anthelmintics
Cestodes			
Raillietina	*Raillietina cesticillus*	Eliminate the intermediate host (beetles) and its breeding habitat	Fenbendazole, praziquantel, niclosamide, and albendazole
Choanotaenia	*Choanotaenia infundibulum*	Eliminate the intermediate host (housefly, beetles) and its breeding habitat	Fenbendazole, praziquantel, niclosamide, and albendazole

References

Abbassi H, Répérant JM (2015) Criptosporidiosis. In: Brugère-Picoux J, Villancourt J-P, Bouzouaia M, Shivaprasad HL, Venne D (eds) Manual de Patologia Aviar. Pag. 418–424. Association française pour l'avancement des sciences (AFAS). Toppan Printing Leefung, China, p 701

Amin A, Bilic I, Liebhart D, Hess M (2014) Trichomonads in birds—a review. Parasitology 141:733–747. https://doi.org/10.1017/S0031182013002096

Atkinson CT, Van Riper C III (1991) Pathogenecity and epizootiology of avian haematozoa: *Plasmodium, Leucocytozoon*, and *Haemoproteus*. In: Bird–parasite interactions: ecology, evolution, and behavior. Oxford University Press, Oxford, pp 19–48

Băcescu B, Petruţ T, Condur D, Iorga S (2011) Observations on nematode infestation in turkeys and guinea fowls in household system. Scientific Works-University of Agronomical Sciences and Veterinary Medicine, Bucharest Series C, Veterinary Medicine 57:96–103

Bangoura B, Alnassan AA, Lendner M, Shehata AA, Krüger M, Daugschies A (2014) Efficacy of an anticoccidial live vaccine in prevention of necrotic enteritis in chickens. Exp Parasitol 145:125–134. https://doi.org/10.1016/j.exppara.2014.08.004

Beer LC, Graham BDM, Barros TL, Latorre JD, Tellez-Isaias G, Fuller AL, Hargis BM, Vuong CN (2022a) Evaluation of live-attenuated *Histomonas meleagridis* isolates as vaccine candidates against wild-type challenge. Poult Sci 101(3):101656. https://doi.org/10.1016/j.psj.2021.101656

Beer LC, Petrone-Garcia VM, Graham BD, Hargis BM, Tellez-Isaias G, Vuong CN (2022b) Histomonosis in poultry: a comprehensive review. Front Vet Sci 9:880738. https://doi.org/10.3389/fvets.2022.880738

Belete A, Addis M, Ayele M (2016) Review on major gastrointestinal parasites that affect chickens. J Biol Agric Healthc 6(11):11–21

Bermudez AJ, Ley DH, Levy MG, Ficken MD, Guy JS, Gerig TM (1988) Intestinal and bursal cryptosporidiosis in turkeys following inoculation with *Cryptosporidium* sp. isolated from commercial poults. Avian Dis 32(3):445–450

Berrilli F, D'Alfonso R, Giangaspero A, Marangi M, Brandonisio O, Kabore Y et al (2012) *Giardia duodenalis* genotypes and *Cryptosporidium* species in humans and domestic animals in Côte d'Ivoire: occurrence and evidence for environmental contamination. Trans R Soc Trop Med Hyg 106:191–195. https://doi.org/10.1016/j.trstmh.2011.12.005

Bilic I, Jaskulska B, Souillard R, Liebhart D, Hess M (2014) Multi-locus typing of *Histomonas meleagridis* isolates demonstrates the existence of two different genotypes. PLoS One 9(3):e92438. https://doi.org/10.1371/journal.pone.0092438

Bowman DD (2019) Georgis' parasitology for veterinarians, 11th edn. Elsevier, St Louis, p 528. https://doi.org/10.1016/C2016-0-02298-2

Boyett T, Crespo R, Vinueza VC, Gaghan C, Mohammed JP, Kulkarni RR (2022) Enumeration and speciation of coccidia affecting turkeys using flow cytometry method. J Appl Poult Res 31(3):100270. https://doi.org/10.1016/j.japr.2022.100270

Brener B, Tortelly R, Menezes RC, Muniz-Pereira LC, Pinto RM (2006) Prevalence and pathology of the nematode *Heterakis gallinarum*, the trematode *Paratanaisia bragai*, and the protozoan *Histomonas meleagridis* in the turkey, *Meleagris gallopavo*. Memorias do Instituto Oswaldo Cruz 101:677–681

Chapman HD (2008) Coccidiosis in the turkey. Avian Pathol 37(3):205–223. https://doi.org/10.1080/03079450802050689

Chapman HD (2018) Applied strategies for the control of coccidiosis in poultry. In: Perspectives in agriculture, veterinary science, nutrition and natural resources. CABI Reviews, pp 1–11. https://doi.org/10.1079/PAVSNNR201813026

Chapman HD, Matsler PL, Chapman ME (2004) Control of coccidiosis in turkeys with diclazuril and monensin: effects upon performance and development of immunity to *Eimeria* species. Avian Dis 48(3):631–634. https://doi.org/10.1637/7136

Chapman HD, Roberts B, Shirley MW, Williams RB (2005) Guidelines for evaluating the efficacy and safety of live anticoccidial vaccines and obtaining approval for their use in chickens and turkeys. Avian Pathol 34(4):279–290. https://doi.org/10.1080/03079450500178378

Clipsham R (1995) Avian pathogenic flagellated enteric protozoa. Semin Avian Exotic Pet Med 4(3):112–125

Collins JB, Andersen EC (2022) The Turkey ascarid, *Ascaridia dissimilis*, as a model genetic system. Int J Parasitol 7519(22):177–181. https://doi.org/10.1016/j.ijpara.2022.10.005

Cooper GL, Shivaprasad HL, Bickford AA, Nordhausen R, Munn RJ, Jeffrey JS (1995) Enteritis in turkeys associated with an unusual flagellated protozoan (*Cochlosoma anatis*). Avian Dis 39(1):183–190

Cupo KL, Beckstead RB (2019) *Heterakis gallinarum*, the cecal nematode of gallinaceous birds: a critical review. Avian Dis 63(3):381–388. https://doi.org/10.1637/0005-2086-63.3.381

de Graaf DC, Vanopdenbosch E, Ortega-Mora LM, Abbassi H, Peeters JE (1999) A review of the importance of cryptosporidiosis in farm animals. Int J Parasitol 29(8):1269–1287. https://doi.org/10.1016/s0020-7519(99)00076-4

Dixon CF, Hansen MF (1965) Helminths of poultry in Kansas. Poult Sci 44(5):1307–1315. https://doi.org/10.3382/ps.0441307

Edgar SA, Flanagan C (1979) Efficacy of Stenorol® (Halofuginone): III. For the control of coccidiosis in turkeys. Poult Sci 58(6):1483–1489. https://doi.org/10.3382/ps.0581483

El-Dakhly KM, Mahrous LN, Mabrouk GA (2016) Distribution pattern of intestinal helminths in domestic pigeons (*Columba livia domestica*) and turkeys (*Meleagris gallopavo*) in Beni-Suef province, Egypt. J Vet Med Res 23(1):112–120. https://doi.org/10.21608/jvmr.2016.43226

El-Sherry S, Ogedengbe ME, Hafeez MA, Sayf-Al-Din M, Gad N, Barta JR (2019) Cecal coccidiosis in turkeys: comparative biology of *Eimeria* species in the lower intestinal tract of turkeys using genetically typed, single oocyst—derived lines. Parasitol Res 118:583–598. https://doi.org/10.1007/s00436-018-6147-5

El-Wahab AA, Visscher C, Haider W, Dimitri R (2021) A case study of histomoniasis in fattening turkeys identified in histopathological investigations. German J Vet Res 1(3):13–18. https://doi.org/10.51585/gjvr.2021.3.0015

Evans NP, Evans RD, Fitz-Coy S, Pierson FW, Robertson JL, Lindsay DS (2006) Identification of the new morphological and life-cycle stages of *Cochlosoma anatis* and experimental transmission using pseudocyst. Avian Dis 50(1):22–27. https://doi.org/10.1637/7360-040405R.1

Farhang HH (2012) The survey of gastrointestinal parasites in turkeys of Tabriz Iran. Life Sci J 9(4):4341–4343

Garton W (2014) Recognition and treatment of coccidiosis in turkeys. Vet Times 44(41):6–7. http://www.vbd.co.uk

Gharagozlou MJ, Dezfoulian O, Rahbari S, Bokaie S, Jahanzad I, Razavi ANE (2006) Intestinal cryptosporidiosis in turkeys in Iran. J Vet Med A Physiol Pathol Clin Med 53(6):282–285. https://doi.org/10.1111/j.1439-0442.2006.00843.x

Gómez-Muñoz MT, Gómez-Molinero MÁ, González F, Azami-Conesa I, Bailén M, García Piqueras M, Sansano-Maestre J (2022) Avian oropharyngeal trichomonosis: treatment, failures and alternatives, a systematic review. Microorganisms 10:2297. https://doi.org/10.3390/microorganisms10112297

Goodwin MA, Steffens WL, Russell ID, Brown J (1988) Diarrhea associated with intestinal cryptosporidiosis in turkeys. Avian Dis 32(1):63–67

Grabensteiner E, Hess M (2006) PCR for the identification and differentiation of *Histomonas meleagridis, Tetratrichomonas gallinarum* and *Blastocystis* spp. Vet Parasitol 142(3–4):223–230. https://doi.org/10.1016/j.vetpar.2006.07.011

Hafez HM (2008) Poultry coccidiosis: prevention and control approaches. Archiv Fur Geflugelkunde 72(1):2–7

Hafez HM, Hauck R, Lüschow D, McDougald L (2005) Comparison of the specificity and sensitivity of PCR, nested PCR (nPCR), and real-time PCR (qPCR) for the diagnosis of histomoniasis. Avian Dis 49(3):366–370. https://doi.org/10.1637/7341-020805R.1

Hafez HM, Hauck R, Gad W, De Gussem K, Lotfi A (2010) Pilot study on the efficacy of paromomycin as a histomonostatic feed additive in turkey poults experimentally infected with *Histomonas meleagridis*. Arch Anim Nutr 64(1):77–84. https://doi.org/10.1080/17450390903478851

Hassanain MA, Rahman A, Khalil AM (2009) New scanning electron microscopy look of *Ascaridia galli* (Schrank, 1788) adult worm and its biological control. Res J Parasitol 4:94–104

Hauck R, Balczulat S, Hafez HM (2010) Detection of DNA of *Histomonas meleagridis* and *Tetratrichomonas gallinarum* in German poultry flocks between 2004 and 2008. Avian Dis 54:1021–1025. https://doi.org/10.1637/9261-012910-Reg.1

Helmy YA, Hafez HM (2022) Cryptosporidiosis: from prevention to treatment, a narrative review. Microorganisms 10(12):2456. https://doi.org/10.3390/microorganisms10122456

Helmy YA, Krücken J, Abdelwhab EM, von Samson-Himmelstjerna G, Hafez HM (2017) Molecular diagnosis and characterization of *Cryptosporidium* spp. in turkeys and chickens in Germany reveals evidence for previously undetected parasite species. PLoS One 12(6):e0177150. https://doi.org/10.1371/journal.pone.0177150

Hemsley RV (1971) Fourth stage *Ascaridia* spp. larvae associated with high mortality in turkeys. Can Vet J 12(7):147–149

Hess M, Liebhart D, Grabensteiner E, Singh A (2008) Cloned *Histomonas meleagridis* passaged in vitro resulted in reduced pathogenicity and is capable of protecting turkeys from histomonosis. Vaccine 26(33):4187–4193. https://doi.org/10.1016/j.vaccine.2008.05.071

Hurst GA, Turner LW, Tucker FS (1979) Capillariasis in penned wild turkeys. J Wildl Dis 15(3):395–397. https://doi.org/10.7589/0090-3558-15.3.395

Kadykalo S, Roberts T, Thompson M, Wilson J, Lang M, Espeisse O (2018) The value of anticoccidials for sustainable global poultry production. Int J Antimicrob Agents 51(3):304–310. https://doi.org/10.1016/j.ijantimicag.2017.09.004

Karaye PG, Ola-Fadunsin SD, Dogo GA (2018) Diversity of gastrointestinal parasites affecting some domestic animals in plateau state, North Central Nigeria. Sci World J 13(1):82–86

Kurt M, Acici M (2008) Cross-sectional survey on helminth infections of chickens in the Samsun region, Turkey. DTW Deutsche Tierarztliche Wochenschrift 115(6):239–242

Landim de Barros T, Vuong CN, Tellez-Isaias G, Hargis BM (2022) Uncontroversial facts and new perspectives on poultry histomonosis: a review. Worlds Poult Sci J 78(4):913–933. https://doi.org/10.1080/00439339.2022.2119915

Liebhart D, Ganas P, Sulejmanovic T, Hess M (2017) Histomonosis in poultry: previous and current strategies for prevention and therapy. Avian Pathol 46(1):1–18. https://doi.org/10.1080/03079457.2016.1229458

Lindsay DS, Larsen CT, Zajac AM, Pierson FW (1999) Experimental *Cochlosoma anatis* infections in poultry. Vet Parasitol 81(1):21–27. https://doi.org/10.1016/s0304-4017(98)00227-1

Marx M, Reiner G, Willems H, Rocha G, Hillerich K, Masello JF et al (2017) High prevalence of *Trichomonas gallinae* in wild columbids across western and southern Europe. Parasites Vectors 10:242. https://doi.org/10.1186/s13071-017-2170-0

McDougald LR (2020) Internal parasites. In: Swayne DE, Boulianne M, Logue CM, McDougald LR, Nair V, Suarez DL, de Wit S, Grimes T, Johnson D, Kromm M, Prajitno TY, Rubinoff I, Zavala G (eds) Diseases of poultry, 14th edn. Wiley, Hoboken, pp 1157–1192

Nadakal AM, Mohandas A, John KO, Muraleedharan K (1973) Contribution to the biology of the fowl cestode *Raillietina echinobothrida* with a note on its pathogenicity. Trans Am Microsc Soc 92(2):273–276

Norton RA, Hopkins BA, Skeeles JK, Beasley JN, Kreeger JM (1992) High mortality of domestic turkeys associated with *Ascaridia dissimilis*. Avian Dis 36(2):469–473

Peirce MA (2005) A checklist of the valid avian species of Babesia (Apicomplexa: Piroplasmorida), *Haemoproteus, Leucocytozoon* (Apicomplexa: Haemosporida), and *Hepatozoon* (Apicomplexa: Haemogregarinidae). J Nat Hist 39:3621–3632

Pramual P, Tangkawanit U, Kunprom C, Vaisusuk K (2020) Seasonal population dynamics and a role as natural vector of *Leucocytozoon* of black fly, *Simulium chumpornense* Takaoka & Kuvangkadilok. Acta Trop 211(6):105617. https://doi.org/10.1016/j.actatropica.2020.105617

Purple KE, Humm JM, Kirby RB, Saidak CG, Gerhold R (2015) *Trichomonas gallinae* persistence in four water treatments. J Wildl Dis 51:739–742. https://doi.org/10.7589/2014-05-137

Qaid MM, Al-Mufarrej SI, Azzam MM, Al-Garadi MA (2021) Anticoccidial effectivity of a traditional medicinal plant, *Cinnamomum verum*, in broiler chickens infected with *Eimeria tenella*. Poult Sci 100(3):100902. https://doi.org/10.1016/j.psj.2020.11.071

Rebolloso SL, Salas Westphal AI, Scott Morales LM (2006) Primer informe de *Metroliasthes lucida* (Cestoda: Dilepididae) en guajolote silvestre Río Grande de Nuevo León, México. Veterinaria México 37(2):263–267

Reid WM, Nugara D (1961) Description and life cycle of *Raillietina georgiensis* n. sp., a tapeworm from wild and domestic turkeys. J Parasitol 47(6):885–889

Seddiek SA, Ali MM, Khater HF, El-Shorbagy MM (2011) Anthelmintic activity of the white wormwood, *Artemisia herba-alba* against *Heterakis gallinarum* infecting turkey poults. J Med Plants Res 5(16):3946–3957

Shehata AA, Attia Y, Khafaga AF, Farooq MZ, El-Seedi HR, Eisen-reich W, Tellez-Isaias G (2022) Restoring healthy gut microbiome in poultry using alternative feed additives with particular attention to phytogenic substances: challenges and prospects. German J Vet Res 2(3):32–42. https://doi.org/10.51585/gjvr.2022.3.0047

Sigrist B, Ng TC, Albini S, Wolfrum N (2022) A new duplex real-time PCR for simultaneous detection and differentiation of *Tetratrichomonas gallinarum* and *Trichomonas gallinae*. J Vet Diagn Invest 34(4):631–637. https://doi.org/10.1177/10406387221098069

Smith T (1895) An infectious disease among turkeys caused by protozoa (infectious enterohepatitis). USDA Bureau Anim Ind Bull 8:3–27

Steele EJ, Noblet GP (1993) Gametocytogenesis of *Leucocytozoon smithi*. J Eukaryot Microbiol 40(3):384–391. https://doi.org/10.1111/j.1550-7408.1993.tb04932.x. PMID: 8508175

Sulaiman IM, Monis P, Lal AA, Fayer R, Pavlasek I (2003) A redescription of *Cryptosporidium galli* Pavlasek, 1999 (Apicomplexa: Cryptosporidiidae) from birds. J Parasitol 89:809–813. https://doi.org/10.1645/GE-74RI

Taylor MA, Coop RL, Wall RL (2016) Veterinary parasitology, 4th edn. Wiley-Blackwell, Oxford, p 1032

Udoh NA, Luka SA, Audu PA (2014) Prevalence of gastrointestinal parasites of domestic turkey (*Meleagris gallopavo* Linnaeus 1758) slaughtered in Kaduna metropolis, Kaduna state, Nigeria. J Nat Sci Res 4(7):105–109

Valkiūnas G, Iezhova TA (2022) Keys to the avian Haemoproteus parasites (Haemosporida, Haemoproteidae). Malar J 21(1):269. https://doi.org/10.1186/s12936-022-04235-1

Valkiūnas G, Duc M, Iezhova TA (2022) Increase of avian plasmodium circumflexum prevalence, but not of other malaria parasites and related haemosporidians in northern Europe during the past 40 years. Malar J 21:105

Villeneuve A, Brugère-Picoux J (2015) Parasitos internos. In: Brugère-Picoux J, Villancourt J-P, Bouzouaia M, Shivaprasad HL, Venne D (eds) Manual de Patologia Aviar. Pag. 428–440. Association française pour l'avancement des sciences (AFAS). Toppan Printing Leefung, China, p 701

Vrba V, Pakandl M (2014) Coccidia of turkey: from isolation, characterization and comparison to molecular phylogeny and molecular diagnostics. Int J Parasitol 44(13):985–1000. https://doi.org/10.1016/j.ijpara.2014.06.004

Wages DP, Ficken MD (1989) Cryptosporidiosis and turkey viral hepatitis in turkeys. Avian Dis 33(1):191–194

Youssefi MR, Mahmoudi P, Ebrahimi A (2018) Prevalence of intestinal parasites in turkeys from Gilan, Mazandarn and Golestan provinces 2015. J Vet Clin Res 9(1):25–29

Zajac AM (2021) Fecal examination for the diagnosis of parasitism. In: Zajac AM, Conboy GA, Little SE, Reichard MV (eds) Veterinary clinical parasitology, 9th edn. Wiley Blackwell, Hoboken, pp 1–190

Ectoparasites Affecting Turkeys

16

Xochitl Hernandez-Velasco, Guillermo Tellez-Isaias, Daniel Hernandez-Patlan, Bruno Solis-Cruz, Víctor M. Petrone-García, Inkar Castellanos-Huerta, Jesús A. Maguey-González, Juan D. Latorre, Saeed El-Ashram, Wolfgang Eisenreich, Hafez M. Hafez, and Awad A. Shehata

X. Hernandez-Velasco (✉)
Departamento de Medicina y Zootecnia de Aves, Facultad de Medicina Veterinaria y Zootecnia, UNAM, Cd. de Mexico, Mexico City, Mexico
e-mail: xochitlh@fmvz.unam.mx

G. Tellez-Isaias · I. Castellanos-Huerta · J. A. Maguey-González · J. D. Latorre
Department of Poultry Science, University of Arkansas, Fayetteville, AR, USA
e-mail: gtellez@uark.edu; icastell@uark.edu; jm201@uark.edu; jl115@uark.edu

D. Hernandez-Patlan · B. Solis-Cruz
Laboratory 5: LEDEFAR, Multidisciplinary Research Unit, National Autonomous University of Mexico-Superior Studies Faculty at Cuautitlan (UNAM-FESC), Cuautitlan Izcalli, Mexico State, Mexico

Nanotechnology Engineering Division, Polytechnic University of the Valley of Mexico, Tultitlan, Mexico State, Mexico
e-mail: danielpatlan@comunidad.unam.mx; bruno_sc@comunidad.unam.mx

V. M. Petrone-García
Departamento de Ciencias Pecuarias, FESC, UNAM, Cuautitlán, Estado de Mexico, Mexico

S. El-Ashram
College of Life Science and Engineering, Foshan University, Foshan, China

Faculty of Science, Kafrelsheikh University, Kafr El-Sheikh, Egypt

W. Eisenreich · A. A. Shehata
Bavarian NMR Center, Structural Membrane Biochemistry Department of Chemistry, Technische Universität München, Garching, Germany
e-mail: wolfgang.eisenreich@mytum.de; awad.shehata@tum.de

H. M. Hafez
Faculty of Veterinary Medicine, Institute of Poultry Diseases, Free University of Berlin, Berlin, Germany
e-mail: hafez.mohamed@fu-berlin.de

H. M. Hafez, A. A. Shehata (eds.), *Turkey Diseases and Disorders Volume 2*,
https://doi.org/10.1007/978-3-031-63322-5_16

Abstract

Parasitosis refers to associations between two living beings of different species, where an individual known as the host (from the Latin hospitator-oris = host) provides a source of food, shelter, or transport to the other called parasite. The types of relationships between the host and the parasite can be diverse; in some cases, the parasite can feed or live only occasionally (facultative parasite) or be totally dependent on its host (permeant parasite). The degree of affectation that the parasite causes to its host is variable and depends on many factors; it can be null or minimal (parasitism), or cause indirect and/or direct damage, and in the presence of complicating factors, it can significantly affect well-being, and the health of its host (parasitosis), or even cause death in severe cases. In general, the economic importance lies in the fact that they affect the well-being and rest of the birds by causing restlessness, nervousness, discomfort, damage to the plumage, and itching, but they can also affect the health of the bird by causing weakness, depression, anemia, loss of body weight, dermatitis, hyperkeratosis, and decreased egg production in commercial laying and breeding birds, when present in large numbers. In addition, the ectoparasites that affect domestic birds can be vectors of pathogenic viruses and/or bacteria and even act as paratenic hosts or intermediaries for other parasites such as protozoa and helminths. Its diagnosis can be made by conventional methods (macroscopic, microscopic) or molecular techniques; the latter allow analyzing the phylogenetic and evolutionary relationships between species. Although many methods have been developed to determine the size of ectoparasite populations, visual inspection is the easiest, fastest, and most practical way. The control and treatment of these ectoparasites is difficult due to the restrictions on the use of some products as well as the presence of species resistant to one or more of them. The increase of free-range rearing form does not allow the use of insecticidal products. The ectoparasites that can parasitize turkeys are mostly not specific to this avian species, in addition to the fact that their prevalence may vary with the geographical region, climate, and the production system in which they are raised. In the present chapter, we will address the most common and important ectoparasites in turkeys.

Keywords

Ectoparasites · *Arthropoda* · Fleas · Flies · Ticks

16.1 Introduction

Ectoparasites are a group of arthropods or invertebrates that have a segmented body and legs divided into articulated segments (arthron = joint, and podos = foot) in addition to an exoskeleton made of chitin, a hard, brown-colored structure that is highly resistant to environmental conditions. In the case of domestic birds, transmission is generally carried out by direct contact between birds, but flies, rodents, or other types of harmful fauna can participate in the transmission of ectoparasites such as fleas, flies, and ticks from one bird to another (Taylor et al. 2016; Mullen and Durden 2002). The main ectoparasites that affect turkeys will be discussed in this chapter.

16.2 Mites

Mites constitute the most diverse group of arachnid symbionts associated with birds (Proctor and Owens 2000). Mites have adapted to various bird tissues such as skin, feathers, subcutaneous tissue, and respiratory tract as well as their habitats (Skoracki et al. 2012). They have a body composed of a cephalothorax and a rounded abdomen with no apparent segments. They are grayish when fasting and dark red when fed since they feed mainly on blood, skin fluids, lipids, and feather keratin (Philips 2000). However, due to their small size (0.3–1.5 mm in length), little knowledge about them, and the fact that most are not highly pathogenic, they can often go undetected (Domrow 1991).

The reproductive cycle consists of several phases: egg, larva (with six legs), protonymph, deutonymph, and adult. Both nymphs and adults have eight legs. They have a wide global distribution and host range, often associated with large numbers of domestic and wild birds (Walter and Proctor 1999). Around 20 species of mites infest domestic birds, but only 12 genus can cause severe infestations and affect commercial poultry.

16.2.1 Skin Mites

The ectoparasitoses caused by the red mite or *Dermanyssus gallinae* (De Geer 1778; Acari: *Dermanyssidae*); the tropical mite, *Ornithonyssus bursa* (Berlese 1888; Acari: Macronyssidae); and the northern mite u *Ornithonyssus sylviarum* (Canestrini and Fanzao 1877; Acari: *Macronyssidae*) are the most common in poultry in different parts of the world (Axtell and Arends 1990; Harrington et al. 2011; Mullens et al. 2009); however, there are reports of other species that, although less common, also come to affect intensive poultry farming such as *Megninia ginglymura* or *Dermatophagoides pteronyssinus* (Quintero et al. 2007, 2010; Camacho-Escobar et al. 2010; Mironov 2013).

The genera *Dermanyssus* and *Ornithonyssus* are frequently found in laying hens because there are several factors that favor their permanence, such as the presence of birds of multiple ages, facilities that offer multiple shelters and longer production cycles, and high production pressure in birds; therefore, the highest infestations occur around the peak of laying in flocks on commercial farms and during the incubation phase in backyard birds. The distribution within the flock is heterogeneous; in cases of severe infestation, they become a nuisance both for the birds and for the workers (Krinsky 1983; Ribeiro et al. 1992).

They are parasites originating from birds of the Order Passeriformes, mainly sparrows, and it is precisely they who favor their recurrent presence in commercial farms and in birds raised under grazing conditions, as well as the presence of cracks and crevices in the facilities and the flow of personnel, equipment, and harmful fauna (Walter and Proctor 1999). However, they can also affect mammals, including humans, which represent an important occupational health issue that is often not reported and has been undervalued. Bird attendants are the most susceptible population (Sparagano et al. 2020; Alves et al. 2023).

16.2.1.1 *Ornithonyssus sylviarum* (Northern Fowl Mite)

It is an obligate blood-sucking parasite with a wide number of possible hosts among domestic, ornamental, and wild birds. It is a mite of economic importance for commercial poultry farming in many countries, and in the American continent, it is the most common and damaging mite in commercial birds, mainly in laying hens and breeding turkeys.

Description

They are small mites (less than 1 mm in length) of a grayish or dark reddish color; if they have already fed, their body is oval with a large number of setae compared to *Dermanyssus*, and they have long legs that allow them to move quickly. The anal plate is teardrop-shaped and has three setae in a triangular arrangement. The chelicerae are elongated and stylelike. The female has two pairs of setae on the sternal plate (Taylor et al. 2016; Di Palma et al. 2012).

Life Cycle

In theory, it carries out its entire production cycle (7 days long) on the bird. After each blood feeding, it deposits from one to five eggs, which is why it can form very large populations in a short time (McCrea et al. 2005; Taylor et al. 2016). However, mite populations and their distribution can vary greatly between birds in a flock, house, or farm (Murillo and Mullens 2017).

Epidemiology

It is distributed worldwide, mainly found in temperate climates, and is particularly important in North America, where it is considered the most common and harmful ectoparasite of poultry (Murillo and Mullens 2017); its prevalence is high and similar in intensive production, such as cage-free or free grazing (Mullens and Murillo 2018). *O. sylviarum* parasitizes a wide variety of wild and commercial birds, mainly in laying hens or breeding turkeys. The infection is transmitted by direct contact with infected birds, through vectors, or material contaminated with mites (Mullens et al. 2009).

Pathogenesis

Being a blood-sucking parasite, in addition to causing weakness in the bird, it can also transmit other infectious diseases such as fowlpox, St. Louis encephalitis, Newcastle disease, chlamydiosis, and western equine encephalomyelitis. According to DeLoach and DeVaney (1981) at high levels of infestation, a Leghorn hen can lose about 6% of blood per day, and in turkeys, there are no studies in this regard.

Fig. 16.1 *Ornithonyssus sylviarum* infestation of the skin and feathers below the cloaca of a bird (X. Hernandez-Velasco)

Signs and Lesions

In moderate to severe infestations, dirty and matted feathers with a blackish material due to the presence of the parasite, traces of dried blood, and feces can be seen on the feathers around the vent and ventral region (Fig. 16.1). They cause irritation and anemia and can affect feed conversion rate, egg production, and the reproductive capacity of the male birds. When there are large populations, they can also be observed in the facilities (Hinkle and Hickle 2008; Murillo and Mullens 2017).

Diagnosis

The main distinctive habits and characteristics of this mite that help us to identify it are finding it present in the bird during the day and night. It is an obligate blood-sucking parasite. To differentiate it from other genera of mites, it is necessary to clarify it and observe it under a light microscope and look for the main distinguishing characteristics of this species, such as the dorsal plate that extends to two-thirds of the total length of the body, the anal plate that is shaped like a teardrop, with the anus in its anterior half (Fig. 16.2), the chelicerae in the form of tweezers with mobile and fixed last digits, and the genitoventral plate that narrows in its caudal portion. The sternal plate narrows sharply and has two pairs of setae, unlike O. bursa, which has three (Di Palma et al. 2012), (Fig. 16.3). It is also possible to differentiate between species by genetic analysis (Takehara et al. 2019).

Control and Treatment

In commercial birds, powdered or liquid pesticide spray products are used and placed directly on the skin in the peri-cloacal or ventral region of the bird for the best effect. Control of these mites can be enhanced by applying liquid spray pesticides to litter and facilities or by placing resin strips in the facilities. Among the most widely used products are pyrethrins, synthetic pyrethroids (e.g., permethrin), organophosphates (e.g., tetrachlorvinphos/dichlorvos), isoxazolines (e.g., fluralaner), and carbamates (e.g., carbaryl, Sevin dust); however, the availability and

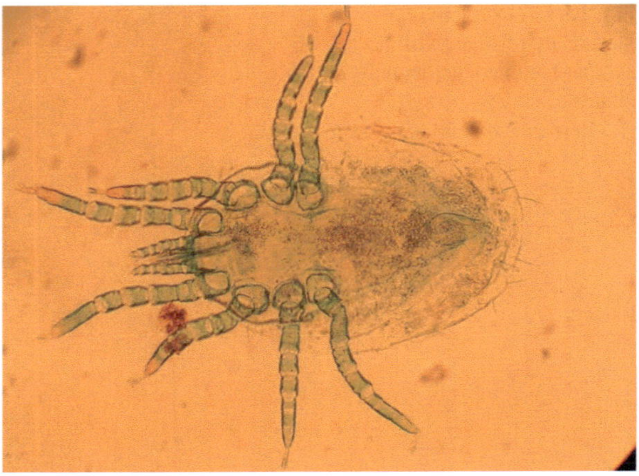

Fig. 16.2 Ventral view of a cleared female *Ornithonyssus sylviarum*, showing the teardrop-shaped anal plate, distinctive to this genus (X. Hernandez-Velasco)

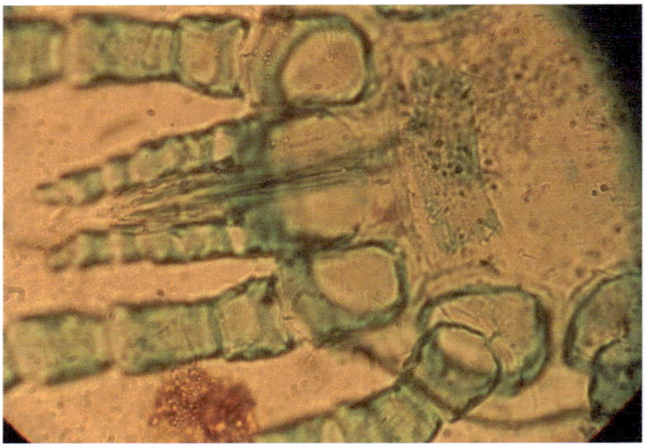

Fig. 16.3 Enlarged view of gnatosoma of the ventral side of *Ornithonyssus sylviarum*, showing the chelicerae and the sternal plate with only two pairs of bristles (X. Hernandez-Velasco)

approval of its use may change according to the geographic region and the available products. The products must be rotated to favor their effectiveness for a longer period of time. However, mites can persist in the environment for weeks or months. To prevent infestations or reinfestations by mites, it is necessary to improve biosecurity measures, mainly to prevent wild birds from having contact with farm birds, which is difficult in free-range production systems (McCrea et al. 2005; Mullens et al. 2017).

16.2.1.2 *Ornithonyssus bursa* (Tropical Fowl Mite)

Ornithonyssus bursa is an ectoparasite of commercial and wild birds, it feeds on blood and can be found in birds, facilities, or nest boxes (if applicable). Unlike *O. sylviarum, O. bursa* occurs mainly in tropical and subtropical climates but has also been reported from southern Europe.

Description

In general, it closely resembles *O. sylviarum* in habits, life cycle, and appearance (Mullen and O'Connor 2009; Takehara et al. 2019). It differs from *O. sylviarum* mainly in that the sternal plate has three pairs of setae, unlike *O. sylviarum* and *D. gallinae*, which have only two pairs.

Life Cycle

Very similar to that of *O. sylviarum*.

Epidemiology

With worldwide distribution, mainly in tropical and subtropical areas, it occurs in domestic birds, is a common parasite in wild birds, and occasionally can also affect humans (Denmark and Cromroy 2003; Waap et al. 2020).

Pathogenesis

Similar to *O. sylviarum*, it causes stress, but it can also be a potential vector for diseases such as Western Equine Encephalitis and Saint-Louis Encephalitis virus (Valiente Moro et al. 2005).

Signs and Lesions

Similar to *O. sylviarum.*

Diagnosis

It is extremely similar to *O. sylviarum*, but it can be differentiated by observing it under a light microscope after processing and rinsing. The most distinctive morphological differences of *O. bursa* are the sternal plate, which has three pairs of bristles, whereas *O. sylviarum* and *D. gallinae* only have two as well as the dorsal plate that tapers toward the caudal end and ends less sharply than *O. sylviarum* (Figs. 16.4 and 16.5).

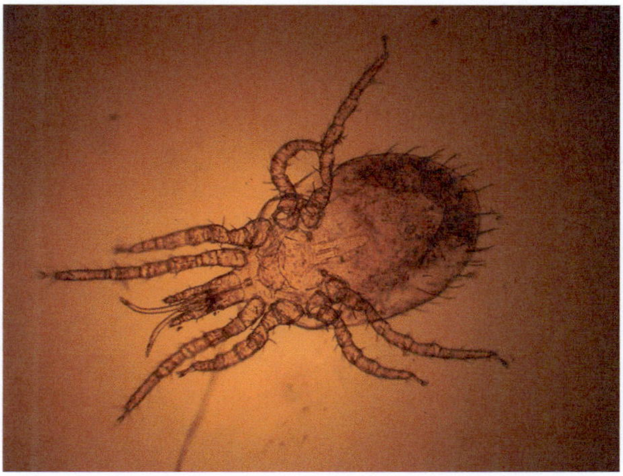

Fig. 16.4 *Ornithonyssus bursa*, ventral view (X. Hernandez-Velasco)

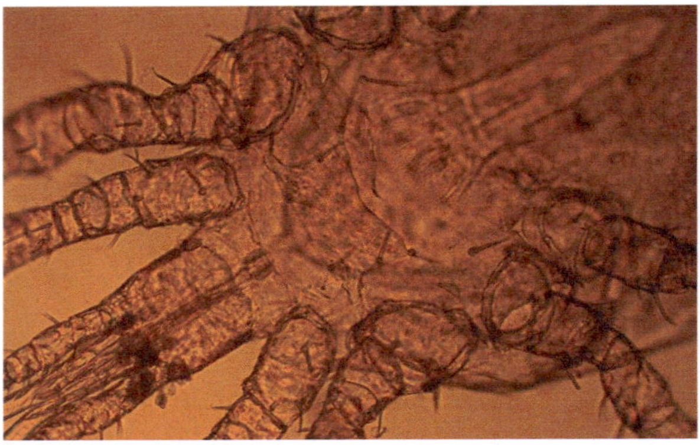

Fig. 16.5 Placa esternal de *Ornithonyssus bursa* (X. Hernandez-Velasco)

Control and Treatment

Similar to *O. sylviarum* (Takehara et al. 2019).

16.2.1.3 *Dermanyssus gallinae* (Poultry Red Mite)

It is one of the most common blood-sucking parasitic mites in commercial and wild birds and can occasionally affect mammals as well. *D. gallinae* is considered the most frequent and important ectoparasite in commercial poultry farms for laying hens in Europe, Brazil, and Japan, where it causes significant economic losses to 80% of poultry farms (Sparagano 2009; Sparagano et al. 2014). In turkeys, there are no studies in this regard.

Description

They are slightly larger and less hairy than *O. sylviarum*, an adult measuring just over 1 mm. Its body is light gray to dark red or black when it has fed. Among the microscopic characteristics that allow their differentiation is the end of the dorsal plate that tapers caudally, but does not reach the posterior end of the body: The anal plate is the width of the genital plate and has the shape of a angular stone, and the anus is located in its back half. The chelicerae are long and stylet-shaped (McCrea et al. 2005; Taylor et al. 2016) (Fig. 16.6).

Life Cycle

D. gallinae feed mainly at night on a regular basis, so to rule out their presence, inspection of the birds and facilities must be carried out at night. During the day, the parasite hides in groups in cracks and fissures in cages, aviaries, or houses, under clods of manure, or in nests. The female deposits from three to seven eggs, 12–24 h after the blood meal, and its incubation takes 2 days. They can survive more than 34 weeks without consuming feed, so they can withstand facility rest days between flocks. Under favorable conditions, the reproductive cycle is completed in 7–14 days and requires more days in cold weather (McCrea et al. 2005; Taylor et al. 2016).

Fig. 16.6 Adult female of *Dermanyssus gallinae*, ventral view (X. Hernandez-Velasco)

Epidemiology

It has a worldwide distribution, endemic in Europe, Japan, and Brazil, and prevalent in the western region of North America (Sparagano et al. 2014; Takehara et al. 2019). Chickens and hens are its most common hosts; however, they can also be found in turkeys (Hinkle and Hickle 2008; Rezaei et al. 2016), pigeons, sparrows, canaries, and wild birds (Tarazona and Cordero del Campillo 1999). It is a more serious problem in winter and in breeding birds with floor nests, stackable litter, or free-range systems (Alabama Cooperative Extension System (Alabama A&M and Auburn University) 2019; Taylor et al. 2016). The mites are easily transmitted by direct contact between birds or through fomites or vectors.

Pathogenesis

D. gallinae feeds on blood and is responsible for causing skin irritation and anemia and affects weight gain, egg production, and quality. It could also be a vector of different pathogens such as fowl pox, St. Louis encephalitis, Eastern equine encephalitis, West Nile viruses, Venezuelan equine encephalitis, Western equine encephalitis, Newcastle disease viruses, *Salmonella gallinarum*, *S. enteritidis*, *Pasteurella multocida*, *Erysipelothrix rhusiopathiae*, *Bacillus thuringiensis*, *Listeria monocytogenes*, *Borrelia burgdorferi*, and *Coxiella burnetii* (Valiente Moro et al. 2009; Sparagano 2009; Mullen and O'Connor 2009).

Signs and Lesions

In affected birds, dirty and matted feathers are observed with remains of excrement from the mites, desquamation, and remains of the bird's blood. In addition, they cause skin irritation, restlessness, paleness, anemia, weight loss, irritability and nervousness, low production, and dirty the egg with spots of the same crushed mites and blood (Proctor and Owens 2000).

Diagnosis

The diagnosis is based on the observation under a microscope of the specimens, previously clarified and based on the previously established identification keys (Di Palma et al. 2012). *Dermanyssus* has retracted, whiplike chelicerae and a distinctive, capstone-shaped anal plate, the anus being on the caudal portion of this plate. The differential diagnosis between these species of mites can also be made by genetic analysis (Takehara et al. 2019).

Both genders may be present in the same bird or flock. The differentiation between these genera of mites is extremely important, because due to their unequal habits, they require different mechanisms for their control.

Control and Treatment

Due to the habits of *D. gallinae*, control should be done with spray miticide products that are applied directly to help the product penetrate their daytime hiding places, such as pyrethroid, carbaryl, coumaphos, malathion, or stirophos. Dimethoate and fenthion can be applied to the facility when it is empty. Fluralaner, ivermectin, or moxidectin can also be administered to birds (McCrea et al. 2005; Taylor et al. 2016). Other nonchemical alternatives may be useful in organic productions and include plant products, essential oils, and diatomaceous earth (Sparagano et al. 2014; George et al. 2015; Murillo and Mullens 2017; Mullens and Murillo 2018; Alves et al. 2023); however, some of these products are not regulated, nor is their effectiveness as dewormers or pesticides sufficiently proven.

16.2.1.4 *Knemidocoptes mutans* (Scaly Leg Mite)

Knemidocoptes mutans is a microscopic sarcoptic mite that affects commercial and wild birds, mainly affecting the skin of the feet, where it builds tunnels or burrows under the scales that deform the legs. The presence of these mites is uncommon in birds raised in intensive systems.

Description

It is a very small mite with an almost circular body, very short legs, and a pair of posterior bristles and penetrates deeply, forming caverns in the skin of the hocks of chickens and turkeys, causing thickening formation of thick scabs with intense keratinization. They are extremely small mites with marked sexual dimorphism. Males measure 220–250 × 140–160 μm, while females measure 445–495 × 340–440 μm in their adult stage (Tarazona and Cordero del Campillo 1999) (Figs. 16.7 and 16.8).

Life Cycle

The reproductive cycle lasts 10–14 days and takes place entirely in the bird. It is transmitted between birds by close contact with infected birds (Hinkle and Hickle 2008).

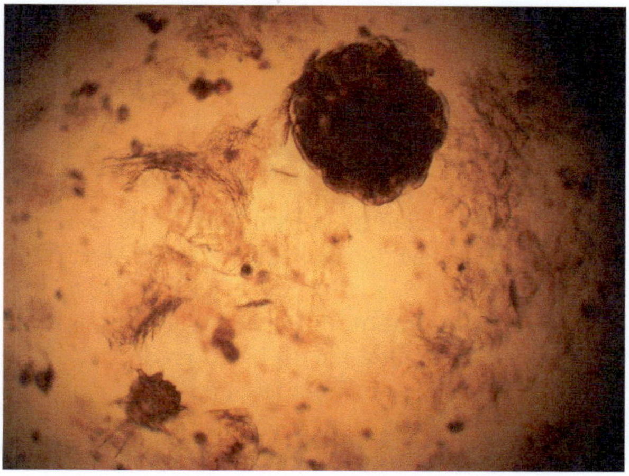

Fig. 16.7 Female (top) and male (bottom left) of *Knemidocoptes mutans* from a sample taken directly from the bird. Note the difference in size between the sexes and the two distinctive longer setae in the male (X. Hernandez-Velasco)

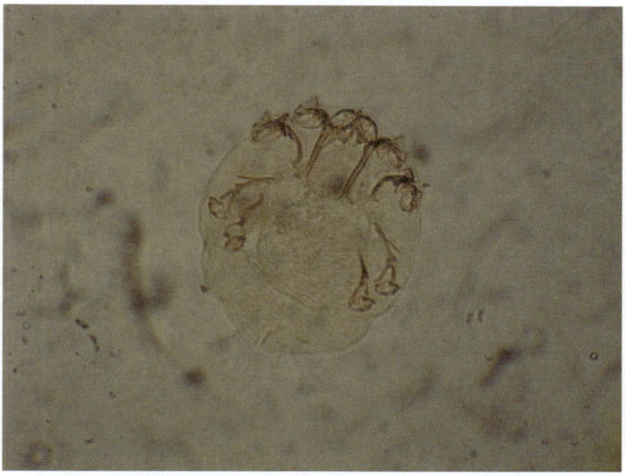

Fig. 16.8 Adult clarified female of *Knemidocoptes mutans,* ventral view (X. Hernandez-Velasco)

Epidemiology

The distribution of *K. mutans* is worldwide. All stages of these mites occur on the hosts. It is more common in chickens, turkeys, pheasants, and partridges raised in semi-freedom or backyard conditions and in ornamental birds such as pheasants (Lawal et al. 2019).

Pathogenesis

The eggs are laid in tunnels formed by the mites when penetrating under the scales of the tarsi, where it produces irritation, inflammation of the tissue and together with the accumulation of granular residues (excretions), and grayish, loosen, and lift the scales. They can also cause hypertrophy, tissue cornification, thickening and deformation of the legs, and lameness in severe cases (Proctor and Owens 2000).

Signs and Lesions

Serious infections can cause difficulty walking or limping. Severe lesions generally occur in adult birds because it is a chronic process (McCrea et al. 2005; Camacho-Escobar et al. 2014).

Diagnosis

The diagnosis is confirmed by observing the parasite under a microscope from a smear of the damaged tissue (Proctor and Owens 2000; Taylor et al. 2016).

Control and Treatment

If bird numbers are low, an oil-based product can be applied to the entire affected area, or by dipping the bird's feet in a container containing petroleum jelly, 50:50 kerosene and cooking oil mix, or blue ribbon once daily for 2 weeks; hexachlorocyclohexane (0.1%), sulfur solution (10%), or sodium fluoride (0.5%) can also be used. Ivermectin treatment may also be effective (McCrea et al. 2005; Taylor et al. 2016).

16.2.2 Ticks

16.2.2.1 Argas

Soft ticks of the genus Argas are distributed in regions with a hot dry climate, and they feed at night, and during the day, they live in crevices or nests near birds' shelters. In addition to the affectation in the host, by the presence of the parasite and the consumption of its blood, it can also transmit bacterial diseases caused by *Rickettsiae* or *Spirochetes*, as well as viral and parasitic diseases in poultry.

Description

They are 5–10 mm long ticks; have an oval and dorsoventrally flattened body, without apparent segments; are brownish yellow to reddish brown in color; and do not present sexual dimorphism, and mouthparts adapted for skin piercing and sucking (chelicerae and pedipalps) are found ventrally (Taylor et al. 2016). When it reaches its full development and is filled with blood, the female can reach a length of up to 6 × 10 mm, and the male is a little smaller (Hinkle and Hickle 2008; Tarazona and Cordero del Campillo 1999) (Fig. 16.9).

Life Cycle

They are nocturnal, feeding on blood for short periods of only a few minutes at night and during the day; due to their flat shape, they can penetrate deep crevices near the flock, where the female deposits her eggs in variable quantities in groups of four to five series one load after each feeding period (Tarazona and Cordero del Campillo 1999; Rassouli et al. 2016). In 10–15 days in hot climates or up to 3 months in cold climates, ticks hatch (nymphs), climb onto the bird's body, and later (in nine more days) mature into adults. Ticks can live for long periods without food (Taylor et al. 2016).

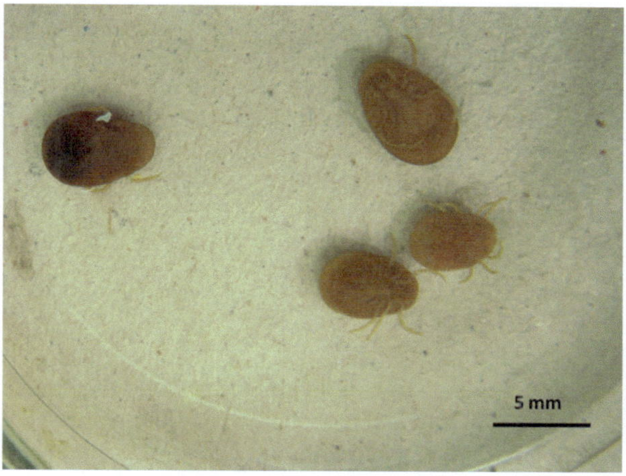

Fig. 16.9 Dorsal and ventral view of *Argas persicus* (X. Hernandez-Velasco)

Epidemiology

Argas persicus is a soft tick with a worldwide distribution, which occurs most frequently in tropical and subtropical areas, mostly in birds that are reared in semi-intensive or backyard facilities and in poor sanitation management. It is the most important tick in domestic birds (chickens, turkeys, ducks, geese, guinea fowl, pigeons, and quail, among others), since they are not species-specific (Hadi and Hind 2015; Rezaei et al. 2016; Lawal et al. 2019). *A. persicus* can also affect humans, mainly in rural areas and when they are in close proximity to domestic birds (Rezaei et al. 2016). *A. persicus* usually visits the bird briefly to feed on blood, usually at night (Haider Shah et al. 2004). Rodents, flies, birds, and bird-caring personnel can carry or store these parasites and transmit them to the flock.

Pathogenesis

They cause skin irritation, anemia, extreme general weakness, emaciation, decreased weight gain, and egg production and, in severe cases, can cause death (Permin et al. 2002). The saliva of this tick is toxic and can cause partial paralysis or paresis in birds. In addition, it has been reported that they can transmit pathogens such as rickettsia (Hinkle and Hickle 2008), *Borrelia gallinarum* (*B. anserina*), *Aegyptianella* spp., and *Pasteurella multocida* (Tarazona and Cordero del Campillo 1999; Haider Shah et al. 2004; Salifou et al. 2008; Rassouli et al. 2016).

Signs and Lesions

They are not characteristic. In cases of severe infestations, they can cause fatal flaccid paralysis in young birds.

Diagnosis

It is unlikely to find them on the bird during the day. The diagnosis is carried out by locating the ticks in the cracks, crevices, or facilities that provide them with shelter and are close to the places where the birds sleep.

Control and Treatment

In infestations with *Argas persicus*, which remain for a short time in their host, control is based on treating the facilities, walls, beds, floors, and ceilings by sprinkling, nebulizing, and pulverizing, ensuring that the product penetrates the cracks. In general, pyrethrins and pyrethroids are preferred for their residual effects. Carbaryl, malathion, pyrethroids, or coumaphos could also be used.

16.2.3 Lice

They are small insects with six legs and do not have wings, and their body is dorsoventrally flattened and divided into a head, which have antennae with three to five segments, a thorax and abdomen with sclerotized plates, and numerous setae. Only *Mallophaga* lice (chewing lice) infect birds. Although more than 40 species associated with domestic birds have been reported, they are less frequent than mites and are only occasionally found in birds in intensive production. Turkeys can harbor many species of chewing lice (*Phthiraptera*: *Ischnocera* and *Amblycera*) (Price et al. 2003). Lice that affect commercial birds are 1–6 mm in length and are generally characterized by prominent rounded mandibles, making the head wider than the thorax (Taylor et al. 2016). They feed mainly on dry skin flakes, scabs, and part of the feathers; however, they can also feed on blood (Price and Graham 1997; Kumar et al. 2017). They can be found both on the skin or the feathers of the bird, they are generally very active and move quickly, and they pass easily from one bird to another when they are in close contact as well as the hands of the workers when they hold an infested bird.

They are obligate parasites, spending their entire cycle (about 3 weeks) on the bird's body. Females have a structure with projections called genitalia in the caudal part of their body that serves to accommodate the eggs to the host's feathers (Johnson et al. 2004). The eggs are subcylindrical with rounded tips and a terminal cover called the operculum. They are attached to the bird's feathers and require an average incubation period of 5 days. However, each genus has specific reproductive habits and cycles.

The most common lice in chickens and turkeys are the so-called chicken body louse or *Menacanthus stramineus* and the feather spine louse or *Menopon gallinae* (Price and Graham 1997; Hinkle and Corrigan 2013). Although its economic importance has not been determined, infestations by lice can cause poor condition of the skin and feathers. Infestations can be more severe in young birds that have had their beaks trimmed, as well as in flocks that are in the season of greatest laying and during autumn and winter (Galloway and Lamb 2021; Clayton et al. 2005). They cause restlessness, irritation, and nervousness in birds and, in severe infestations, loss of body weight and decreased egg production; they can also cause discomfort to workers tending or handling these birds (Price and Graham 1997; Hinkle and Corrigan 2013).

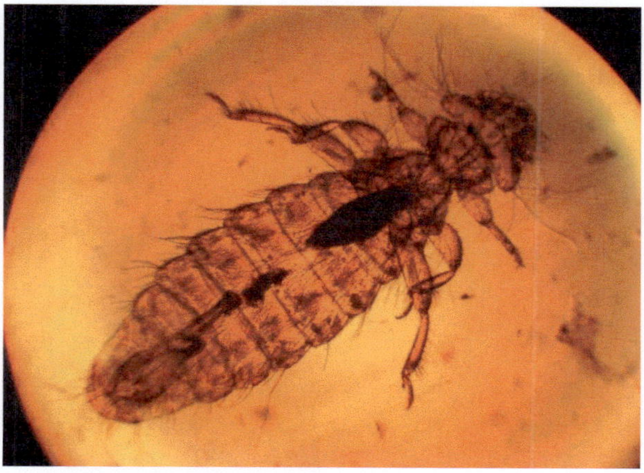

Fig. 16.10 Ventral view of a male *Menacanthus stramineus* (X. Hernandez-Velasco)

16.2.3.1 *Menacanthus stramineus* (Chicken Body Louse or Yellow Body Louse)

It has a wide world distribution and is the most economically important genus in commercial poultry farming, especially in hens that produce eggs for consumption, although it has a wide range of avian hosts in the Orders *Galliformes*, *Piciformes,* and *Passeriformes*. It is the most prevalent and damaging species of louse that infests poultry.

Description

It is a 3 mm long, pale-yellow louse. The head is almost triangular, wider than it is long, with inconspicuous antennae, with four segments. The terminal segment of the antenna is not divided and has numerous ridges. The thorax is narrower than head, with elongated oval body that ends narrower and with a rounded tip, and each abdominal segment has two rows of setae in the dorsal region, unlike *Menopon*, which only has one (Price and Graham 1997; Prastowo et al. 2020). Male faint chitinized characteristic genitalia (Adly et al. 2022), (Fig. 16.10).

Life Cycle

The eggs are deposited by the female at the base of the feathers, especially in the feathers that cover the lower part of the head and throat, and if there are more severe infestations, they can be found at the base of the feathers of the abdomen, head, neck, legs, and underwing. The eggs take 4–7 days to become nymphs with the same habits and characteristics as the adults, and only they are smaller. Molting from nymph to adult occurs in 6 days (Hinkle and Hickle 2008). *M. stramineus* eggs are distinguished by having filaments in the anterior half of the egg, and on the operculum, it has a thicker and longer bristle (Fig. 16.11) (Price and Graham 1997).

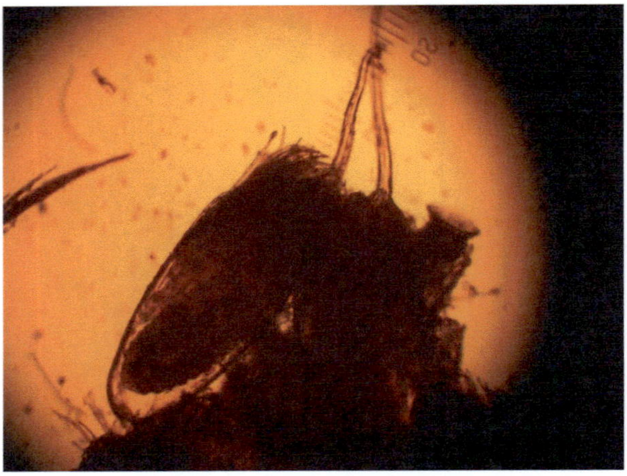

Fig. 16.11 Egg of *Menacanthus stramineus* (X. Hernandez-Velasco)

Epidemiology

It is the most common louse in chickens, turkeys, guinea fowl, quail, pheasants, and canaries. It is found mainly on the skin, especially around the vent, chest, underwings, and thighs (Hadi and Hind 2015; Prastowo et al. 2020).

Pathogenesis

In general, they do not produce important pathogenic pictures, but their presence in the bird can cause irritation, inflammation, and itching on the skin due to their movement and chewing activity (Axtell 1999), and it is the species that most affects weight gain and infestations severe can cause anemia, dermatitis, scabbing, loss of some feathers, and reduced posture (Rezaei et al. 2016). In general, malnourished and immunosuppressed birds have a higher number of lice. *M. stramineus* can be a vector of various pathogenic viruses and bacteria for birds, such as equine encephalomyelitis virus (Taylor et al. 2016) or Pasteurella multocida (Mateo 1999; Lane et al. 2006).

Signs and Lesions

M. stramineus can cause lethargy, sluggishness, drooping wings, nervous tension, and broken and shed feathers. In addition to the change in skin color. The bird's constant scratching can cause more severe injuries. The most affected areas may present scabs in the form of patches on the skin, where the greatest number of lice are found (Hadi and Hind 2015).

Diagnosis

The *Menacanthus* egg has a distinctive operculum that appears to be sculpted and has four to eight tufted processes attached to the tip of the operculum. *Menacanthus* can also be differentiated by having two short spinelike processes on the underside of the head and male faint chitinized characteristic genitalia (Adly et al. 2022).

Control and Treatment

Birds must be inspected frequently, and the presence of a few lice on several birds or eggs is sufficient to indicate treatment. Bird grooming has been reported to be important in controlling avian lice populations, and the level of infestation has been lower in birds with intact beaks (Brown 1972). Treatment is similar for different species of lice. It is recommended to treat at intervals of 7–10 days until population levels decrease. Organophosphorus (malathion, tetrachlorvinphos, and coumaphos), trichorphan (neguvon), pfispray, pyrethroids (permethrin, cypermethrin, and deltamethrin), dichlorvos, phosalone, and carbamates (carbaryl) have proven highly effective in controlling lice populations; however, the use of some chlorinated hydrocarbon insecticides (DDT, chlordane, methoxychlor, heptachlor) has been restricted or prohibited due to their prolonged persistence, toxicity, and low specificity. The effectiveness of some products of natural origin, such as pestoban and neem seed extract, and spores and endotoxin of *Bacillus thuringiensis* has also been reported (Khan et al. 2003; Al-Quraishy et al. 2012; Alves et al. 2023).

16.2.3.2 *Menopon gallinae* (Feather Shaft Louse)

It is a louse with worldwide distribution, small in size, and light yellow in color. *M. gallinae* is found primarily on the abdomen, chest, back, and thigh feathers of chickens, turkeys, and other domestic and wild birds. It moves quickly. It is a common louse but a little pathogenic, except when severe infestations develop in young birds.

Description

It is a pale-yellow louse and is 1.5–2.5 mm long (Rezaei et al. 2016). It is frequently found near the axis of the feathers on the back, chest, and abdomen and not so much on the skin, as in the case of *Menacanthus* (Price and Graham 1997). It has short antennae hidden in the slits behind the eyes. The abdomen narrows posteriorly in the female and ends rounded in the male (Adly et al. 2022) (Fig. 16.12).

Life Cycle

Eggs are deposited at the base of the shaft. These require 4–7 days of incubation and 10–15 days to go from nymphs to adults. An adult female can lay 50 or more eggs, and her life cycle lasts 3 weeks.

Epidemiology

It is a louse of chickens and turkeys, including domestic and wild birds (Garcia-Rejon et al. 2021). Like *Menacanthus, Menopon* feeds on sloughing skin feathers, and adult stages may also feed on blood (Kumar et al. 2017).

Pathogenesis

In general, *Menopon gallinae* is less damaging than *Menacanthus* because it remains longer on the root of the feathers rather than on the skin of the bird; however, in severe infestations, it can be irritating and cause weakness of the birds (Price and Graham 1997; Rezaei et al. 2016). *M. gallinae* feeds by gnawing through the

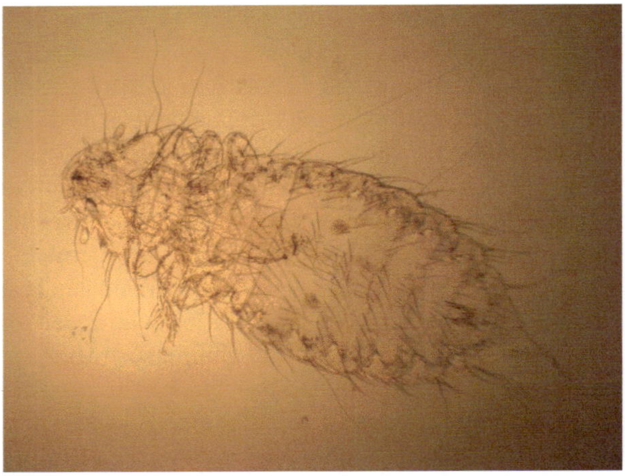

Fig. 16.12 Male *Menopon gallinae*, ventral view (X. Hernandez-Velasco)

epidermis or puncturing the quills of young feathers. Like *Menacanthus*, it can also transmit *Pasteurella multocida, Salmonella gallinarum*, *Escherichia coli*, and *Streptococcus* sp. (Kumar et al. 2017).

Signs and Lesions

If the degree of infestation is low, it can go unnoticed, while in severe infestations and mainly in young birds, it can cause restlessness, anemia, and weakness and damage feathers and even the skin, and in laying birds, it can reduce egg production (Camacho-Escobar et al. 2014).

Diagnosis

It basically consists of the detection of lice and eggs with the naked eye and is confirmed by the microscopic identification of a row of setae in each abdominal segment and distinctive genitalia in the male (Adly et al. 2022).

Control and Treatment

Treatment is similar to that used for *Menacanthus stramineus.*

16.2.3.3 *Lipeurus caponis* (Wing Lice)

It is a louse with worldwide distribution and parasitizes several species of *Galliformes* birds, including turkeys. It is mainly found on the barbules of the wing and tail feathers and generally does not cause severe infestations or serious damage to the health of the birds.

Description

It has a long and narrow gray body (2.5 mm in length and 0.3 mm in width). The head is longer than it is wide, rounded in the frontal region, with rounded, slightly extended temporal lobes. The antennae present marked sexual dimorphism, in the male the first segment is the largest and thickest, while the third segment is shorter

Fig. 16.13 Male *Lipeurus caponis,* dorsal view. Note the first segment of the antenna enlarged, and the genitalia (X. Hernandez-Velasco)

than the second and has an apical process. The front legs are very small, the middle legs are intermediate in size, and the hind legs are larger and longer than the middle legs. The abdomen is elongated. Male genitalia are distinctive and heavily chitinized (Mateo 1999; Adly et al. 2022) (Fig. 16.13).

Life Cycle

It deposits its eggs between the barbules of the bird's wing feathers, and the incubation period is 4–7 days (Adly et al. 2022).

Epidemiology

It affects gallinaceae, including domestic and wild turkeys (Canul et al. 2014; Gómez 2019; Prastowo et al. 2020; Zarith et al. 2017). In turkeys, they are found more often in the rachis of flight and tail feathers and less often in those of contour feathers (Al-Waaly and Jasim 2018; Salifou et al. 2008).

Pathogenesis

It causes slight damage in healthy birds and with an adequate state of nutrition, basically causing discomfort and irritation. Severe infestations are not frequent, so they generally do not affect the general health state of the bird (Price and Graham 1997; Mateo 1999).

Signs and Lesions

Because it is inactive and feeds mainly on the bird's feathers, in most cases, it causes only mild discomfort and irritation (Al-Waaly and Jasim 2018).

Diagnosis

The diagnosis is made by observing the specimens under a stereoscopic or light microscope, by observing the characteristics above mentioned, in the description of the parasite and the previously reported identification keys, among which stand out the sexual dimorphism in the antennae and the distinctive genitalia of the male (Mateo 1999; Adly et al. 2022).

Control and Treatment

Similar to that used for other lice.

16.2.3.4 *Cuclotogaster heterographus* (Chicken Head Louse)

It is a genus of chewing lice that are obligate parasites and are distributed worldwide. They parasitize gallinaceae such as chickens, peafowl, pheasants, chukar, and turkeys, as well as wild birds. *C. heterographus* is the most common species in this genus. It is a louse with an elongated body 2–2.5 mm in length and an almost triangular head. They tend to live along the base of the head and neck feathers. The female lays her eggs singly at the base of the feathers.

Description

It is a grayish-colored louse that is 2–3 mm long. It has a large and rounded head at the front, exposed antennae, and five segments. It has a small and broad body with an oval abdomen (Al-Waaly and Jasim 2018; Price et al. 2003) (Fig. 16.14).

Fig. 16.14 Female *Cuclotogaster heterographus* (X. Hernandez-Velasco)

Life Cycle

Their life cycle from egg to adult lasts 32–36 days. Eggs are laid singly at the base of small feathers on the bird's head. It feeds mainly on barbules in the softest part of the feather (Price and Graham 1997).

Epidemiology

It is a louse with worldwide distribution in most domestic birds, including turkeys (Lozoya et al. 1986; Ali et al. 2012). In the bird, it is found mainly near the base of the head and neck feathers, but in severe infestations, it can spread to other parts (Al-Waaly and Jasim 2018; Price et al. 2003).

Pathogenesis

It can cause restlessness, irritation, weakness, and decreased production. In severe infestations, it can cause further damage or even death in young birds. They can also be vectors of pathogens that cause fowl cholera, fowl typhoid, and toxoplasmosis (Kakarsulemankhel et al. 2010).

Signs and Lesions

In general, it causes restlessness, irritation, and weakness and can cause low production. (Wall and Shearer 1997; Kakarsulemankhel et al. 2010).

Diagnosis

By direct observation under an optical microscope. For the identification of the ectoparasites, previously reported keys and descriptions were used (Martín 2002).

Control and Treatment

Similar to other lice.

16.2.3.5 *Goniocotes gallinae* (Poultry Fluff Louse)

Goniocotes gallinae is a louse with worldwide distribution that affects domestic birds, mainly chickens, turkeys, pheasants, and pigeons. Its presentation frequency is much lower compared to other lice. It is present on the entire bird's body but is less common on the head and wings. It feeds mainly on the rest of the feathers, so in general, it does not cause serious damage to birds.

Description

Goniocotes gallinae is the smallest louse (1–1.5 mm) that parasitizes commercial birds, it is pale to dark yellow in color, and its body is broad. The head has a rounded anterior edge, and the preantenal area is small. The antennae are exposed, filiform, and without sexual dimorphism. The abdomen is oval, and in the female, it ends in a rounded shape, while in the male, it is concave with a rounded central bulge, the presence of simple, thick chitinized male genitalia, which is typical to the genus (Taylor et al. 2016; Adly et al. 2022) (Fig. 16.15).

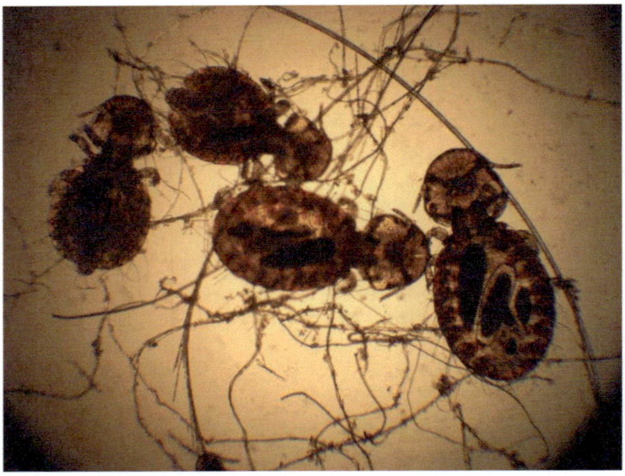

Fig. 16.15 Dorsal view of a female *Goniocotes gallinae* (X. Hernandez-Velasco)

Life Cycle

According to in vivo and in vitro studies, *G. gallinae* lives up to 35 days, and an adult female is capable of producing about 13 eggs in her lifetime. Hatching occurs 12 days after laying (Saxena et al. 2007).

Epidemiology

It is found in low quantities mainly in the down at the base of the feathers on the back or abdomen (Mateo 1999; Fabiyi et al. 2017).

Pathogenesis

It is an inactive louse, and it is considered less harmful than other lice, and it can cause irritation and loss of feathers and affect egg production (Mateo 1999; Al-Waaly and Jasim 2018).

Signs and Lesions

It causes discomfort, irritation, feather loss, and egg production in hosts (Fabiyi et al. 2017).

Diagnosis

It is performed based on the anatomical characteristics; the head is rounded, without expansion in the temporal region. It has a small, oval prothorax and equal size pleural ribs in all segments. The presence of simple, thick, chitinized male genitalia is a typical character of the genus (Kakarsulemankhel et al. 2010; Shaikh et al. 2022).

Control and Treatment

Similar to other lice.

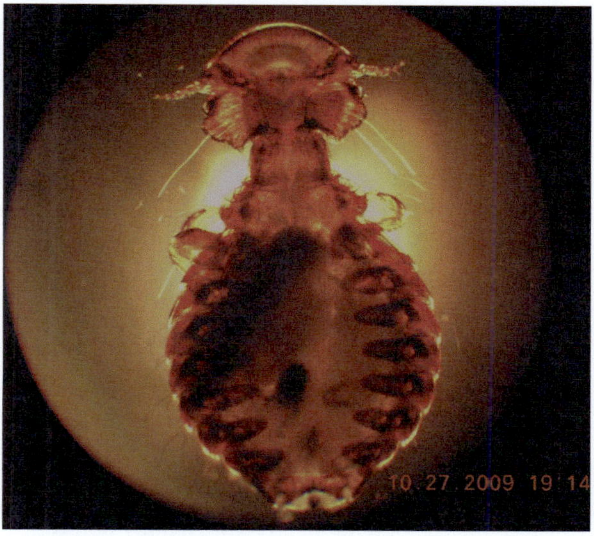

Fig. 16.16 *Goniodes gigas* (X. Hernandez-Velasco)

16.2.3.6 *Goniodes gigas* (Large Chicken Louse)

Goniodes gigas is a louse with worldwide distribution, and it is distinguished by being the largest louse that parasitizes domestic birds, mainly gallinaceous. Its frequency of presentation is lower than that of other lice. In general, it does not cause severe infestations or serious injuries since it feeds on feathers or their remains. It is easy to diagnose by its distinctive features.

Description

Goniodes gigas is dark brown in color, measuring 3–4 mm long (Adly et al. 2022). The head is broad, almost round at the front, and concave at the back; the antennae are filiform; and the temporal lobes are moderately prominent and end with three long bristles. The thorax is trapezoidal in shape. It has a wide body and an oval abdomen, with triangular areas on the sides of the segments, the abdomen ending in a concave shape (Adly et al. 2022; Kakarsulemankhel et al. 2010; Taylor et al. 2016) (Fig. 16.16).

Life Cycle

A female can produce a maximum of 14 eggs in her life, and her life cycle lasts one month (Mateo 1999). The eggs hatch 4–7 days after being laid, and the nymphs molt three times in a period of 2–3 weeks. Nymphs and adults feed on feather remains (Taylor et al. 2016).

Epidemiology

Goniodes gigas can affect domestic and wild turkeys (Lozoya et al. 1986; Hadi and Hind 2015; Rassouli et al. 2016). Although it is distributed worldwide, its prevalence is higher in places with warm climates. In birds, it is mainly found in the feathers of the neck, chest, or back (Adly et al. 2022).

Pathogenesis

Goniodes gigas infestations are not usually serious, but they can affect the health and welfare of the birds, causing irritation and inflammation of the skin (Mateo 1999; Adly et al. 2022).

Signs and Lesions

Nervousness, restlessness, wings sagging, discomfort, and damaged feathers, especially in the back area.

Diagnosis

Goniodes gigas are distinctive for their large size and head shape.

Control and Treatment

Like other lice.

16.2.3.7 *Chelopistes meleagridis* (Large Turkey Louse)

This louse is cosmopolitan, affecting mainly domestic and wild turkeys, although it can occasionally parasitize other bird species. Its prevalence is higher in places with warm climates. It is transmitted by close contact from bird to bird of the same or different species. The economic losses caused have not been analyzed in turkeys but may be underestimated.

Description

It is dark in color, measures 3.2–3.5 mm in length, and has a broad body, dorsoventrally flattened, and a rounded abdomen, with triangular areas on the sides of the segments. The head is almost round in the anterior part and concave in the caudal part. It is characterized by having temporal lobes extended caudally, with the tip forming a long stylet-like process (Naz et al. 2003). Each tarsus has two nails (Prastowo et al. 2020). The pleural plates are highly chitinized, arching superiorly beyond the anterior segment. The last abdominal segment of the female is bifurcated (Mateo 1999) (Fig. 16.17).

Life Cycle

According to in vitro studies carried out by Maturano and Daemon (2014), the development time of males and females is 30 days, and they can live up to 46 days. A female can produce up to 32 eggs, and the population of *C. meleagridis* grows at 12.5 times per generation.

Epidemiology

It is a cosmopolitan parasite that can parasitize several species of domestic birds; it is found mainly in domestic and wild turkeys (Lane et al. 2006; Fabiyi et al. 2017) and guinea fowl that inhabit tropical regions. *Meleagris gallopavo* wild is reported as the original host-parasite (Camacho-Escobar et al. 2014). In turkeys, it is found

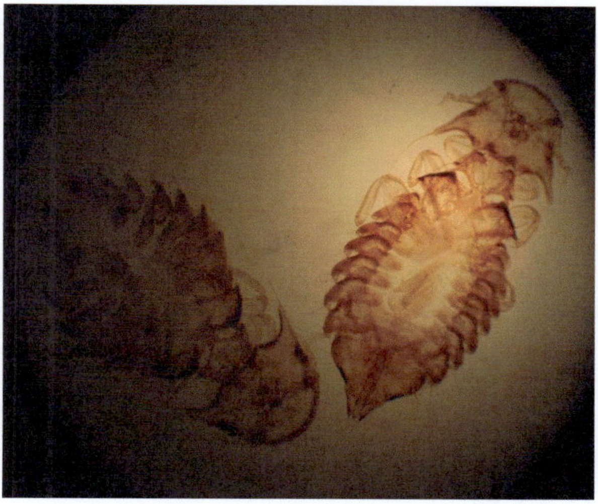

Fig. 16.17 Female and male *Chelopistes meleagridis* (X. Hernandez-Velasco)

mainly on the breast and neck feathers but also on the skin in these areas, as it feeds on both feathers and skin flaking (Johnson and Clayton 2003; Prastowo et al. 2020).

Pathogenesis

It can occur in small numbers in adult birds; in general, infestations are more severe in young birds, where it causes weight loss, weakness, and sometimes death. Therefore, it generally does not produce severe infestations (Mateo 1999; Taylor et al. 2016).

Signs and Lesions

It causes nervousness and restlessness, affects the rest of the birds, and can cause irritation with a decrease in weight gain and egg production (Camacho-Escobar et al. 2014).

Diagnosis

The diagnosis is relatively simple due to its morphological characteristics and is confirmed by observation under an optical or stereoscopic microscope.

Control and Treatment

Similar to other lice. It is recommended to avoid risk factors (e.g., crowding of hosts and high humidity) (Maturano and Daemon 2014).

Fig. 16.18 The sticktight flea, *Echidnophaga gallinacea* (X. Hernandez-Velasco)

16.2.4 Fleas

16.2.4.1 *Echidnophaga gallinacea* (Sticktight or Southern Chicken Flea)

There are various genera of fleas or *Siphonaptera* that can cause infestations in birds; however, they are rare in turkeys, and when they do occur, it is mainly in birds raised in free range in tropical areas or warm places. The most common species in commercial birds worldwide is *Echidnophaga gallinacea*; however, it can also affect mammals, including humans. Affected birds are seen with small black spots on the skin of the head and around the eyes.

Description

It is 1–2 mm long, dark brown in color, with a laterally flattened body, a narrow thorax, and legs adapted for jumping. The head is angular, devoid of a genal and pronotal comb (Taylor et al. 2016) (Fig. 16.18).

Life Cycle

The female adheres to the skin of the face and head of the host at a right angle to the ground, mainly on the skin of the head, and remains so for a period of 4–19 days. The female produces one to four eggs per day while she is attached to the host. Under favorable conditions, the incubation period varies from 4 to 14 days. The larvae feed on organic matter present in the soil; after 14–30 days, they go into a pupal phase that lasts between 9 and 16 days until an adult emerges from it. The complete cycle can vary between 1 and 2 months, but it can remain in the environment for several months as a larva (Taylor et al. 2016; Mullens and Murillo 2018).

Epidemiology

It is more common in areas with a tropical climate, although it is a parasite of backyard birds; however, it can also occur in mammals, including humans, in close association with domestic birds. It is more prevalent in cage-free and free-grazing production systems (Mullens and Murillo 2018).

Pathogenesis

Female fleas adhere to the bird's skin for days, mainly on the head and around the eyes, causing restlessness, restlessness, irritation, anemia, and blood loss. They are not considered important vectors of other diseases (Rezaei et al. 2016).

Signs and Lesions

It causes itching and inflammation due to the constant biting. It causes severe skin irritation at the adhesion site or ulcerations in severe cases. It can also cause decreased growth and production, anemia, and death in young and immunocompromised birds (Hinkle and Hickle 2008; Mateo 1999).

Diagnosis

Their diagnosis is relatively easy due to their color and location on the bird, in addition to the fact that they tend to be found in dense groups and because the mouthparts are embedded in the skin of the bird due to their continuous feeding, unlike other fleas that they feed only intermittently (Rezaei et al. 2016). It is difficult to separate them from the skin by brushing or scraping or even with tweezers. Its physical characteristics facilitate its diagnosis under the microscope (Taylor et al. 2016).

Control and Treatment

They can be removed with tweezers by holding and pulling them firmly, applying directly to the fleas, topical insecticides, or petroleum jelly with a Q-tip. Biosecurity measures should also be strengthened as well as treatment of facilities, removal, and surface treatment of litter with carbaryl, coumaphos, malathion, or pyrethroids and monitoring for reinfestations (Nair et al. 2021).

References

Adly E, Alkhalaf AA, Nasser M, Al Ashaal S (2022) Contribution to the knowledge of chewing lice of Turkey *Meleagris gallopavo domesticus* Linnaeus, 1758 (Galliformes: Phasianidae) encountered in Egypt and Saudi Arabia with new records and identification key. Int J Trop Insect Sci 42:2693–2700. https://doi.org/10.1007/s42690-022-00798-3

Alabama Cooperative Extension System (Alabama A&M and Auburn University) (2019) Poultry pest management. https://www.aces.edu/wp-content/uploads/2019/05/ANR-0483-Poultry-Pest-Management_051719L-copy.pdf

Ali WK, Koyee QM, Ahmed RK, Abdullah SMA (2012) *Cuclotogaster heterographus* (Phthiraptera: Philopteridae) infestation on the body feathers of Turkey *Meleagris gallopavo* as a new host from Erbil City, Kurdistan region, Iraq

Al-Quraishy S, Abdel-Ghaffar F, Al-Rasheid KA, Mehlhorn J, Mehlhorn H (2012) Effects of a neem seed extract (MiteStop®) on mallophages (featherlings) of chicken: in vivo and in vitro studies. Parasitol Res 110(2):617–622. https://doi.org/10.1007/s00436-011-2533-y

Alves LFA, Johann L, Oliveira DGP (2023) Challenges in the biological control of pests in poultry production: a critical review of advances in Brazil. Neotrop Entomol 52:292–301. https://doi.org/10.1007/s13744-022-01021-1

Al-Waaly ABM, Jasim DN (2018) A comparative study of ectoparasites infestation in domestic chickens and Turkey in Al-Diwaniya Province, Iraq. Biochem Cell Arch 18(1):219–224. http://www.connectjournals.com/bca

Axtell RC (1999) Poultry integrated pest management: status and future. Integr Pest Manag Rev 4:53–73. https://doi.org/10.1023/A:1009637116897

Axtell RC, Arends JJ (1990) Ecology and managment of arthropod pests of poultry. Annu Rev Entomol 35:101–126. https://doi.org/10.1146/annurev.en.35.010190.000533

Brown NS (1972) The effect of host beak condition on the size of *Menacanthus stramineus* population of domestic's chickens. Poult Sci 51:162–164. https://doi.org/10.3382/ps.0510162

Camacho-Escobar MA, Pérez-Lara E, Arroyo-Ledezma J, Sánchez-Bernal EI, García-López JC (2010) Parasitic mites in backyard turkeys. Trop Subtrop Agroecosystems 12:675–679. https://www.redalyc.org/articulo.oa?id=93915170025

Camacho-Escobar MA, Arroyo-Ledezma J, Ávila-Serrano NY, Jerez-Salas MP, Sánchez-Bernal EI, García-López JC (2014) Ectoparasites and their damage in backyard turkeys in Oaxaca's coast, Mexico. Eur J Vet Med 2014:7. http://scik.org

Canul M, Sierra A, Azcorra G, Nava F, Amaya S (2014) Contribution to the identification of mallophaga turkeys in natives in the state of Yucatan. Actas Iberoam Conserv Anim 4:279–281

Clayton DH, Moyer BR, Bush SE, Jones TG, Gardiner DW et al (2005) Adaptive significance of avian beak morphology for ectoparasite control. Proc R Soc B 272:811–817. https://doi.org/10.1098/rspb.2004.3036

DeLoach JR, DeVaney JA (1981) Northern fowl mite, *Ornithonyssus sylviarum* (Acari: Macronyssidae), ingests large quantities of blood from white Leghorn hens. J Med Entomol 18(5):374–377. https://doi.org/10.1093/jmedent/18.5.374. PMID: 7299791

Denmark HA, Cromroy HL (2003) Tropical fowl mite, *Ornithonyssus bursa* (Berlese) (Arachnida: Acari: Macronyssidae). The University of Florida's Institute of Food and Agricultural Sciences Extension. EENY-297. http://entnemdept.ifas.ufl.edu/creatures

Di Palma A, Giangaspero A, Cafiero MA, Germinara GS (2012) A gallery of the key characters to ease identification of *Dermanyssus gallinae* (Acari: Gamasida: Dermanyssidae) and allow differentiation from *Ornithonyssus sylviarum* (Acari: Gamasida: Macronyssidae). Parasit Vectors 5:104. https://doi.org/10.1186/1756-3305-5-104

Domrow R (1991) Acari Prostigmata (excluding Trombiculidae) parasitic on Australian vertebrates: an annotated checklist, keys and bibliography. Invertebr Taxon 4:1283–1376. https://doi.org/10.1071/IT9901283

Fabiyi JP, Alayande MO, Akintule AO, Lawal MD, Mahmuda A, Usman M (2017) Prevalence and seasonal fluctuations of ectoparasites infesting backyard turkeys, *Meleagris gallopavo*, in Sokoto, northwestern Nigeria. Rev Elev Med Vet Pays Trop 70(1):21–24

Galloway TD, Lamb RJ (2021) Population dynamics of chewing lice (Phthiraptera) infesting birds (Aves). Annu Rev Entomol 66:209–224. https://doi.org/10.1146/annurev-ento-041420-075608

Garcia-Rejon JE, Tzuc-Dzul JC, Cetina-Trejo R, Madera-Navarrete MI, Cigarroa-Toledo N, Chan-Perez JI, Ortega-Pacheco A, Torres-Chable O, Pietri JE, Baak-Baak CM (2021) Identification of parasitic arthropods collected from domestic and wild animals in Yucatan, Mexico. Ann Parasitol 67(4):647–658. https://doi.org/10.17420/ap6704.381

George DR, Finn RD, Graham KM, Mul MF, Maurer V, Moro CV, Sparagano OAE (2015) Should the poultry red mite *Dermanyssus gallinae* be of wider concern for veterinary and medical science? Parasit Vectors 8:1–10. https://doi.org/10.1186/s13071-015-0768-7

Gómez V (2019) Identificación de parásitos externos e internos en el *Meleagris gallopavo* del estado de Yucatán. (Tesis de pregrado) Tecnológico Nacional de México, Instituto Tecnológico de Conkal, pp 42

Hadi MHA, Hind AA (2015) Ectoparasites of domestic Turkey (*Meleagris gallopavo*) in Al-Diwaniya City/Iraq. Int J Curr Microbiol Appl Sci 4(10):669–667

Haider Shah A, Nisar Khan M, Iqbal Z, Sohail Sajid M (2004) Tick infestation in poultry. Int J Agric Biol 6:1162–1165

Harrington DWJ, George DR, Guy JH, Sparagano OAE (2011) Opportunities for integrated pest management to control the poultry red mite, *Dermanyssus gallinae*. Worlds Poult Sci J 67:83–93. https://doi.org/10.1017/S0043933911000079

Hinkle N, Corrigan RM (2013) External parasites and poultry pests. In: Swayne DE, Glisson JR, McDougald LR, Nolan LK, Suarez DL, Nair V (eds) Diseases of poultry, 13th edn. Wiley-Blackwell, Ames, pp 1099–1116

Hinkle NC, Hickle L (2008) External parasites and poultry pests. In: Saif YM, Fadly AM, Glisson JR, McDougald LR, Nolan LK, Swayne DE (eds) Diseases of poultry, 12th edn. Blackwell Publishing, Ames, pp 1011–1024

Johnson KP, Clayton DH (2003) The biology, ecology, and evolution of chewing lice. In: Price RD, Hellenthal RA, Palma RL, Johnson KP, Clayton DH (eds) The chewing lice: world checklist and biological overview. Illinois natural history survey special publication 24. Champaign, pp 449–475

Johnson KP, Yoshizawa K, Smith V (2004) Multiple origins of parasitisms in lice. Proc R Soc Lond B 271:1771–1776. https://doi.org/10.1098/rspb.2004.2798

Kakarsulemankhel JK, Kakarsulemankhel MN, Yasinzai MI, Kakarbarakzai B, Kakarsulemankhel G, Malghani MAK (2010) Morpho-taxonomy and new record of *Cuclotogaster heterographus* (Nitzsch, (in Giebel), 1866) (Phthiraptera: Ischnocera: Philopteridae) from Balochistan, Pakistan. Pak Entomol 32(1):65–75

Khan MN, Nadeem M, Iqbal Z, Sajid MS, Abbas RZ (2003) Lice infestation in poultry. Int J Agric Biol 5:213–216

Krinsky WL (1983) Dermatoses associated with the bites of mites and ticks (Arthropoda: Acari). Int J Dermatol 22:75–91. https://doi.org/10.1111/j.1365-4362.1983.tb03319.x

Kumar S, Ahmad A, Ali R, Kumar V (2017) A note on the haematophagous nature of poultry shaft louse, *Menopon gallinae* (Amblycera: Phthiraptera). J Parasit Dis 41(1):117–119. https://doi.org/10.1007/s12639-016-0760-y

Lane RS, Kucera TF, Barrett RH, Mun J, Wu C, Smith VS (2006) Wild Turkey (*Meleagris gallopavo*) as a host of ixodid ticks, lice and Lyme disease spirochetes (*Borrelia brugdorferi* sensu Lato) in California state parks. J Wildl Dis 42(4):759–571. https://doi.org/10.7589/0090-3558-42.4.759

Lawal JL, Mustapha M, Adamu L, Dauda L, Biu AA (2019) Ectoparasitosis in domesticated turkeys (*Meleagris gallopavo*) in Jere area, Borno State, Nigeria. Int J Curr Microbiol App Sci 5(1):11–22. https://doi.org/10.18488/journal.110.2019.51.11.22

Lozoya SA, Quiñones LS, Aguirre ULA, Guerrero RE (1986) Malófagos de las aves domésticas en 4 municipios del sureste del estado de Coahuila. Agraria 2(2):203–221

Martín MMP (2002) Mallophaga: Amblycera, In: Ibérica F, Ramos MA et al (eds) Museo Nacional de Ciencias Naturales, vol 20. CSIC, Madrid, pp 187

Mateo MP (1999) Malofagidosis y pulicosis. En: Martínez FAR, Sánchez AC, Hernández RS, Navarrete LCI, Díez BP, Quiroz RH, Carvalho VM (eds), Parasitología Veterinaria, McGraw-Hill Interamericana, Madrid, España, pp 833–843

Maturano R, Daemon E (2014) Reproduction, development and habits of the large turkey louse *Chelopistes meleagridis* (Phthiraptera: Ischnocera) under laboratory conditions. Braz J Biol 74(3):712–719. https://doi.org/10.1590/bjb.2014.0085

McCrea B, Jefrey S, Ernst A (2005) Common lice and mites of poultry: identification and treatment. Oakland California: Division of Agriculture and Natural Resources, University of California. Publication 8162. http://anrcatalog.ucdavis.edu

Mironov SV (2013) *Allopsoroptoides galli* n. g., n. sp., a new genus and species of feather mites (Acari: Analgoidea: Psoroptoididae) causing mange in commercially raised domestic chicken in Brazil. Syst Parasitol 85(3):201–212

Mullen G, Durden L (2002) Medical and veterinary entomology. Academic, San Diego

Mullen GR, O'Connor BM (2009) Mites (Acari). In: Mullen GR, Durden LA (eds) Medical and veterinary entomology, 2nd edn. Academic, San Diego, pp 433–492

Mullens BA, Murillo AC (2018) The future of poultry pest management. In: Mench JA (ed) Woodhead Publishing Series in Food science, technology and nutrition, advances in poultry welfare. Woodhead Publishing, Duxford, pp 295–321. https://doi.org/10.1016/B978-0-08-100915-4.00014-2

Mullens BA, Owen JP, Kuney DR, Szijj CE, Klingler KA (2009) Temporal changes in distribution, prevalence and intensity of northern fowl mite (*Ornithonyssus sylviarum*) parasitism in commercial caged laying hens, with a comprehensive economic analysis of parasite impact. Vet Parasitol 160:116–133. https://doi.org/10.1016/j.vetpar.2008.10.076

Mullens BA, Murillo AC, Zoller H, Heckeroth AR, Jirjis F, Flochlay-Sigognault A (2017) Comparative in vitro evaluation of contact activity of fluralaner, spinosad, phoxim, propoxur, permethrin and deltamethrin against the northern fowl mite, *Ornithonyssus sylviarum*. Paras Vectors 10(1):358. https://doi.org/10.1186/s13071-017-2289-z

Murillo AC, Mullens BA (2017) A review of the biology, ecology, and control of the northern fowl mite, *Ornithonyssus sylviarum* (Acari: Macronyssidae). Vet Parasitol 246:30–37. https://doi.org/10.1016/j.vetpar.2017.09.002

Nair S, Gouge DH, Murillo A (2021) Backyard chickens and ectoparasites: introduction and management. The University of Arizona, Cooperative Extension. AZ1878

Naz S, Rizvi SA, Ahmad Z (2003) Redescription of *Chelopistes meleagridis* (Linnaeus) (Phthiraptera: Ischnocera: Philopteridae) from Pakistan with reference to its morphotaxonomical and genitalial studies. Pak J Entomol 18:29–35

Permin A, Esmann JB, Hoj CH, Hove T, Mukatirwa S (2002) Ecto-, endo- and haemoparasites in free range chicken in the Gomoronzi District in Zimbabwe. Prev Vet Med 54:213–224. https://doi.org/10.1016/s0167-5877(02)00024-7

Philips JR (2000) A review and checklist of the parasitic mites (Acarina) of the Falconiformes and Strigiformes. J Raptor Res 34:210–231

Prastowo J, Priyowidodo D, Nurcahyo W, Chusnaifah DL, Wusahaningtyas LS, Firdausy LW, Sahara A (2020) Lice infestation and diversity in turkeys (*Meleagris gallopavo*) in the special region of Yogyakarta and Central Java, Indonesia. Vet World 13:782–788. https://doi.org/10.14202/vetworld.2020.782-788

Price MA, Graham OH (1997) Chewing and sucking lice as parasites of mammals and birds. United States Department of Agriculture. Technical bulletin number 1849, pp 7–30

Price RD, Hellenthal RA, Palma RL, Johnson KP, Clayton DH (2003) The chewing lice: world checklist and biological overview. Illinois Natural History Survey Special Publication 24

Proctor H, Owens I (2000) Mites and birds: diversity, parasitism and coevolution. Trends Ecol Evol 15(9):358–364

Quintero MT, Juárez VG, Gutiérrez RS, Grifaldo AF, Cisneros FV, Nieves GE, Vega PJL (2007) Sobre un caso de infestación masiva de acaros *Dermatophagoides pteronyssinus* (Acari: Astigmata: Pyroglyphidae) en una granja de gallinas ponedoras en el estado de Sonora México y sus implicaciones en salud pública. Trabajo publicado en las memorias del XII Congreso Iberoamericano de Parasitología, Madrid, España

Quintero MT, Itza M, Juarez G, Eleno A (2010) Seasonality of *Megninia ginglymura*: a one-year study in a hen farm in Yucatan, Mexico. In: Sabelis M, Bruin J (eds) Trends in acarology. Springer, Dordrecht, pp 537–538. https://doi.org/10.1007/978-90-481-9837-5_92

Rassouli M, Darvishi MM, Lima SRR (2016) Ectoparasite (louse, mite and tick) infestations on female turkeys (Galliformes, Phasianidae. Meleagris gallopavo) in Iran. J Parasit Dis 40(4):1226–1229

Rezaei F, Hashemnia M, Chalechale A, Seidi S, Gholizadeh M (2016) Prevalence of ectoparasites in free-range backyard chickens, domestic pigeons (*Columba livia domestica*) and turkeys of Kermanshah province, west of Iran. J Parasit Dis 40(2):448–453. https://doi.org/10.1007/s12639-014-0524-5

Ribeiro VLS, Moojen V, Telles APD (1992) *Ornithonyssus bursa*: parasito de aves causando acaríases cutâneas em humanos no Rio Grande do Sul, Brasil. Anis Bras Dermatol 67:31–34

Salifou S, Nattay A, Odjo AM, Pangui LJ (2008) Arthropodes ectoparasites du dindon (*Meleagris gallopavo*) dans le nord-oust du Béenin. Rev Elev Med Vet Pays Trop 61(3–4):185–189. https://doi.org/10.19182/remvt.9987

Saxena AK, Kumar S, Gupta N, Singh R (2007) Population expansion of the poultry fluff louse, *Goniocotes gallinae* (De Geer, 1778) (Ischnocera, Phthiraptera). Zool Sci 24:327–330. https://doi.org/10.2108/zsj.24.327

Shaikh F, Naz S, Ali Birmani S, Akbar S, Dahri S, Akbar S (2022) Morpho-taxonomy of new host and locality record of *Goniocotes gallinae* (De Geer, 1778) (Phthiraptera: Ischnocera) from Hyderabad, Sindh, Pakistan. Pak J Parasitol 73:17–23

Skoracki M, Zabludovskaya SA, Bochkov AV (2012) A review of Prostigmata (Acariformes: Trombidiformes) permanently associated with birds. Acarina 20:67–107

Sparagano O (2009) Control of poultry mites (*Dermanyssus*). Springer Netherlands, Dordrecht. https://doi.org/10.1007/978-90-481-2731-3

Sparagano OA, George DR, Harrington DW, Giangaspero A (2014) Significance and control of the poultry red mite, *Dermanyssus gallinae*. Annu Rev Entomol 59:447–466. https://doi.org/10.1146/annurev-ento-011613-162101

Sparagano O, George DR, Finn RD, Giangaspaero A, Bartley K, Ho J (2020) *Dermanyssus gallinae* and chicken egg production: impact, management, and a predicted compatibility matrix for integrated approaches. Exp Appl Acarol 82:441–453. https://doi.org/10.1007/s10493-020-00558-3

Takehara M, Murata S, Katakura K, Fujisawa S, Hmoon MM, Win SY, Bawm S, Htun LL, Aung YH, Win MM, Isezaki M, Maekawa N, Okagawa T, Konnai S, Ohashi K (2019) Haematophagous mites on poultry farms in the Republic of the Union of Myanmar. Heliyon 5(4):e01544. https://doi.org/10.1016/j.heliyon.2019.e01544

Tarazona VJM, Cordero del Campillo M (1999) Parasitosis cutáneas y oculares. En: Martínez FAR, Sánchez AC, Hernández RS, Navarrete LCI, Díez BP, Quiroz RH, Carvalho VM (eds), Parasitología Veterinaria. McGraw-Hill Interamericana, Madrid, pp 824–843

Taylor MA, Coop RL, Wall RL (2016) Veterinary parasitology, 4th edn. Blackwell Publishing, Oxford. https://doi.org/10.1002/9781119073680

Valiente Moro C, Chauve C, Zenner L (2005) Vectorial role of some dermanyssoid mites (Acari, Mesostigmata, Dermanyssoid). Parasite 12:99–109. https://doi.org/10.1051/parasite/2005122099

Valiente Moro C, De Luna CJ, Tod A, Guy JH, Sparagano OAE, Zenner L (2009) The poultry red mite (*Dermanyssus gallinae*): a potential vector of pathogenic agents. Exp Appl Acarol 48:93–104. https://doi.org/10.1007/s10493-009-9248-0

Waap H, Aguin-Pombo D, Maia M (2020) Case report: human dermatitis linked to *Ornithonyssus bursa* (Dermanyssoidea: Macronyssidae) infestation in Portugal. Front Vet Sci 7:567902. https://doi.org/10.3389/fvets.2020.567902

Wall R, Shearer D (1997) Veterinary entomology. Chapman & Hall, New York

Walter DE, Proctor HC (1999) Mites: ecology, evolution and behaviour. CAB International

Zarith M, Suhaila AH, Izzauddin NHN, Khadijah S (2017) Parasites prevalence in poultry: focusing on free range turkeys (*Meleagris gallopavo*). Malaysian J Vet Res 8(1):1–9

Part III

Nutritional Disorders

Nutritional Disorders in Fattening Turkeys

17

Amr Abd El-Wahab, Bussarakam Chuppava, Awad A. Shehata, Shereen Basiouni, Wolfgang Eisenreich, and Hafez M. Hafez

Abstract

The feed industry has long known that inadequate dietary nutrient levels may not support optimum poultry health, performance, or welfare. However, the research on the effects of additional nutrients (amino acids, minerals, and vitamins) on turkeys is sparse and partly outdated. Turkeys are omnivores, but for economic reasons, they are fed largely vegetarian diets supplemented with purified sources of nutrients to meet their nutritional requirements (Klasing and Korver 2020). The growth rate of turkeys has increased significantly during the past decade due to genetic selection. Nutrient deficiencies or toxicities often result from diet formulation and/or milling errors. In this chapter, we will discuss the main nutritional disorders in turkeys.

A. A. El-Wahab · B. Chuppava
Foundation Institute for Animal Nutrition, University of Veterinary Medicine Hannover, Hannover, Germany
e-mail: amr.abd.el-wahab@tiho-hannover.de

A. A. Shehata · W. Eisenreich
TUM School of Natural Sciences, Bavarian NMR Centre (BNMRZ), Structural Membrane Biochemistry, Technical University of Munich, Garching, Germany
e-mail: awad.shehata@tum.de; wolfgang.eisenreich@mytum.de

S. Basiouni
Institute of Molecular Physiology, Johannes-Gutenberg University, Mainz, Germany
e-mail: sbasioun@uni-mainz.de

H. M. Hafez (✉)
Faculty of Veterinary Medicine, Institute of Poultry Diseases, Free University of Berlin, Berlin, Germany
e-mail: hafez.mohamed@fu-berlin.de

H. M. Hafez, A. A. Shehata (eds.), *Turkey Diseases and Disorders Volume 2*, https://doi.org/10.1007/978-3-031-63322-5_17

Keywords

Amino acids · Vitamins · Minerals · Nutritional disorders

17.1 Protein and Amino Acids

Protein is one of the major cost components in the turkeys' feed and has major effects on performance and the overall cost of the final product (Table 17.1). Thus, commercial diets usually are formulated using a "least cost" approach to meet the protein requirements. Generally, there is a need for ten essential amino acids (arginine, histidine, isoleucine, leucine, lysine, methionine, phenylalanine, threonine, tryptophan, and valine). Arginine, lysine, and methionine benefit gastrointestinal function and the gut-associated immune system. It was found that a diet with higher levels of arginine, lysine, and methionine can positively impact turkeys' performance and immune system. It also enhances the specific indicators that help maintain the integrity of the gut, even under challenging conditions (Konieczka et al. 2022).

Turkey poults are severely protein-depleted at hatch since the protein is preferentially used as an energy source during hatching. This is because gluconeogenesis from protein does not require oxygen, which is limited during the transition from chorioallantois to pulmonary respiration (Uni and Ferket 2004). The transition from the yolk sac to feed is compromised because of the immature digestive tract in young poults, which leads to a reduced ability to digest and absorb protein (Halevy et al. 2000; Halevy et al. 2003; Moore et al. 2005).

When used as the sole protein source in pre-starter diets, soybean meal with high levels of anti-nutritional factors can have adverse effects (Palliyeguru et al. 2011). Anti-nutritional factors such as protease inhibitors impair protein digestion, lectins affect carbohydrate digestion and can damage the intestinal wall, phytic acid interferes with mineral and protein absorption, and non-starch polysaccharides cause fluid retention and poor nutrient absorption (Cowieson et al. 2004; Stein et al. 2008). Although industrial processing reduces the levels of protease inhibitors and lectins, high inclusion rates of soybean meal may still have adverse effects. Additionally, high potassium levels in soybean meals have been associated with increased water intake and excreta moisture.

It was found that combining soybean meal with canola protein concentrate, fish meal, porcine meal, or corn gluten meal improves poult performance during the first 14 days of life in comparison to feeding SBM alone (Ross et al. 2019).

Amino acids are the building blocks of protein, the major dry matter component of growth in poultry and their hatching eggs. Adequate intakes of dietary amino acids are crucial for the optimum efficiency of poultry production. An excess or a deficiency of amino acids can negatively impact the health and productivity of birds. A study revealed that diets with a 90% arginine-to-lysine ratio and high methionine content can help reduce oxidative stress, regulate metabolic parameters, and affect markers of intestinal barrier integrity in turkeys with necrotic enteritis (Ognik et al. 2020).

Table 17.1 Protein requirement (%) of fattening male and female turkeys

Reference	Growing of fattening turkeys (male/female) per week of age							
NRC (1994)[a]	**0–4**	**4–8**	**8–12/8–11**	**12–16/11–14**	**16–20/14–17**	**20–24/17–20**	**–**	**–**
	28	26	22	19	16.5	14	–	–
Aviagen (2017)	**0–4**	**5–6**	**7–9/7–8**	**10–12/9–10**	**13–14/11–12**	**15–16/13–14**	**17–18/15–16**	**19–21/17–20**
	26–28	24–26	23–25	20–22	18–20	16–18	15–17	14–16

[a] Current National Research Council (NRC)

Deficiencies of individual essential amino acids usually have the same effect as when protein is deficient; however, additional symptoms may appear that characterize certain amino acid deficiencies. The effects of essential amino acid deficiencies are nonspecific and may lead to decreased body weight and feed intake in adults. The disorders associated with amino acid deficiencies are summarized as follows: (1) A lack of methionine reduces growth performance and may predispose foot pad dermatitis (FPD) (Abd El-Wahab et al. 2016). (2) Low contents of some essential amino acids (lipotropic factors), such as methionine, are a potential predisposing factor to hepatic lipidosis (Abd El-Wahab et al. 2021) (Fig. 17.1). (3) Inadequate lysine is known to cause depigmentation of the wing feathers in Bronze turkey poults, as well as can result in stunting and retarded development in birds (Biesiadecki et al. 2004). (4) Muscular dystrophy is accompanied by cysteine and methionine deficiencies. (5) Arginine, lysine, and methionine deficiency can impair turkeys' performance, gut integrity, and immune status (Konieczka et al. 2022).

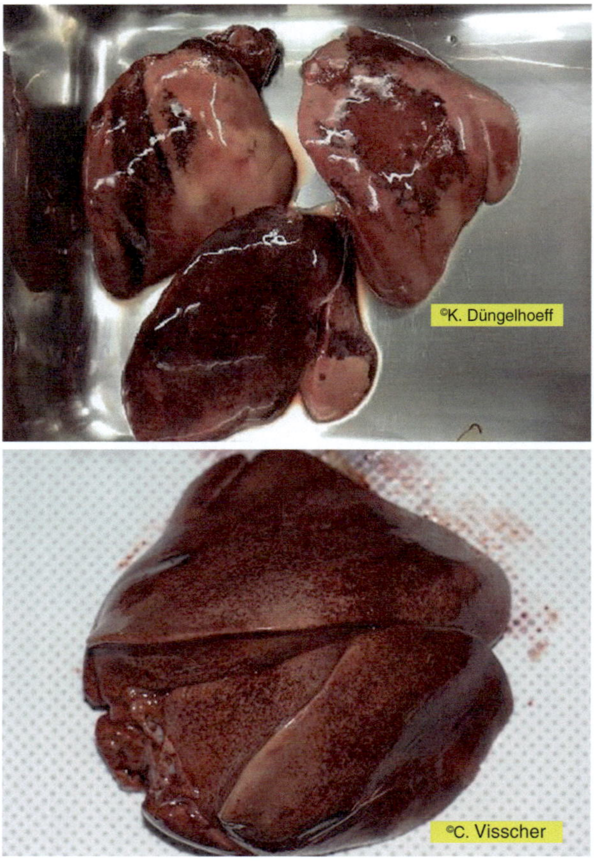

Fig. 17.1 Liver with hepatic lipidosis. Enlarged, mottled liver with large, pale, scattered, well-demarcated foci, and dark red areas (Abd El-Wahab et al. 2021)

17.2 Vitamins

Vitamins are a heterogeneous group of fat-soluble (A, D, E, and K) and water-soluble compounds essential in poultry nutrition. Generally, the vitamins are cofactors for enzymes, hormones (e.g., vitamins A, D), or antioxidants (pre-vitamin A and vitamin E). The amounts of vitamins needed in poultry diets are very low, yet moderate deficiencies are often more debilitating than those of protein, energy, or fat.

Vitamin deficiency reduces the growth rate of young birds and affects feather follicles, the epithelial surfaces of skin, hematopoietic tissues, and the growth plate of bones. Common symptoms that arise from defects in these tissues are dyschondroplasia, chondrodystrophy, dermatitis, poor feathering, anemia, and increased susceptibility to infectious diseases (Klasing and Korver 2020). In practice, poultry diets should be formulated to contain a large margin of safety for all vitamins to compensate for possible losses during feed processing, transportation, storage, variations in feed composition, and environmental conditions.

Birds with abundantly tissue stores of the fat-soluble vitamins A, D, and E can withstand long periods of depletion before deficiency symptoms can be manifested. Also, birds with marginal stores of vitamins can withstand feed restriction without deficiency symptoms because tissue catabolism liberates vitamins from muscle, fat, and other tissues.

17.2.1 Fat-Soluble Vitamins

17.2.1.1 Vitamin A

From a nutritional perspective, vitamin A is the most challenging of the vitamins because it is deficient in many feedstuffs and is among the most likely to become toxic upon over-supplementation. Vitamin A refers to a group of fat-soluble and brightly pigmented molecules, including retinol, retinal, retinoic acid, and several pro-vitamin A carotenoids converted to retinoids within the intestine and in other tissues (Koutsos et al. 2003). Among the 700 known carotenoids, only 50 have pro-vitamin A activity, with ß-carotene, ß-cryptoxanthin, and a-carotene being the major pro-vitamin A molecules (Romanchik et al. 1995).

Carotenoids have several functions, including immune-regulatory and immune-stimulatory functions, besides antioxidant, antimutagenic, and anticarcinogenic properties (Mora et al. 2008). The vitamin A requirement of the fattening turkey ranged between 8000 and 12,000 IU/kg feed.

Deficiency Signs

Vitamin A deficiency is more common in young birds. The main clinical signs are: (1) general depression, ruffled feathers, stunted growth, poor feathering, drowsiness, unsteady gait, and postural imbalance (Fig. 17.2); (2) a watery discharge from the nostrils and eyes and eyelids with thick exudate; (3) hyperkeratosis on the mucous membrane of the mouth and esophagus; and (4) drop in egg production in

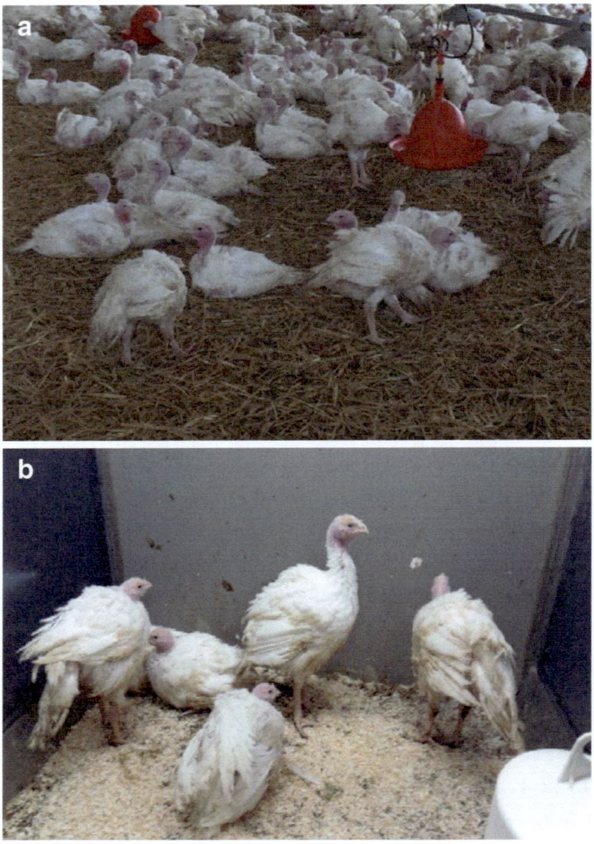

Fig. 17.2 (**a**) Turkeys on a commercial farm with ruffled feathers, an inability to stand, and abnormal gait. (**b**) Turkeys in the diagnostic trial were drowsy and moved uncoordinatedly when agitated, with a droopy head, ruffled feathers, and undeveloped feather tail (Abd El-Wahab et al. (2017)

breeder flocks, decreased hatchability, and early embryonic mortality during the first week of incubation.

Histopathology

Vitamin A-deficiency lesions first appear in the esophagus (Fig. 17.3) and pharynx and are confined largely to mucous glands and their ducts. Small, white nodules may be found in the nasal passages, mouth, esophagus, and pharynx and may extend into the crop. The original epithelium is replaced by a keratinizing epithelium (i.e., squamous metaplasia) that blocks ducts of the mucous glands, causing them to become distended with secretions and necrotic materials (Klasing and Korver 2020).

Hypovitaminosis A is characterized by thin diphtheritic membranes and nasal plugs that are usually limited to the cleft palate and its adjacent epithelium (Klasing and Korver 2020). Exudate also may fill sinuses and other nasal cavities, causing swelling of one or both sides of the face. As the condition progresses, the mucous membrane is covered with a dry, dull, fine, slightly uneven film, whereas a normal membrane is even and moist.

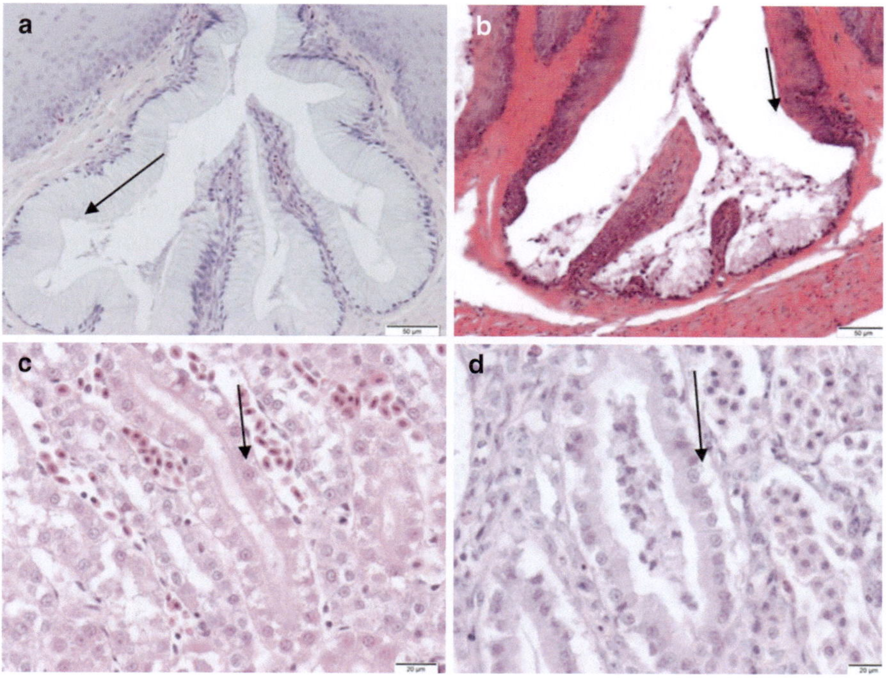

Fig. 17.3 Microscopic lesions of tissues from control (**a**, **c**) and experimental birds (**b**, **d**). (**a**) Normal esophageal mucosa and mucus glands (arrow); (**b**) esophagus with hyperplastic, thickened mucosa and submucosal squamous metaplastic mucus glands (arrow); (**c**) normal renal tubules (arrow); and (**d**) mild degeneration in renal tubules (arrow) (Abd El-Wahab et al. 2017)

Chronic vitamin A deficiency causes damage to the kidney tubules (Fig. 17.3), which leads to azotemia and visceral urate deposits (e.g., "visceral gout") in severe cases. A severe deficiency results in a grossly abnormal cardiovascular system, characterized by an absence of vascular networks and by a ballooned, non-compartmentalized, randomly positioned heart without an inflow tract at the posterior site of the heart.

Toxicity

Mistakes in the formulation of vitamin premixes can result in toxicity at vitamin A levels between 35,000 and 60,000 IU/kg. Signs of toxicity include slow growth, an unsteady gait, reluctance to walk, reduced bone mineralization, and a higher incidence of tibial dyschondroplasia. Severe toxicity results in anorexia, conjunctivitis, adhesions of the eyelids, encrustations around the mouth, and thinning of the frontal bones of the skull with thickened osteoid seams (Degernes et al. 2011). Additionally, hypervitaminosis A interferes with calcium absorption and vitamin E.

Diagnosis

The diagnosis of vitamin A deficiency depends on the clinical signs and lesions, the amount of vitamin A in the ration, and/or the vitamin A level in the liver. Respiratory diseases should be taken into consideration for the differential diagnosis.

Treatment

Poultry severely deficient in vitamin A should be given a stabilized vitamin A preparation at approximately 10,000 IU vitamin A/kg of ration. Absorption of vitamin A is rapid; therefore, chickens or turkeys not in advanced stages of deficiency should respond promptly, except for blindness, which may be permanent (Klasing and Korver 2020).

17.2.1.2 Vitamin D

Vitamin D is a fat-soluble vitamin obtained by either producing in the skin upon exposure to sunlight or by supplementation in the feed. Dietary vitamin D is absorbed by the small intestine, bound by vitamin D-binding protein or albumin in the blood, and then rapidly taken up by the liver.

Two main forms of vitamin D are known: ergocalciferol (Vitamin D_2) and cholecalciferol (Vitamin D_3). Poultry can efficiently utilize vitamin D3. Vitamin D undergoes two enzymatic hydroxylations in the body before it becomes metabolically active because vitamin D-binding protein cannot bind to vitamin D_2 effectively. The first hydroxylation occurs in the liver resulting in 25-dihydroxy vitamin D_3 ($25(OH)_2D_3$), which is followed by the second hydroxylation in the kidneys, resulting in $1,25(OH)2D_3$ with the aid of 1a-hydroxylase (Christakos et al. 2016). In poultry, it is inadequate for biological function in chickens.

The major and conventional function of $1,25(OH)_2D_3$, also called calcitriol, assists the formation of bone and eggshell by regulating calcium and phosphorus homeostasis in the body by influencing intestinal and renal and renal homeostasis absorption. The optimal dose of vitamin D_3 in turkey diets is 2500–4000 IU/kg feed (Klasing and Korver 2020).

Deficiency Signs

Vitamin D_3 deficiency is associated with inadequate dietary calcium and phosphorus, leading to rickets (growing birds) and osteomalacia (in adults). **Field rickets** in turkey poults is mostly caused by impairment of intestinal absorption of minerals and vitamin D3 (Huff et al. 1999).

The first signs of vitamin D_3 deficiency in growing chicks or poults are slower growth and an awkward gait. As the deficiency advances, rickets becomes evident as severe fragility and bending of long bones caused by poor mineralization. Beaks and claws become soft and pliable (Fig. 17.4), feathering is poor, and birds walk with obvious effort and take a few unsteady steps before squatting on their hocks, which they rest upon while swaying slightly from side to side (Klasing and Korver 2020). Enlargement of ribs at their junctions with vertebrae is also a common postmortem lesion (Fig. 17.5).

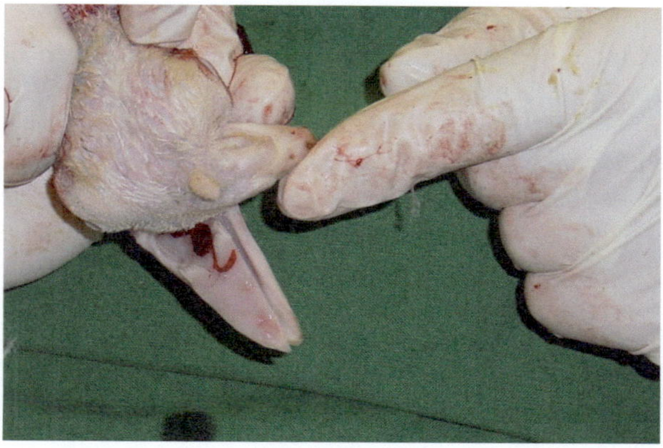

Fig. 17.4 Rickets in turkeys, Beak is soft and pliable. (Hafez-Berlin)

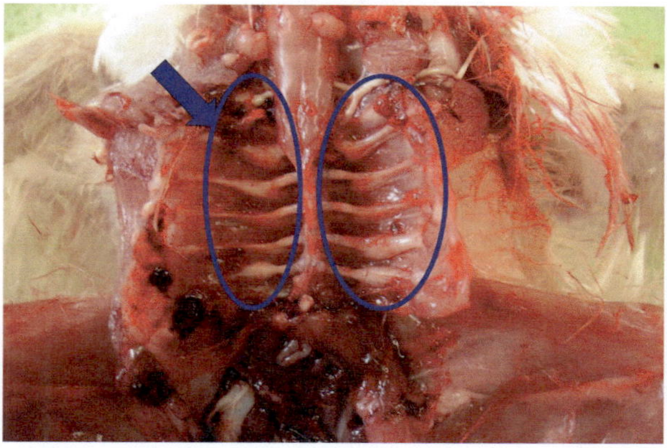

Fig. 17.5 Rickets in turkeys. Enlargement of ribs at their junctions with vertebrae. (Hafez-Berlin)

A moderate deficiency of vitamin D_3 results in an increased incidence of tibial dyschondroplasia, especially when calcium or phosphorus levels are not optimal. Many of the signs of a vitamin D_3 deficiency may be similar to calcium deficiency, and analysis of levels of these two nutrients in the diet, or D_3 in blood plasma, can confirm the cause (Bowes et al. 1988).

Hatchability is reduced markedly by vitamin D_3 deficiency, mainly caused by increased embryonic mortality late in incubation. Chicks and poults that do not hatch have a high incidence of chondrodystrophy, in which the upper or lower mandible is shortened or deformed (Klasing and Korver 2020).

In layers and breeders, signs of deficiency begin to occur as soon as 2 weeks after they are deprived of vitamin D_3. The first sign is a marked increase in thin-shelled and soft-shelled eggs, followed by a marked decrease in egg production. Biochemical indicators include a rapid decrease in the concentrations of 25-(OH)D_3 in the blood, followed by a decrease in blood calcium concentration.

Egg production and eggshell strength may vary in a cyclic manner. During periods of extreme leg weakness, hens show a characteristic posture described as a "penguin-type squat" (Breeding et al. 1994).

Histopathology

In turkeys receiving deficient vitamin D_3, characteristic changes observed on necropsy are confined to bones and parathyroid glands. The latter becomes enlarged from hypertrophy and hyperplasia. Bones are soft and break easily. Well-defined knobs are present on the inner surface of the ribs at the costochondral junction. The most characteristic internal signs of vitamin D_3 deficiency poults are bending the ribs at their juncture with the spinal column and bending them downward and posteriorly (Klasing and Korver 2020).

Toxicity

Vitamin D toxicity in turkeys is not well documented. However, the toxicity is exacerbated by high dietary calcium or phosphorus levels, especially in the growing chick. In broiler chicks, pathology can be detected at 30,000 IU/kg of vitamin D_3 when fed throughout the growing period. Lesions include atrophy of the parathyroid gland and calcium deposits in renal tubular lumina. Hens are generally more resistant to vitamin D toxicity than growing chicks, but toxicity can be transferred to the egg, causing excessive mobilization of eggshell calcium. Very high levels of vitamin D up to four million IU/kg diet rapidly induce renal damage.

Diagnosis

The diagnosis of hypovitaminosis D depends on clinical signs and lesions and the diet's contents of calcium, phosphorus, and vitamin D_3.

Treatment and Prevention

Feeding a single dose of 15,000 IU of vitamin D_3 cured the affected birds better than when generous levels of the vitamin were added to the feed. However, it should be remembered that excess vitamin D can be harmful. Hypovitaminosis D can be prevented by providing a balanced diet with adequate calcium, phosphorus, and vitamin D levels.

17.2.1.3 Vitamin E

Vitamin E is a fat-soluble antioxidant, with four different tocopherols, alpha, beta, gamma, and delta, as the functional forms of this vitamin. The alpha-tocopherol is the most commonly found form in nature and is considered to be the most biologically active. In poultry production, vitamin E supplementation is essential for maintaining fertility and hatchability in parent stocks. It also has a crucial role in preventing nutritional encephalopathy and myopathies in chickens and turkeys (Klasing and Korver 2020).

Supplementing poultry feed with vitamin E (or other antioxidants) is essential, especially when oxidizable fats are included in the feed because, upon oxidation, these fats release metabolically harmful free radicals that affect poultry health and production (Engberg et al. 1996; Lu et al. 2014). Vitamin E supplementation in chicken feed also prevents the oxidation of lipids in unsaturated fatty acids (Rama Rao et al. 2011); because of this, the amount of active vitamin E that reaches the intestine for absorption can be reduced in poultry diets high in unsaturated fat

(Villaverde et al. 2008). Under these circumstances, the antioxidant status in poultry can be decreased due to increased lipid peroxidation (Rama Rao et al. 2011). Generally, the dietary requirement of vitamin E for fattening turkeys ranged from 30 to 50 mg/kg feed.

In a study, supplementation of layer hen feed with 30 IU/kg of vitamin E resulted in a significant increase in serum superoxide dismutase and glutathione peroxidase, which are enzymes with antioxidant activities (Liu et al. 2019).

Deficiency Signs

Hypovitaminosis E is usually seen in young birds. Vitamin E is very unstable in diets rich in oxidized products. Most cases are observed in birds fed rations high in polyunsaturated fatty acids (cod liver oil, soya bean oil) that oxidize and become rancid.

Exudative diathesis is an edema of subcutaneous tissues characterized by a subdermal deposit of viscous green-blue-colored exudate from endothelial failures in sections of the circulatory system caused by a deficiency in vitamin E and selenium (Klasing and Korver 2020). This colored viscous fluid is seen easily through the skin because it usually contains some blood components from slight hemorrhages that appear in the breast and leg muscles. Exudative diathesis can be eliminated, and most myopathies can be greatly relieved when increased selenium alone.

Encephalomalacia is a nervous syndrome characterized by ataxia or paresis, backward or downward retractions of the head (sometimes with lateral twisting), forced movements, incoordination, tremors and relaxation of the legs, and finally death (Klasing and Korver 2020). Poults with paresis usually do not have brain lesions but have poliomyelomalacia (Klasing and Korver 2020).

Nutritional myopathy in chickens, turkeys, waterfowl, and ostriches is attributed to vitamin E/selenium deficiency, characterized by pale foci or light-colored streaks of breast muscles. Vitamin E and selenium deficiency, especially in turkeys, may result in an extreme myopathy of the proventriculus and heart muscles (Klasing and Korver 2020).

Diagnosis

Hypovitaminosis E is diagnosed based on signs, lesions, and histopathological examination. Analysis of vitamin E and selenium levels in the ration is also recommended.

Treatment

If not too far advanced, exudative diathesis is readily reversed by administering proper levels of vitamin E and selenium supplementation either by injection and/or by oral dosing or in feed. A single dose of 300 IU/bird is required for treatment. A dose of 30–150 mg of vitamin E/Kg feed is recommended as a preventive dose. The 150 ppm vitamin E diet did reduce the vulnerability of the liver and red blood cells to in vitro peroxidation (Applegate and Sell 1996). Supplementing turkeys with 50 mg of vitamin E per kg significantly reduced thiobarbituric acid reactive compounds in the liver, demonstrating antioxidative protection of this tissue that was not seen with increasing selenium levels (Mueller et al. 2009). However, when the turkeys were healthy (not diseased), a desirable performance with 20 mg vitamin E per kg diet is enough (Sell et al. 1997).

17.2.1.4 Vitamin K

Vitamin K is a cofactor in the liver and bone proteins to synthesize carboxyglutamate residues from glutamic acid. In the absence of vitamin K, an abnormal prothrombin lacking a-carboxyglutamic acid is secreted in the blood by the liver since prothrombin is an important part of the blood-clotting mechanism. Deficiency of vitamin K results in markedly prolonged blood-clotting time (Jin and Sell 2001); an affected poults may bleed to death from a slight bruise or other injuries.

There are three possible sources of vitamin K_1–K_3 in the diet: Vitamin K_1, known as phylloquinone, is found mostly in the leafy sections of plants. Vitamin K_2, also known as menaquinone, is produced by bacteria, particularly those found in the large intestine. Vitamin K_3, or menadione, is a synthetic vitamin that does not occur naturally. Adults usually have well-developed intestinal microflora, and vitamin K inadequacies are unusual. Vitamin K_2 is not readily absorbed from the large intestine, but it is digested after coprophagy of cecal excreta. A 1.8 mg menadione/kg is necessary for turkey poults for adequate prothrombin time during the first 4 weeks of life. These recommendations were also adopted by the NRC (1994). Generally, a 2–3 mg/kg diet is considered acceptable for fattening turkeys.

Deficiency Signs

Turkey poults with vitamin K deficiency have a prolonged prothrombin time in the blood, and damage in the gizzard has been observed. Oral antibiotic administration did not affect the derived vitamin K requirement. This indicates that the contribution from intestinal microbial synthesis is relatively small in young poults (Jin and Sell 2001). According to Klasing (2008), signs of vitamin K deficiency occur most frequently between second and third weeks after the birds are placed on a vitamin K-deficient diet. Large hemorrhages appear on the breast, legs, wings muscles, and/or abdominal cavity. Generally, 4–6 h after vitamin K is given to deficient birds, blood clots normally, but recovery from anemia or disappearance of hemorrhages cannot be expected.

17.2.2 Water-Soluble Vitamins

Generally, except for biotin, the scientific information related to vitamin B requirements for turkeys is limited (GfE 2004). For example, in very few cases, systematic increase trials over several doses allow reliable derivation of supply recommendations (GfE 2004). The validity of some studies is also limited concerning the duration of the trials, as the experiments often do not cover an entire fattening period but only a relatively short period.

17.2.2.1 Thiamin (Vitamin B_1)

Thiamin is a cofactor for several enzymes catalyzing decarboxylation and transmetalation-type reactions. Based on these findings and following the suggestions of the NRC (1994), a supply recommendation of a 2 mg/kg diet is established.

Deficiency Signs

Thiamin deficiency leads to ataxia, muscle weakness, and muscle degeneration. Backward movement and bending of the head and neck are typical nervous symptoms. **"Stargazing"** in young turkey poults, thiamin deficiency also leads to changes in amino acid levels in the liver and some brain regions (NRC 1994; Klasing and Korver 2020). **Polyneuritis** is observed in adult birds approximately 3 weeks after being placed on a thiamin-deficient diet. Turkeys suffering from thiamin deficiency respond in a few hours to oral administration of the vitamin.

17.2.2.2 Riboflavin (Vitamin B_2, Lactoflavin)

Riboflavin is a cofactor in many enzyme systems in the body, such as NAD- and NADP-cytochrome reductases, succinic dehydrogenase, acyl dehydrogenase, etc.

Deficiency Signs

Vitamin B_2 deficiency in turkey poults resulted in a very slow growth rate, and birds become weak and emaciated, with diarrhea developing between the first and second weeks of age. Birds do not walk except when forced to walk on their hocks with the help of their wings. Toes are curled inward, when walking and resting (**curled-toe paralysis**) (Klasing and Korver 2020). Riboflavin deficiency in young turkeys is characterized by poor growth, poor feathering, leg paralysis, and encrustations in the corners of the mouth and eyelids (Ruiz and Harms 1989). Severe dermatitis of the feet and shanks, marked by edematous swelling, desquamation, and deep fissure of the feet. In severe cases of riboflavin deficiency, birds show marked swelling and softening of sciatic and brachial nerves (Cai et al. 2006). Sometimes, the sciatic nerves reach a diameter four to five times thicker than normal.

Histopathology

Histology examination of affected nerves shows degenerative changes in myelin sheaths of the main peripheral nerve trunks. This may be accompanied by axis cylinder swelling and fragmentation. Schwann cell proliferation, myelin changes, gliosis, and chromatolysis occur in the spinal cord. The sciatic nerve exhibits myelin degeneration in one or more branches. Fine structural examination of the sciatic nerve reveals that redundant folds and loops of myelin form symmetry or asymmetric expansions of the sheath resulting in segmental demyelination (Cai et al. 2006). In cases of curled-toe paralysis, degeneration of the neuromuscular end plate and muscle tissues is often found. Riboflavin is probably also essential for the myelin metabolism of the main peripheral nerve trunks.

Treatment

Two 100-μg doses of riboflavin should be sufficient for treating riboflavin-deficient poults, followed by adding an adequate level to the diet. However, when the curled-toe deformity is long-standing, irreparable damage occurs; in this case, riboflavin administration cannot cure the condition (Klasing and Korver 2020).

17.2.2.3 Pyridoxine (Vitamin B_6)

Vitamin B_6 acts as a cofactor in amino acid decarboxylation and transamination processes. The three active forms of vitamin B_6 are pyridoxine, pyridoxal, and pyridoxamine.

In fattening turkey, trials by Jeroch et al. (1978) showed that 5.7 mg/kg levels in the starter feed and 4.0–4.2 mg/kg in the individual fattening phases are sufficient. However, according to the NRC, a dose level of 4.5 mg/kg in the starter phase and 3.5–3.0 mg/kg in the subsequent fattening phase are sufficient (NRC 1994). However, the INRA (1984) recommends a much lower supply (2 mg/kg).

Deficiency Signs

Severely pyridoxine-deficient birds show low appetite, poor growth, chondrodystrophy, and characteristic nervous signs (Klasing and Korver 2020). During these convulsions, poults may run aimlessly about, flapping their wings and falling to their sides or rolling completely over on their backs, where they perform rapid jerking motions with their feet and heads (Klasing and Korver 2020). A pyridoxine deficiency causes a defect in collagen fibers in cortical bone and articular cartilage matrix and increased solubility of proteoglycans and collagen (Massé et al. 1998). These structural defects cause chondrodystrophy and osteoarthritis in deficient birds. In adult birds, vitamin B_6 causes a marked reduction of egg production and hatchability, decreased feed intake, loss of BW, and deaths.

17.2.2.4 Cobalamin (Vitamin B_{12})

Vitamin B_{12} is a cofactor for enzymes that transfer one-carbon units and catalyze carbon skeleton rearrangements. In contrast, the interconversion of methyl malonyl coenzyme A to succinyl coenzyme. Although plant diets are deficient in vitamin B_{12}, it is readily available in animal products and the cecal microbiota following coprophagy, making its deficiency rare (Watanabe et al. 2013). While the NRC (1994) recommends a supply of 0.003 mg/kg for all life stages of fattening turkeys, the INRA recommends 0.015 mg/kg for chicks and 0.01 mg/kg for fatteners (1984). The committee recommends 0.01 mg/kg for the entire fattening period.

Deficiency Signs

Deficiencies in laying hens also in laying turkey have been induced by dramatically increasing the amount of protein in their diet, increasing carbon rearrangement enzymes' activity (Ward et al. 1985). Klasing and Korver (2020) stated that vitamin B_{12} deficiency results in slow growth, decreased FCR, increased the mortality rate, and reduced egg size and hatchability in turkey. Vitamin B_{12} deficiency causes myelin degeneration in birds. Chondrodystrophy may occur in vitamin B_{12}-deficient chicks or poults, when their diets lack choline or methionine as sources of methyl groups.

Treatment

Adding a 4 mg vitamin B_{12}/ton breeding diet is sufficient to maintain maximum hatchability and produce poults or chicks, with sufficient vitamin stores to prevent any deficiency during the first few weeks of life. Similar injections with the vitamin B_{12} under field condition of young chicks followed by diet supplementation also will correct the deficiency. Halle et al. (2011) found that 20 μg of vitamin B_{12}/kg feed meets the requirements of growing chickens and ducks for fattening. And NRC (1994) stated that 0.003 mg/kg diet is enough for growing turkeys during the whole fattening period.

17.2.2.5 Pantothenic Acid (Vitamin B_5)

Pantothenic acid is a component of coenzyme A, which is involved in the formation of citric acid in the citric acid cycle, synthesis and oxidation of fatty acids, oxidation of keto acids, and acetylation of choline. In studies by Jeroch et al. (1978), the content of pantothenic acid in feed mixtures was sufficient for optimal growth in fattening turkeys. Therefore, a supply of 13 mg/kg in the first 6 weeks and 10 and 8 mg/kg in the following fattening period is sufficient. The NRC (1994) recommends 10 mg/kg for turkey poults (up to 4 weeks of age) and 9 mg/kg for fattening turkeys. However, the INRA (1984) recommends a supply of 10 mg/kg for turkey poults and 5 mg/kg for fattening turkeys.

Deficiency Signs

Signs of pantothenic acid deficiency are difficult to differentiate from those of biotin deficiency. Generally, signs are dermatosis, broken feathers (retard, rough feather growth), chondrodystrophy, poor growth, and mortality. A viscous exudate mostly sticks eyelids together, and vision is impaired. Outer layers of skin between the toes and on the bottoms of the feet sometimes peel off, and small cracks and fissures appear at these points (Klasing and Korver 2020). Supplementing pantothenic acid in the poultry diet can improve the health and performance of chickens.

A pasty substance in the mouth and an opaque gray-white exudate in the proventriculus could be detected at necropsy. The liver is hypertrophied and may vary in color from a faint to dirty yellow. The spleen is slightly atrophied. Kidneys are enlarged. Nerves and myelinated fibers of the spinal cord show myelin degeneration. These degenerating fibers occur in all cord segments down to the lumbar region.

Treatment

Pantothenic acid deficiency appears to be completely reversible, if not too far advanced, by oral treatment or injection with the vitamin followed by restoration of an adequate level in the diet (Klasing and Korver 2020).

17.2.2.6 Niacin (Nicotinic Acid, Nicotinamide, Vitamin B_3)

Niacin is a component of two important coenzymes: nicotinamide adenine dinucleotide (NAD) and nicotinamide adenine dinucleotide phosphate (NADP), which are mostly involved in carbohydrate, fat, and protein metabolism. Although tryptophan can be converted to niacin, the efficiency is low, and it is not advised as a replacement for dietary supplementation (Ruiz et al. 1990). Niacin availability in grains and grain by-products is frequently low (Yen et al. 1977). The NRC (1994) recommends 60 mg/kg in the starter feed and 40 mg/kg in the fattening phase.

Deficiency Signs

The main lesion in young birds is an enlargement of the hock joint (chondrodystrophy) and bowing of the legs. The main difference between this condition and the chondrodystrophy caused by manganese or choline deficiency is that the Achilles tendon rarely slips from its condyles in nicotinic acid deficiency. In some cases, inflammation of the mouth, diarrhea, and poor feathering may be seen. Additionally, a decrease in feed intake, BW, rate of egg production, and hatchability could be noted too.

Treatment

Supplementing a deficient ration with required amounts of nicotinic acid has little or no effect on cases that have progressed to the extent that the tendon has slipped from its condyles (chondrodystrophy) or on advanced cases of enlarged hock disorder in adult tom turkeys. Excessive supplementation should be avoided because levels above 0.75% dietary niacin cause decrease of bone thickness dimensions and bone strength (Johnson et al. 1995).

17.2.2.7 Folic Acid

Folic acid is a part of the enzyme system involved in single-carbon metabolism. It is involved in the synthesis of purines and the methyl of choline, methionine, and thymine. Thus, it is necessary for nucleic acid metabolism. Froehlich (1987) suggests folic acid supplements in starter feed of 1–2 mg/kg in chicks and for fattening turkeys of 1–1.5 mg/kg. In agreement with NRC (1994), a 1.0 mg/kg folic acid concentration is recommended during the first 8 weeks in turkeys. During the remainder of fattening, 0.8 mg/g is adequate.

Deficiency Signs

Folic acid deficiency leads to impaired growth rate, poor feathering, anemia, perosis, and reduced hatchability (Klasing and Korver 2020). Insufficient folic acid supply of the turkey breeder hens led to an impairment of the young turkey poults' live weight development and cervical dislocation. The most obvious symptom of inadequate folic acid is cervical paralysis with turkey poults (Miller and Balloun 1967). Poults show only a slight anemia.

Folic Acid-Choline Interrelationship

Folic acid has a central role in methyl group metabolism. Young et al. (1955) observed that, when a diet for birds is deficient in folic acid, an increase in the dietary level of choline reduces, but does not completely prevent the incidence and severity of chondrodystrophy. A growth depression has been observed in chicks fed a practical diet low in folic acid and marginally deficient in methionine and choline. Supplementing the diet with folic acid or methionine and choline stimulated the growth of poults under these conditions (Pesti et al. 1991).

Treatment

A single intramuscular injection (IM) of 50–100 μg pure pteroylglutamic (folic) acid causes a peak reticulocyte response within 4 days in severely anemic folic acid-deficient birds (Robertson et al. 1947). Adding 500 μg folic acid/100 g feed caused recovery comparable to that obtained with vitamin injection (Klasing and Korver 2020).

17.2.2.8 Biotin (Vitamin B_7)

Biotin is a cofactor in carboxylation and decarboxylation reactions involving carbon dioxide fixation. These reactions have important roles in anabolic processes and in nitrogen metabolism. Biotin is concentrated in the liver, kidney, and bone, which are the principal areas of activity for enzymes that require it.

According to the results of Jeroch et al. (1978), 0.24 mg/kg in the first 6 weeks of life and 0.18–0.11 mg/kg in the subsequent fattening period of meat turkeys are sufficient for optimal supply. Misir and Blair (1988), who conducted studies on the bioavailability of biotin from various diets with young turkey poults, found a linear increase in growth rate in pre-depleted birds up to supplements of 0.2 mg/kg. According to INRA (1984), biotin levels can be lowered from 0.3 mg/kg in fattening turkey poults to 0.05 mg/kg. NRC (1994) supply recommendations are similar at 0.1–0.25 mg/kg for fattening poults. Data from NRC (1994) or GfE (1999) on biotin requirements vary in the range of 150–300 μg/kg (usual) complete feed. In the abovementioned experimental studies, considerably higher biotin contents/additions were used (400/800/1600/2000 μg/kg), so it must be assumed that the dosages and effects are "exceeding requirements" or "supra-nutritive."

Biotin bioavailability for chickens and turkeys varies greatly among practical feed ingredients (Frigg 1984; Misir and Blair 1988). Biotin is no more than 12% available in some grains but almost completely available in others (Bryden et al. 1991). This is an important consideration in formulating diets to satisfy the biotin requirements of poultry. Sun and Kim (2017) showed that increasing dietary biotin (1521 μg/kg) improves the performance and well-being of broilers stocked at high densities (16 bird/m^2) in litter-independent and litter-dependent manners, respectively.

Deficiency Signs

In turkey poults, a condition characterized by hock dysfunction, broken feathers, dermatitis, and diarrhea was found; when biotin supplementation was significantly reduced, the prevalence and severity were increased (NRC 1994; GfE 2004). Feeding with folic acid and calcium pantothenate had little effect, but when combined with biotin, almost disorder-free poults could be generated.

Among the nutrients with potential importance for footpad health frequently examined in more detail in experimental studies, biotin is probably the most important, and publications on this topic have been published for decades (Abd El-Wahab et al. 2013a). In turkeys, severe pododermatitis is a prevalent condition.

Foot pad lesions have been linked to the nutrition factors, and poor litter conditions. Turkey poults fed a semi-purified, low-biotin diet developed severe foot pad dermatitis. It could be linked to the skin's physiology.

In contrast, biotin supplementation reduced the severity of foot pad dermatitis in poults reared on dry litter. However, it did not affect the severity of the lesions in poults reared on wet litter (Harms and Simpson 1977). Nevertheless, Abd El-Wahab et al. (2013a) observed that high biotin supplementation (2000 μg/kg) reduced foot pad lesions by about 18% and 30% in two repetition trials in broiler chickens exposed experimentally to wet litter.

Treatment

Injection or oral administration of biotin was sufficient to prevent biotin deficiency signs in chicks and turkey poults.

17.2.2.9 Choline

Choline is present in acetylcholine and phospholipids. It acts as a methyl source in synthesis within the body of methyl-containing compounds such as methionine, creatine, and carnitine. Choline per se does not act as a methyl donor but first must be oxidized to betaine. Although poultry can synthesize choline, the amount produced is limited, and supplementation is required, when demand exceeds biosynthetic capacity (NRC 1994; GfE 2004). Jeroch et al. (1978) suggested levels in the starter phase of 1590 mg/kg and the fattening phase between 990 and 675 mg/kg for an adequate supply for turkeys.

The INRA (1984) recommends an allowance of choline chloride of 800–500 mg/kg of choline chloride for turkey poults. The NRC (1994) supply recommendations decrease from 1600 mg/kg for chicks to 800 mg/kg by the end of fattening.

Deficiency Signs

In addition to poor growth, chondrodystrophy is the most consistent lesion of choline deficiency in chicks and young poults (Klasing and Korver 2020). Chondrodystrophy is first characterized by pinpoint hemorrhages and a slight swelling about the hock joint, followed by an apparent flattening of the tibio-metatarsal joint caused by rotation of the metatarsus (Klasing and Korver 2020).

Treatment

If choline deficiency is noted in chicks and/or poults before severe signs of chondrodystrophy have developed, the deficiency can be cured by supplementing the diet with sufficient choline to meet the requirements. If the tendon slipped, the damage is irreparable.

17.2.2.10 Vitamin C (Ascorbic Acid)

Vitamin C, known as L-ascorbic acid, is a water-soluble vitamin synthesized from glucose (Sahin et al. 2003). It has notable antioxidant properties, because of its ability to donate electrons, and it protects the integrity of many cells, including lymphocytes, against damage from free radicals generated in response to infection or toxins (Nimse and Pal 2015). Unlike fat-soluble vitamins, vitamin C is not stored in the body, and elevated dietary intake of vitamin C results in decreased absorption and rapid excretion by the kidneys (Johnston et al. 2006). Poultry, unlike humans, can synthesize vitamin C endogenously, thanks to the L-gluconolactone oxidase enzyme, which is present in the renal tissue where it converts L-gulono-g-lactone into ascorbic acid (Hooper et al. 2001). However, the requirements for vitamin C are increased under stressful conditions such as beak trimming, vaccination, transportation, thermal stress, or infection.

Turkeys may be exposed to stress factors (high stocking density, temperature, etc.) that increase vitamin C requirements, so positive effects were found after supplementation (Nagórna-Stasiak et al. 1999). However, in studies by Sell et al. (1997), ascorbic acid supplements of 300 mg/kg up to the 41st day of age of turkey poults, did not affect gain, feed expenditure, and viability.

17.3 Minerals

Essential mineral elements are as important as amino acids and vitamins in maintaining life, well-being, and production in poultry. The minerals essential for the maintenance of well-being are calcium (Ca), phosphorus (P), sulfur (S), magnesium (Mg), potassium (K), sodium (Na), and chlorine (Cl) and the trace elements such as manganese (Mn), iron (Fe), copper (Cu), zinc (Zn), iodine (I), molybdenum (Mo), chromium (Cr), and selenium (Se).

17.3.1 Major Elements

The elements Ca, P, Mg, Na, K, Cl, and S are understood as major elements. In the feeds used in poultry nutrition, the content of K is so high that an insufficient supply is not to be expected. Nevertheless, recommendations on the supply of K are given below to provide farms with appropriate data for nutrient balancing. Since the supply of S-containing amino acids is adequate, no specific recommendations are given for this element. Recommendations for the supply of major elements to broiler turkeys have been published by the World's Poultry Science Association (WPSA) (1985) and the NRC (1994).

17.3.1.1 Recommendations for the Supply of Major Elements

For element P, data are given based on total P and available P. This is to consider that the P is not all available in the complete feed. As soon as measurement data on the usability of P from a sufficiently large number of individual components are available, a more differentiated calculation of the contents of total P will be possible (Table 17.2).

For all elements, the contents in the feed can be significantly reduced during fattening. This results from changes in the increment composition and the feed input (feed/growth). Some experiments carried out with variation in P supply show that with the P contents recommended here, no negative effects occur even at a high growth level (Hocking et al. 2002a, b; Rodehutscord et al. 2003). All in all, it becomes clear that extensive research still has to be done to validate the data on the contents in the live weight increment, the factors influencing the unavoidable losses, and the usability.

17.3.1.2 Calcium and Phosphorus

In general, dietary allowances for poultry and all nutrients are, for example, provided by the German Society of Nutrition Physiology (GfE) or the WPSA. These recommendations are based on calculations of the factorial approached requirement data (Jeroch et al. 2019). For turkey parent stock, no recommendations have been published for Ca and P, which indicates a poor database of requirements for maintenance and growth for turkey breeders. Other concentration norms for minerals are estimated based on evaluations of dose and response. This system is used by the NRC (1994), Jeroch et al. (2019), and in trials by Leeson and Summers (2005). All these publications focused on heavy breeds only (Table 17.3). Calcium or phosphorus imbalances in poultry feed can cause significant losses in production. Rickets, a disturbance of normal ossification, is marked by distortion of the bones (Klasing 2008). Some birds mobilize large amounts of Ca from their skeleton during this period, and the bones may become so demineralized that the birds are unable to stand and appear paralyzed. The sternum and rib bones are frequently deformed, and all bones are easily broken (NRC 1994).

Deficiencies of Ca and phosphorus in growing chicks cause rickets that differ in histopathology from those of vitamin D deficiency. Tibiae from chicks fed a diet containing 0.3% Ca from the time of hatching showed, by 2 weeks, a widening of the proliferating pre-hypertrophic zone of epiphyseal cartilage and irregular contours in the boundary between the zones of proliferating and hypertrophic cartilage. Irregular cartilage columns and elongated epiphyseal vessels were present. By 4 weeks, the epiphyseal growth plate had widened and sometimes extended as a cartilaginous plug into the metaphysis. Histologically, the proliferating and hypertrophic zones were irregular and often contained areas of nonviable cells. The hypertrophied zone was markedly widened in some chicks by 4 weeks. Metaphyseal blood vessels invaded along the lateral, but not the apical, region of the cartilaginous plug; cartilage columns of the metaphysis were thickened and irregular. The investigators note that the pathology is similar to that of tibial dyschondroplasia.

Table 17.2 Requirements of major elements for turkeys as percentages or units per kilogram of diet on 90% dry matter according to NRC (1994)

		Growing turkeys (male/female) age per week					
Nutrient	Unit	0–4	4–8	8–12/8–11	12–16/11–14	16–20/14–17	20–24/17–20
Calcium	%	1.2	1.0	0.85	0.75	0.65	0.55
Nonphytate phosphorus	%	0.6	0.5	0.42	0.38	0.32	0.28
Magnesium	mg	500	500	500	500	500	500
Sodium	%	0.17	0.15	0.12	0.12	0.12	0.12
Chlorine	%	0.15	0.14	0.14	0.12	0.12	0.12
Potassium	%	0.7	0.6	0.5	0.5	0.4	0.4

Table 17.3 Recommendations for Ca, available P for heavy turkey breeder hens

	Age (weeks)	Ca (g/kg)	Available P (g/kg)
Leeson and Summers (2005)	0–3	1.40	0.80
	4–7	1.30	0.70
	8–11	1.10	0.60
	12–14	1.00	0.50
	15-lighting	0.90	0.45
	Breeder	2.60–2.80	0.35–0.40
Jeroch et al. (2019)	0–3	1.35	0.75
	4–6	1.25	0.72
	7–11	1.10	0.65
	12–15	1.00	0.55
	16–28	0.90	0.45
	29–30	2.50–3.20	0.45–0.50
	Breeder	2.50–3.20	0.45–50

Calcium Toxicity

In leghorn pullets, nephrosis and visceral urate deposition (e.g., "visceral gout") were observed in the high Ca treatment by 16 weeks of age. Phosphorus deficiency (0.2% available dietary P) and Ca excess (2.24% Ca and 0.45% available P) resulted in similar abnormalities of the tibia. Several histologic abnormalities were observed, but the most conspicuous was a marked lengthening of the cartilage columns of the degenerating hypertrophied epiphyseal cartilage and metaphyseal primary spongiosa. Some chicks could not stand at 4 weeks, displaying a spraddle-legged posture. Folding fractures and bowing or rotation of the tibiotarsus were frequently observed. However, Ca toxicity in turkeys is still not widely studied or presented.

17.3.1.3 Magnesium

Magnesium is essential for carbohydrate metabolism and activating many enzymes, especially those involved in phosphorylation reactions. It is essential for bone formation, about two-thirds being present in bone chiefly as a carbonate. Eggshells contain about 0.4% mg.

Deficiency and Signs

Chicks fed a magnesium-deficient diet grew slowly for approximately one week, then ceased growing, and became lethargic. When disturbed, these birds frequently passed into a brief convulsion accompanied by gasping and finally into a comatose state, sometimes ending in death. Magnesium deficiency signs of turkey poults are similar to those of chicks. Hypomagnesemia and hypocalcemia are associated with severe Mg deficiency in birds (Weaver and Welsh 1993). Tibiae exhibit abnormalities, including thickening of trabeculae, increased retention of cartilage cores, and the occurrence of elongated and inactive osteocytes in the metaphysis.

17.3.1.4 Sodium and Chlorine

Sodium chloride, carbonate, or phosphate is found mainly in blood and body fluids. Sodium is connected intimately with maintaining membrane potentials, cellular transport processes, and regulating the hydrogen ion concentration of blood. Chloride, the major mineral anion in extracellular fluid, plays a role in fluid, ionic, and acid-base balance. Jankowski et al. (2012a) observed that an increase in the Na content of diets from 0.07 to 0.22% led to an increase in the BW of 8-week-old turkeys, whereas a further increase in dietary Na levels was ineffective (Jankowski et al. 2012b). Generally, young turkeys (6–8 weeks of age) are more sensitive to dietary sodium deficiency, than older birds at the end of rearing.

Deficiency Signs

Animals receiving diets deficient in Na fail to grow and develop bone softening, eye keratinization, gonadal inactivity, adrenal hypertrophy, cellular function changes, feed utilization impairment, and decrease in both plasma and special fluid volumes (Klasing and Korver 2020). Poults fed a diet without salt show retarded growth with decreased efficiency of food utilization (Klasing and Korver 2020). Also, salt deprivation in turkeys impairs egg production and hatchability (Harms et al. 1985). Chloride deficiency in chickens and/or turkeys exhibited extremely poor growth rate, high mortality, dehydration, and reduced blood chloride. In addition, deficient chicks and turkey poults showed nervous signs characteristic of Cl deficiency.

Sodium-calcium interactions also play a vital role in the pathogenesis of skeletal muscle damage in broiler chickens and in turkeys (Sandercock and Mitchell 2004; Klasing and Korver 2020), which may lead to locomotion problems. In turkeys, leg problems often result from genetic selection for a rapid increase in body weight and breast muscle yield (Crespo et al. 2000). Research suggests that the most common reason for insufficient bone mineralization in broiler chickens and growing turkeys is mineral metabolism disorders, in particular a deficiency of Ca and P (Kestin et al. 2001; Rao et al. 2003; Venäläinen et al. 2006; Tatara et al. 2011).

Little attention has been paid to the effects of essential dietary electrolytes (Na, K, and Cl) on bone mineralization in young birds. Murakami et al. (1997) reported that, in broiler chickens, the ash content of bones decreased with increasing dietary Na levels and increased in response to a higher dietary chloride intake. In a study by Jankowski et al. (2011), broiler chickens' lowest and highest dietary Na levels deteriorated tibia composition and elasticity. In the cited study of Jankowski et al. (2011), a Na-deficient diet (0.02%) fed to broilers caused a significant decrease in the crude ash and P content of tibia dry matter, which made the bones more brittle and more prone to break easily.

Toxicity

Large amounts of salt in the ration are toxic to chickens and turkeys. The lethal dose is approximately 4 g/kg BW. Young chicks and poults appear to be more susceptible to salt's toxic effects than older birds. Poultry is much less tolerant to salt supplied via the water than by the diet. Signs of salt intoxication include the inability to stand, intense thirst, pronounced muscular weakness, and convulsive movements preceding death (Klasing and Korver 2020). Matterson et al. (1946) fed day-old poults graded quantities of salt for 23 days and observed 25% edema and 25% mortality at 4.0% salt but none at 2.0%. However, Swayne et al. (1986) described a case of accidental salt poisoning in 5- to 11-day-old poults in which a diet contained 1.85% salt. From another point of view, dietary Na intake is often increased to stimulate the growth of birds, leading to excess litter wetness and excreta moisture, as demonstrated in chicks by many authors (Murakami et al. 2001; Borges et al. 2003; Vieira et al. 2003; Mushtaq et al. 2007). Abd El-Wahab et al. (2013b) also observed a higher incidence of foot pad dermatitis in turkeys fed diets containing 0.31% Na and 1.53% K, compared with birds that received lower dietary electrolytes (0.16% and 0.78%, respectively).

17.3.1.5 Potassium

Potassium is found primarily in the body's cellular compartment; the bird's soft tissues contain over three times as much K as Na. As a major cation in intracellular fluid, K is essential in maintaining membrane potential and cellular fluid balance. Potassium participates directly in numerous biochemical reactions and is necessary for normal heart activity, reducing heart muscle contractility and favoring relaxation (Klasing and Korver 2020). Potassium contents of ingredients used in the formulation of complete diets for poultry are presented in Table 17.4.

Table 17.4 Potassium contents of ingredients used in the formulation of complete diets for poultry according to Spark (2004); Youssef et al. (2008), and Kamphues et al. (2011)

Protein-supplying components	K content (g/kg fresh basis)	Energy supplying components	K content (g/kg fresh basis)
Whey protein powder	48–55	**Grain**	
Soy protein	23.0	Wheat fodder meal	12.0
Soybean meal	19.0–22.1	Tapioca	8.0
Rapeseed meal	13.– 17.0	Rye	6.0
Broad beans	12.8–14.0	Triticale	5.0–5.50
Sunflower extraction	12.1–14.0	Wheat	5.0
meal	13.0	Corn	3.0–3.60
Rapeseed cake	11.0–13.0	Sorghum bicolor	3.0
Peas	11.0	Bread flour	2.50
Peanut meal	11.0	Corn starch	1.20
Hemoglobin powder	10.9	Corn oil	0.01
Dried distillers grains	8.80–10.8	**Other feed materials**	
Lupine	8.74–29.6	Yeasts	22.6–26.0
Microalgae	8.30	Brewer's yeast	22.5
Corn extraction meal	8.0	Kluyveromyces yeast	14.4–24.0
Potato protein	2.90–6.0	Molasses	13.2–35.4
Fish meal	2.90–5.50		
Meat meal	4.80		
Whole egg powder	3.20		
Pea protein	1.70		
Blood meal	0.87–1.23		
Wheat gluten			

Deficiency Signs

The deficiency of K is very rare; however, severely affected individuals may exhibit tetanic seizures followed by death. A low K level in the vital organs of animals may occur during severe stress (Klasing and Korver 2020).

Toxicity

High levels of dietary electrolytes also increase fecal moisture and can cause problems with wet litter. Increasing dietary concentrations of Na, K, or P causes linear increases in the water intake and linear increases in the moisture content of excreta. Each 1 g/kg increase in dietary minerals increased the moisture content of the excreta by 9.0, 12.0, and 5.6 g/kg for Na, K, and P, respectively (Smith et al. 2000). Feeding the high electrolyte diet resulted in much higher litter moisture (37.5% DM) compared with the normal dietary Na and K levels (64.5% DM content of litter) at day 35 of turkey poults. Feeding the high electrolyte diet resulted in significantly higher external foot pad dermatitis scores (3.65 ± 1.03) than birds fed normal electrolyte levels (Abd El-Wahab et al. 2013b).

17.3.2 Trace Elements

17.3.2.1 Summary Recommendations for Supply

The essential trace elements Fe, Cu, Zn, Mn, I, and Se are important in the feed of fattening turkeys. They must be added to the feed to a greater or lesser extent. Table 17.5 compiles supply recommendations for the trace elements mentioned for fattening turkeys.

The data represent supply recommendations derived from various dose-response experiments. However, the data are partly based on older experiments with animals whose growth potential did not reach that of today's breeding products.

A factorial derivation of the requirement is not possible at present because the data basis on the contents in the growth is insufficient, and there is no data on the endogenous losses and the element utilization. It is discussed in the literature that various stresses, such as stress or infections (e.g., coccidiosis), require a higher element intake due to reduced absorption and/or altered metabolic behavior in turkeys (Noy et al. 1994).

17.3.2.2 Iron

Iron is an essential component of heme, the porphyrin nucleus of hemoglobin, and the cytochromes and is a component of several enzymes, including catalase, peroxidase, phenylalanine hydroxylase, tyrosinase, and proline hydroxylase (Klasing and Korver 2020).

Systematic studies on the Fe requirements of fattening turkeys have been largely lacking. The INRA (1984) recommends an allowance of 40 mg/kg feed in the first 8 weeks and 30 or 20 mg/kg in subsequent fattening periods. According to the NRC (1994), supply recommendations range from 80 to 50 mg/kg feed with 90%, depending on age. Given these data, 80 and 60 mg/kg are recommended for an adequate supply. High Fe supplements of up to 500 mg/kg have little effect on the oxidative stability of turkey meat during storage (Bartov and Kanner 1996).

Table 17.5 Requirements of trace elements for turkeys as percentages or units per kg of diet on 90% dry matter according to NRC (1994)

		Growing turkeys (male/female) age per week						Breeders	
Nutrient	Unit	0–4	4–8	8–12/8–11	12–16/11–14	16–20/14–17	20–24/17–20	Holding	Laying
Iron	mg	80	60	60	60	50	50	50	60
Copper	mg	8	8	6	6	6	6	6	8
Zinc	mg	70	65	50	40	40	40	40	65
Manganese	mg	60	60	60	60	60	60	60	60
Iodine	mg	0.4	0.4	0.4	0.4	0.4	0.4	0.4	0.4
Selenium	mg	0.2	0.2	0.2	0.2	0.2	0.2	0.2	0.2

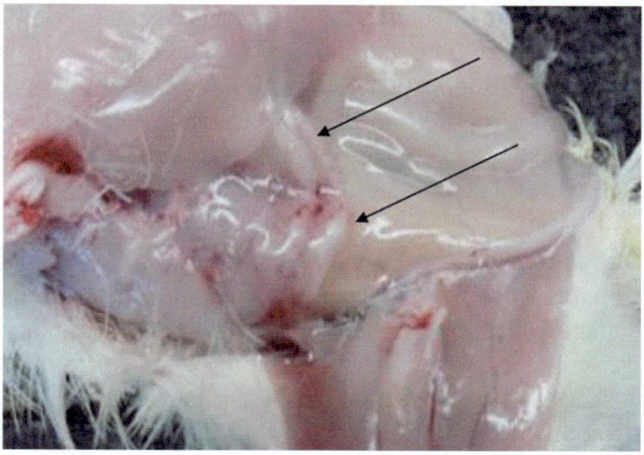

Fig. 17.6 Distortion of ribs (arrows). (Photo by Günther et al. (2019))

Deficiency Signs

Iron deficiency results in hypochromic, microcytic anemia and reduced concentration of non-heme Fe in plasma and prevents normal feather pigmentation in breeds having colored plumage. A deficiency in laying hens and turkey breeder hens also causes anemia in the developing chick embryo and reduced hatchability. Chicks that survive incubation are weak and listless; however, they recover when given supplemental Fe (Morck and Austic 1981; Klasing and Korver 2020).

Toxicity

In a case report by Günther et al. (2019), unintended very high levels of Fe were sevenfold to tenfold higher (557 mg/kg) than the recommendation (60–80 mg/kg, according to Jeroch (2013) in diets of poults at 3 weeks age. The poults started to show ruffled feathers and retarded growth, followed by lameness in about 2% of the flock within a few days. Clinical examination revealed an unwillingness to walk, wing-assisted hobbling, and sporadic, slight swelling of the hock joints. Several birds showed distortion of the ribs (Fig. 17.6) and swollen rib heads. Applying a breaking force test on the long bones of the hindlimb revealed a delayed breakage of long bones.

17.3.2.3 Copper

Copper is essential for the formation of hemoglobin. In the absence of Cu, dietary Fe is absorbed and deposited in the liver and elsewhere, but hemoglobin synthesis does not occur. Copper is a component of several enzymes that participate in redox reactions. Lysyl oxidase is a copper-containing enzyme that catalyzes the oxidation of lysine residues in the formation of the cross-linking structure desmosine in elastin (Rucker et al. 1999).

A dose-response study of trace minerals has been reported (Richards et al. 2010), but a specific systematic dose-response study to derive Cu requirements for broilers and turkeys is not available in the literature. The INRA (1984) provided Cu supplements of 2–4 mg/kg for different ages or fattening stages. The NRC (1994) recommended Cu levels in the first 8 weeks of 8 mg/kg, reduced to 5 mg/kg in subsequent fattening stages. According to European Commission (2013), the Cu content of animal diets should not exceed 35 mg/kg feed, including up to 20 mg Cu/kg feed from cupric chelate of amino acid hydrate. The Cu un-supplemented basal diet containing 11.3 mg Cu/kg is enough for turkeys, and further Cu-supplementations are unnecessary. Further supplementations from inorganic and/or organic sources did not significantly improve turkeys' performance and health. Further, Cu supplementation from inorganic and/or organic sources increases the Cu-excretion (from 808 to 2164 mg per animal) and may cause additional environmental problems (Mikulski et al. 2009).

Deficiency Signs

Copper deficiency has been reported to cause decrease the cross-linking in bone collagen and to increase bone fragility of chick bone (Opsahl et al. 1982). This weakens the elastin structure, leading to an aortic rupture in poultry (Buckingham et al. 1981). Copper deficiency in chicks results in anemia, characterized by reduced numbers of circulating erythrocytes and impaired feather pigmentation in colored breeds of fowl (Grau et al. 1989). A Cu deficiency also causes prolonged prothrombin time in broilers (Kaya et al. 2006).

17.3.2.4 Zinc

Traces of Zn appear to be necessary for life in all animals. It is a constituent of the enzyme carbonic anhydrase and is an activator or a cofactor of more than 299 enzymes (Dewar and Downie 1984).

Dose-response experiments with turkey poults based on semisynthetic diets were conducted by Dewar and Downie (1984). Depending on the criterion covered, requirements ranged from 25 (optimal growth) to 41 mg/kg (plasma Zn concentration). However, these studies were conducted on animals with growth rates well below those of present-day heavy-fattening turkeys. Considering the existing literature, 50 mg/kg in the first 4 weeks and 40 mg/kg during the subsequent fattening period are determined for an adequate supply.

According to Whitehead (1990), an undersupply of Zn (<30 mg/kg complete feed) is also supposed to be associated with disorders of skin health, especially in the form of lesions on the pads of the feet. A special Zn supplementation in the form of Zn-methionine, Zn-lysine, or their combination, so that in each case an additional 40 mg Zinc per kg complete feed was supplied, resulted in significantly lower alterations in the form of foot pad dermatitis (FPD), but (inexplicably) only in female broilers (Hess et al. 2001). Abd El-Wahab et al. (2013a) suggested that to combine the maximum levels of Zn (150 mg/kg, especially of Zn-methionine) and high levels of biotin (2000 μg/kg of diet) in broilers experimentally reared on litter with critical moisture content (35% moisture).

Deficiency Signs

Deficiency of Zn results in retarded growth; poor feathering; enlarged hocks; short, thickened long bones (chondrodystrophy); scaling of the skin and dermatosis, particularly on the feet; and an awkward arthritic gait (Klasing and Korver 2020).

17.3.2.5 Manganese

Manganese is an activator of several enzymes required for normal bones, growth, reproduction, and prevention of chondrodystrophy. Manganese is also essential for pyruvate carboxylase, which contains biotin and controls gluconeogenesis.

Studies by Kealy and Sullivan (1966) on turkey poults up to 4 weeks of age indicated a 37–72 mg/kg requirement for growth and preventing leg damage. According to Noy et al. (1994), 50–60 mg/kg is usually recommended for adequate Mn supply in fattening turkeys. However, some breeding companies recommended 100 mg/kg concentrations (Noy et al. 1994). Manganese deficiency impairs endochondral bone growth and the synthesis of glycosaminoglycan molecules, thus affecting the cartilage of the epiphyseal growth plate (Leach Jr 1986; Liu et al. 1994).

The main signs of manganese deficiency in newly hatched chicks are ataxia and tetanic spasms. Poor growth, skeletal deformities, decreased egg production, and reduced hatchability (may reach 50%) are the signs in adult turkeys (Klasing and Korver 2020).

17.3.2.6 Iodine (I)

Traces of I are required for the normal functioning of the thyroid gland in poultry and in other animals. Thyroxine contains approximately 65% iodine, an important regulating agent in body metabolism. When the intake of iodine is suboptimal, the thyroid tissue enlarges, and goiter results. For iodine, no dose-response studies are available in the literature to determine the requirements for turkeys. Therefore, certain inferences must be drawn from studies on broiler chickens. From the NRC (1994), the iodine requirement for broiler chicks is 0.35 mg/kg, and for fattening turkeys is 0.4 mg/kg feed. Much higher supply recommendations are sometimes made by breeding companies, especially for young turkey poults in the first 4–8 weeks of life (Noy et al. 1994). The INRA (1984) also recommends an iodine supplementation during rearing and fattening of between 1.0 and 0.5 mg/kg. However, no experimental findings are cited for the necessity of such high dosages.

Iodine deficiency in poultry has been largely prevented by the widespread use of iodine in iodized salt or as part of the trace mineral premix. Some turkey breeders feed high levels of I to enrich eggs and provide added value (Klasing and Korver 2020).

Toxic effects of iodine on hatchability were seen when 350 ppm iodine was fed during the entire 20-week egg production period. Toxicity was also seen in the 35 ppm iodine treatment during the egg laying period from 16 to 20 weeks of lay. Furthermore, the 35 and 350 ppm iodine treatments depressed egg weights. Data from the two trials of the experiment suggest that the levels of dietary iodine that supported optimal reproduction in turkey breeder hens were in the range from 0 to 3.5 ppm of supplemental iodine (0.7–4.2 ppm total dietary iodine). Typical toxicity symptoms observed in chickens when fed diets containing high levels of iodine include decreased egg production, decreased egg size, depressed fertility and hatchability, enlarged thyroids in hatching chicks, and wiry down (Perdomo et al. 1966; Wilson et al. 1968). All of these symptoms were seen in turkeys fed 350 ppm iodine in the present study. A major difference between the present study and previous studies involving chickens was the amount of iodine that resulted in toxicity. Toxic levels of dietary iodine in turkeys were nearly half those levels reported to cause problems in chickens (600 ppm) (Arrington et al. 1967) suggesting that the turkey breeder hen may be more sensitive to high dietary iodine than is the chicken.

17.3.2.7 Selenium

Selenium is an essential mineral element for both chicks and poults. It is a constituent of the enzymes glutathione peroxidase and phospholipid hydroperoxide glutathione peroxidase, which serve to protect tissues against oxidative damage, and it is a component of iodothyronine 5′-deiodinase, an enzyme that is involved in the conversion of thyroxine to its active form (Burk and Hill 1993).

Selenium metabolism interacts with other dietary factors in a variety of ways. The interaction between Se and vitamin E and the sulfur-containing amino acids was described (Underwood and Suttle 1999). Selenium prevents the development of exudative diathesis in young chickens and myopathy of the ventriculus and heart in young turkeys. Scott et al. (1967)* determined that 0.18 mg Se/kg was required to prevent myopathies of the gizzard and heart in turkey poults that were adequately supplied with vitamin E. Sell et al. (1997) also refer to the importance of Se supply in their experiments to increase nutritive vitamin E intake in meat turkeys. In short-term studies by Bartov (1983) on older fattening turkeys adequately supplied with vitamin E, supplements of Se in the range of 0.1–1.1 mg/kg did not affect live weight development or vitamin E concentration in blood plasma. Bonomi (2001) postulates a 0.3 mg/kg requirement. The INRA (1984) supply recommendations are 0.15 mg/kg allowance during the first 8 weeks and 0.1 mg/kg during the remainder of fattening. The NRC (1994) recommends 0.2 mg/kg selenium concentrations throughout fattening.

Deficiency Signs

Exudative diathesis, encephalomalacia, and myopathy of the ventriculus and heart in young turkeys are associated with selenium deficiency. Poults severely deficient in Se exhibit poor growth and feathering, impaired fat digestion, pancreatic atrophy, fibrosis, and reduced Se-dependent glutathione peroxidase activity in the pancreas (Whitacre et al. 1987).

Toxicity

Excess organic Se, usually as selenomethionine, is incorporated readily into proteins because RNA, and does not distinguish selenomethionine from methionine, and selenomethionine is readily incorporated into proteins in place of methionine (Klasing and Korver 2020). Excess inorganic Se interferes with sulfur metabolism due to the formation of sulfur-Se complexes and the substitution of Se for sulfur in cysteine. These aberrations result in impaired protein synthesis, impaired function of proteins, mutagenesis, hepatotoxicity, and/or feather loss. Salyi et al. (1993) reported an incidence of acute Se toxicity in chicks that was evident by watery diarrhea, weakness, and cerebellar edema. A decrease in growth rate occurs with 5 mg/kg Se in chicks (Todorovic et al. 1999).

17.4 Summary

The fat-soluble and water-soluble vitamins, their function, and effects of hypo- and hypervitaminosis in turkeys are shown in Tables 17.6 and 17.7. Additionally, Table 17.8 shows the pathological conditions in turkeys due to the deficiencies and toxicities of the main elements.

Table 17.6 Fat-soluble vitamins: Function and effects of hypo- and hypervitaminosis (Hafez and Jodas 1997)

Vitamin	Function	Hypovitaminosis	Hypervitaminosis	Requirements	
				Poults	Fattening turkeys
A	– Involved in the formation of proteins of many cell types – Epithelial cells – Gland cells – Chondro-, osteoblasts, osteocytes – Immune system cells	– Growth retardation – Immunosuppression – Increased lacrimal flow – Cheese-like, but odorless, discharge from one or more eye – Hyperkeratosis on the mucous membrane of the mouth and esophagus – Disturbance of skeletal and feather development – Disturbance of spermatogenesis – Egg production and hatchability decrease in adults	– Growth retardation – Emaciation – Decreased laying performance – Disturbance of skeletal development	12,000–15,000 IU	8000–12,000 IU
D_3	– Regulation of calcium and phosphorus metabolism – Skeletal development – Eggshell formation – Effect on fertilization and hatching rate	– Bone calcification disorder (rickets (in growing birds), osteomalacia (in adults)) – Growth retardation – Decreased eggshell quality – Decreased hatchability – Ruffled feathers	– Deposition of calcium phosphate in tissues, organs and blood vessels – Release of calcium from bones	3000–5000 IU	2500–4000 IU

(continued)

Table 17.6 (continued)

Vitamin	Function	Hypovitaminosis	Hypervitaminosis	Requirements	
				Poults	Fattening turkeys
E	– Antioxidant effect – Effect on protein metabolism and membranes (function in combination with selenium)	– Encephalomalacia (crazy chick disease; movement coordination, stretched legs (clonic spasms), opisthotonos, torticollis, brain (cerebellum) oedema hemorrhage) – Muscular dystrophy – Exudative diathesis – Hock joint swelling – Embryonic development disorder		40–60 mg	30–50 mg
K	– Involved in prothrombin synthesis	– Impairment of blood coagulation		3–5 mg	2–3 mg

Table 17.7 Water-soluble vitamins: Function and effects of hypovitaminosis (Hafez and Jodas 1997)

Vitamin	Function	Hypovitaminosis	Requirements	
			Poults	Fattening turkeys
Thiamin (Vitamin B_1)	– Carbohydrate metabolism	– Growth retardation – Ataxia, cramps, opisthotonos, specific posture with flexed legs and the head drawn back (stargazing) – Mortality – Decreased hatchability	3–5 mg	2–3 mg
Riboflavin (Vitamin B_2)	– Involved in enzyme systems in fat and amino acid metabolism – Protection of membranes against oxidation	– Curled-toe paralysis due to nerve degeneration (twisting, muscle atrophy) – Growth retardation – Severe dermatitis of the feet and shanks as well as incrustations on the corners of the mouth and eyelids – Increased embryonic mortality	7–9 mg	5–7 mg
Pyridoxine (Vitamin B6)	– Coenzyme in amino acid metabolism	– Growth retardation – Perosis – Ataxia, cramps – High mortality – Decreased laying performance – Increased embryonic mortality	5–7 mg	4–6 mg
Cobalamin (Vitamin B12)	– Coenzyme in amino acid metabolism	– Growth retardation – Weak chicks – Increased embryonic mortality (deficiency symptoms are exacerbated by choline)	0.03–0.04 mg	0.02–0.04 mg

(continued)

Table 17.7 (continued)

Vitamin	Function	Hypovitaminosis	Requirements	
			Poults	Fattening turkeys
Pantothenic acid (Vitamin B_5)	– Component of coenzyme A (fat, carbohydrate, and amino acid metabolism)	– Growth retardation, high mortality – Dermatitis mainly on the corners of the mouth and eyelids – Keratinization of the skin epithelium – Ruffled feathers – Degeneration of the nervous system – Decreased laying performance – Increased embryonic mortality (embryonic malformation)	11–14 mg	10–14 mg
Niacin (Vitamin B3)	– Formation of NAD[a] and NADP[b] (carbohydrate, fat, and protein metabolism)	– Hock joint swelling – Bowing of the legs – Growth retardation – Dermatitis – Ruffled feathers	60–100 mg	50–80 mg
Folic acid	– Nucleic acid synthesis	– Growth retardation – Anemia and atrophy of the lymphatic tissue – Ruffled feathers – Embryo mortality after picking – Cervical paralysis leading to death	1.0–2.0 mg	1.0–1.5 mg
Biotin (Vitamin B7)	– Participation in gluconeogenesis and fat metabolism	– Growth retardation – Dermatitis and keratinization (beak, eye) – Perosis – High mortality – Increased embryonic mortality (embryonic malformation)	0.10–0.25 mg	0.10–0.20 mg

Choline	– Formation of acetylcholine and phospholipids	– Growth retardation – Perosis – Curvature of the long bones – Decreased hatchability	500–700 mg[c]	400–600 mg[c]
Ascorbic acid (Vitamin C)	– Participation in many metabolic processes	– Susceptibility to disease – Bleeding mucous membranes – Impaired eggshell quality	100–150 mg	100–150 mg

[a] NAD: Nicotinamide adenine dinucleotide
[b] NADP: Nicotinamide adenine dinucleotide phosphate
[c] Choline demand depends on the fat content of the feed

Table 17.8 Quantitative elements: Function and effect of a deficiency and toxicity (Hafez and Jodas 1997)

Element	Function	Deficiency	Toxicity
Calcium (Ca)	– Component of the skeletal system – Formation of eggshell – Blood clotting – Nerve excitation – Muscle contraction – Ion passage through cell membrane	– Rickets – Osteomalacia – Osteoporosis – Decreased quality of eggshell	– May interfere with absorption of other minerals and medicines (e.g., tetracyclines)
Phosphorus (P)	– Component of the skeletal system – Energy transfer in cell – Maintaining ion balance	– Disturbance of bone mineralization – Bones bending – Leg weakness	– Dyschondroplasia
Magnesium (Mg)	– Component of the skeletal system and eggshell – Carbohydrate metabolism – Activator for enzyme systems – Nerve excitation – Muscle contraction	– Growth and feathering disorder – Fragility of bones – Twisting of legs – Overexcitability – Tremor – Ataxia – Cramps	– Increased Calcium loss
Sodium (Na) and Chlorine (Cl)	– Osmotic regulation – pH regulation – Nerve excitation – Muscle contraction	– Decreased efficiency of food utilization – Growth retardation – Renal damage – Paralysis – Decreased Laying performance – Decreased hatchability	– Hydropericardium – Ascites – Hydrothorax – Intense thirst – Diarrhea – Sudden deaths
Potassium (K)	– Nerve excitation – Muscle contraction – Maintaining membrane potential and cellular fluid balance	– High mortality – Growth retardation – Muscle weakness – Decreased quality of eggshell	– Not observed

References

Abd El-Wahab A, Radko D, Kamphues J (2013a) High dietary levels of biotin and zinc to improve health of foot pads in broilers exposed experimentally to litter with critical moisture content. Poult Sci 92:1774–1782. https://doi.org/10.3382/ps.2013-03054

Abd El-Wahab A, Visscher C, Beineke A, Beyerbach M, Kamphues J (2013b) Effects of high electrolyte contents in the diet and using floor heating on development and severity of foot pad dermatitis in young turkeys. J Anim Physiol Anim Nutr 97:39–47. https://doi.org/10.1111/j.1439-0396.2011.01240.x

Abd El-Wahab A, Ahmed M, Ibrahim T (2016) Impact of dietary methionine levels and sources on performance and health of foot pad in broilers. Asian J Anim Vet Adv 11:357–362. https://doi.org/10.3923/ajava.2016.357.362

Abd El-Wahab A, Visscher C, Ratert C, Kölln M, Diephaus D, Beineke A, Kamphues J (2017) Outcome of an experimental study in growing turkeys suspected of having a diet related, uncommon and uncoordinated gait. Vet Sci 4:49. https://doi.org/10.3390/vetsci4040049

Abd El-Wahab A, Chuppava B, Radko D, Visscher C (2021) Hepatic lipidosis in fattening turkeys: a review. German J Vet Res 1:48–66. https://doi.org/10.51585/gjvr.2021.3.0022

Applegate TJ, Sell JL (1996) Effect of dietary linoleic to linolenic acid ratio and vitamin E supplementation on vitamin E status of poults. Poult Sci 75:881–890. https://doi.org/10.3382/ps.0750881

Arrington L, Santa Cruz R, Harms R, Wilson H (1967) Effects of excess dietary iodine upon pullets and laying hens. J Nutr 92:325–330

Aviagen (2017) Management guidelines for growing commercial Turkeys. Aviagen Turkeys Limited, Cheshire

Bartov I (1983) Effect of various dietary factors and age on plasma α-tocopherol concentration of Turkeys. Poult Sci 62:635–641. https://doi.org/10.3382/ps.0620635

Bartov I, Kanner J (1996) Effect of high levels of dietary iron, iron injection, and dietary vitamin E on the oxidative stability of Turkey meat during storage. Poult Sci 75:1039–1046. https://doi.org/10.3382/ps.0751039

Biesiadecki BJ, Schneider KL, Yu ZB, Chong SM, Jin JP (2004) An R111C polymorphism in wild Turkey cardiac troponin I accompanying the dilated cardiomyopathy-related abnormal splicing variant of cardiac troponin T with potentially compensatory effects. J Biol Chem 279:13825–13832. https://doi.org/10.1074/jbc.M314225200

Bonomi A (2001) Selenium in the feeding of milk fed veal. J Food Sci Nutr 30:299–311

Borges S, Da Silva AF, Ariki J, Hooge D, Cummings K (2003) Dietary electrolyte balance for broiler chickens under moderately high ambient temperatures and relative humidities. Poult Sci 82:301–308. https://doi.org/10.1093/ps/82.2.301

Bowes VA, Julian RJ, Leeson S, Stirtzinger T (1988) Research note: effect of feed restriction on feed efficiency and incidence of sudden death syndrome in broiler chickens. Poult Sci 67:1102–1104. https://doi.org/10.3382/ps.0671102

Breeding S, McRee W, Ficken M, Ferket P (1994) Effect of protein restriction during brooding on spontaneous Turkey cardiomyopathy. Avian Dis 38:366–370. https://doi.org/10.2307/1591965

Bryden WL, Mollah Y, Gill RJ (1991) Bioavailability of biotin in wheat. J Sci Food Agric 55:269–275. https://doi.org/10.1002/jsfa.2740550212

Buckingham K, Heng-Khoo CS, Dubick M, Lefevre M, Cross C, Julian L, Rucker R (1981) Copper deficiency and elastin metabolism in avian lung. Proc Soc Exp Biol Med 166:310–319. https://doi.org/10.3181/00379727-166-41066

Burk RF, Hill KE (1993) Regulation of selenoproteins. Annu Rev Nutr 13:65–81. https://doi.org/10.1146/annurev.nu.13.070193.000433

Cai Z, Finnie J, Blumbergs P (2006) Avian riboflavin deficiency: an acquired tomaculous neuropathy. Vet Pathol 43:780–781. https://doi.org/10.1354/vp.43-5-780

Christakos S, Dhawan P, Verstuyf A, Verlinden L, Carmeliet G (2016) Vitamin D: metabolism, molecular mechanism of action, and pleiotropic effects. Physiol Rev 96:365–408. https://doi.org/10.1152/physrev.00014.2015

Cowieson AJ, Acamovic T, Bedford MR (2004) The effects of phytase and phytic acid on the loss of endogenous amino acids and minerals from broiler chickens. Br Poult Sci 45:101–108. https://doi.org/10.1080/00071660410001668923

Crespo R, Stover SM, Taylor K, Chin RP, Shivaprasad H (2000) Morphometric and mechanical properties of femora in young adult male turkeys with and without femoral fractures. Poult Sci 79:602–608. https://doi.org/10.1093/ps/79.4.602

Degernes L, Lynch P, Shivaprasad H (2011) Degenerative joint disease in captive waterfowl. Avian Pathol 40:103–110. https://doi.org/10.1080/03079457.2010.541421

Dewar WA, Downie JN (1984) The zinc requirements of broiler chicks and Turkey poults fed on purified diets. Br J Nutr 51:467–477. https://doi.org/10.1079/BJN19840052

Engberg RM, Lauridsen C, Jensen SK, Jakobsen K (1996) Inclusion of oxidized vegetable oil in broiler diets. Its influence on nutrient balance and on the antioxidative status of broilers. Poult Sci 75:1003–1011. https://doi.org/10.3382/ps.0751003

European Commission (2013) Commission regulation (EU) no 68/2013 of 16 January 2013 on the catalogue of feed materials, The Official Journal of the European Union 29:1–64

Frigg M (1984) Available biotin content of various feed ingredients. Poult Sci 63:750–753. https://doi.org/10.3382/ps.0630750

Froehlich DM (1987) Importance offolic acid in Turkey diets explored. Feedstuffs 59:26–29

GfE (1999) Empfehlungen zur Energie-und Nährstoffversorgung der Legehennen und Masthühner (Broiler). DLG-Verlag, Frankfurt (Main)

GfE (2004) Empfehlungen zur Energie-und Nährstoffversorgung der Mastputen. DLG Verlag, Frankfurt (Main)

Grau C, Roudybush T, Vohra P, Kratzer F, Yang M, Nearenberg D (1989) Obscure relations of feather melanization and avian nutrition. Worlds Poult Sci J 45:241–246

Günther R, Lauterbach L, Brunk S, Visscher C (2019) Unusual rickets in young turkeys—a case report. Proceedings of the 12th international symposium on Turkey diseases, Berlin, Germany

Hafez HM, Jodas S (1997) Putenkrankheiten (Turkey diseases). Enke Verlag, Stuttgart

Halevy O, Geyra A, Barak M, Uni Z, Sklan D (2000) Early posthatch starvation decreases satellite cell proliferation and skeletal muscle growth in chicks. J Nutr 130:858–864. https://doi.org/10.1093/jn/130.4.858

Halevy O, Nadel Y, Barak M, Rozenboim I, Sklan D (2003) Early posthatch feeding stimulates satellite cell proliferation and skeletal muscle growth in Turkey poults. J Nutr 133:1376–1382. https://doi.org/10.1093/jn/133.5.1376

Halle I, Henning M, Köhler P (2011) Influence of vitamin B12 and cobalt on growth of broiler chickens and Pekin ducks. Landbauforschung Volkenrode 4:299–306

Harms RH, Simpson CF (1977) Influence of wet litter and supplemental biotin on foot pad dermatitis in Turkey poults. Poult Sci 56:2009

Harms RH, Buresh RE, Wilson HR (1985) Sodium requirement of the Turkey hen. Br Poult Sci 26:217–220. https://doi.org/10.1080/00071668508416806

Hess JB, Bilgili SF, Parson AM, Downs KM (2001) Influence of complexed zinc products on live performance and carcass grade of broilers. J Appl Anim Res 19:49–60. https://doi.org/10.1080/09712119.2001.9706709

Hocking PM, Robertson GW, Nixey C (2002a) Effects of dietary calcium and phosphorus on mineral retention, growth, feed efficiency and walking ability in growing turkeys. Br Poult Sci 43:607–614. https://doi.org/10.1080/0007166022000004525

Hocking PM, Wilson S, Dick L, Dunn LN, Robertson GW, Nixey C (2002b) Role of dietary calcium and available phosphorus in the aetiology of tibial dyschondroplasia in growing turkeys. Br Poult Sci 43:432–441. https://doi.org/10.1080/00071660120103729

Hooper CL, Maurice DV, Lightsey F, Toler E (2001) Factors affecting ascorbic acid biosynthesis in chickens. I. Adaptation of an assay and the effect of age, sex and food deprivation. J Anim Physiol Anim Nutr 84:48–56. https://doi.org/10.1046/j.1439-0396.2000.00286.x

Huff WE, Huff GR, Clark FD, Moore PA, Rath NC, Balog JM, Barnes DM, Erf GF, Beers KW (1999) Research on the probable cause of an outbreak of field rickets in turkeys. Poult Sci 78:1699–1702. https://doi.org/10.1093/ps/78.12.1699

INRA (1984) L'alimentation des animaux monogastriques. Institute National de la Recherche Agronomique, Paris

Jankowski J, Zduńczyk Z, Juśkiewicz J, Kwieciński P (2011) The effect of different dietary sodium levels on the growth performance of broiler chickens, gastrointestinal function, excreta moisture and tibia mineralization. J Anim Feed Sci 20:93–106. https://doi.org/10.22358/jafs/66161/2011

Jankowski J, Juśkiewicz J, Lichtorowicz K, Zduńczyk Z (2012a) Effects of the dietary level and source of sodium on growth performance, gastrointestinal digestion and meat characteristics in turkeys. Anim Feed Sci Technol 178:74–83. https://doi.org/10.1016/j.anifeedsci.2012.09.012

Jankowski J, Zduńczyk Z, Lichtorowicz K, Juskiewicz J (2012b) Effect of different levels of dietary sodium from sodium chloride on gastrointestinal tract response, tibia mineralization, and footpad dermatitis incidence in young turkeys. J Appl Poult Res 21:856–867. https://doi.org/10.3382/japr.2012-00572

Jeroch H (2013) Empfehlungen zu Spurenelementkonzentrationen in Mastputenalleinfutter. Geflügelernährung. Verlag Eugen Ulmer, Stuttgart

Jeroch H, Prinz M, Hennig A (1978) Untersuchungen zum Vitaminbedarf und zur Vitaminversorgung von Mastputen. Arch Anim Nutr 28:53–65. https://doi.org/10.1080/17450397809428214

Jeroch H, Simon A, Zentek J (2019) Geflügelernährung. Verlag Eugen Ulmer, Stuttgart

Jin S, Sell JL (2001) Dietary vitamin K1 requirement and comparison of biopotency of different vitamin K sources for Young Turkeys. Poult Sci 80:615–620. https://doi.org/10.1093/ps/80.5.615

Johnson NE, Qiu XL, Gautz LD, Ross E (1995) Changes in dimensions and mechanical properties of bone in chicks fed high levels of niacin. Food Chem Toxicol 33:265–271. https://doi.org/10.1016/0278-6915(94)00143-C

Johnston CS, Corte C, Swan PD (2006) Marginal vitamin C status is associated with reduced fat oxidation during submaximal exercise in young adults. Nutr Metab 3:35. https://doi.org/10.1186/1743-7075-3-35

Kamphues J, Youssef I, Abd El-Wahab A, Üffing B, Witte M, Tost M (2011) Influences of feeding and housing on foot pad health in hens and turkeys. Übers Tierernährg 39:147–193

Kaya A, Altıner A, Özpınar A (2006) Effect of copper deficiency on blood lipid profile and haematological parameters in broilers. J Vet Med Ser A 53:399–404. https://doi.org/10.1111/j.1439-0442.2006.00835.x

Kealy RD, Sullivan TW (1966) Studies on manganese requirement and interactions in the diet of Young turkeys. Poult Sci 45:1352–1358. https://doi.org/10.3382/ps.0451352

Kestin S, Gordon S, Su G, Sørensen P (2001) Relationships in broiler chickens between lameness, liveweight, growth rate and age. Vet Rec 148:195–197. https://doi.org/10.1136/vr.148.7.195

Klasing KC (2008) Nutritional diseases. In: Saif YM et al (eds) Diseases of poultry. Blackwell Publishing Professional, Ames

Klasing KC, Korver DR (2020) Nutritional diseases. In: Swayne DE (ed) Diseases of poultry. Wiley, Hoboken, pp 1257–1285

Konieczka P, Tykałowski B, Ognik K, Kinsner M, Szkopek D, Wójcik M, Mikulski D, Jankowski J (2022) Increased arginine, lysine, and methionine levels can improve the performance, gut integrity and immune status of turkeys but the effect is interactive and depends on challenge conditions. Vet Res 53:59

Koutsos EA, Clifford AJ, Calvert CC, Klasing KC (2003) Maternal carotenoid status modifies the incorporation of dietary carotenoids into immune tissues of growing chickens (Gallus gallus domesticus). J Nutr 133:1132–1138. https://doi.org/10.1093/jn/133.4.1132

Leach RM Jr (1986). Mn(II) and glycosyltransferases essential for skeletal development. In: Wedler VSAE (ed) Manganese in metabolism and enzyme function. Academic, London, pp 81–91

Leeson S, Summers JD (2005) Feeding programs for broilers chickens. In: Leeson S, Summers JD (eds) Commercial poultry nutrition. University Books, Guelph, pp 229–296

Liu ACH, Heinrichs BS, Leach RM (1994) Influence of manganese deficiency on the characteristics of proteoglycans of avian epiphyseal growth plate cartilage. Poult Sci 73:663–669. https://doi.org/10.3382/ps.0730663

Liu Y, Zhao L, Mosenthin R, Zhang J, Ji C, Ma Q (2019) Protective effect of vitamin E on laying performance, antioxidant capacity, and immunity in laying hens challenged with *Salmonella* Enteritidis. Poult Sci 98:5847–5854. https://doi.org/10.3382/ps/pez227

Lu T, Harper A, Zhao J, Corl B, LeRoith T, Dalloul R (2014) Effects of a dietary antioxidant blend and vitamin E on fatty acid profile, liver function, and inflammatory response in broiler chickens fed a diet high in oxidants. Poult Sci 93:1658–1666. https://doi.org/10.3382/ps.2013-03827

Massé PG, Ziv I, Cole DEC, Mahuren JD, Donovan SM, Yamauchi M, Howell DS (1998) A cartilage matrix deficiency experimentally induced by vitamin B6 deficiency. Proc Soc Exp Biol Med 217:97–103. https://doi.org/10.3181/00379727-217-44210

Matterson LD, Scott HM, Jungherr E (1946) Salt tolerance of turkeys. Poult Sci 25:539–541. https://doi.org/10.3382/ps.0250539

Mikulski D, Jankowski J, Zduńczyk Z, Wróblewska M, Mikulska M (2009) Copper balance, bone mineralization and the growth performance of turkeys fed diet with two types of Cu supplements. J Anim Feed Sci 18:677–688. https://doi.org/10.22358/jafs/66441/2009

Miller DL, Balloun SL (1967) Folacin requirements of Turkey breeder hens. Poult Sci 46:1502–1508. https://doi.org/10.3382/ps.0461502

Misir R, Blair R (1988) Biotin bioavailability of protein supplements and cereal grains for starting Turkey Poults. Poult Sci 67:1274–1280. https://doi.org/10.3382/ps.0671274

Moore D, Ferket P, Mozdziak P (2005) Muscle development in the late embryonic and early post-hatch poult. Int J Poult Sci 4:138–142

Mora JR, Iwata M, von Andrian UH (2008) Vitamin effects on the immune system: vitamins A and D take centre stage. Nat Rev Immunol 8:685–698. https://doi.org/10.1038/nri2378

Morck TA, Austic RE (1981) Iron requirements of White Leghorn hens. Poult Sci 60:1497–1503. https://doi.org/10.3382/ps.0601497

Mueller AS, Fischer J, Most E, Pallauf J (2009) Investigation into selenium requirement of growing turkeys offered a diet supplemented with two levels of vitamin E. J Anim Physiol Anim Nutr 93:313–324. https://doi.org/10.1111/j.1439-0396.2008.00849.x

Murakami A, Watkins S, Saleh E, England J, Waldroup P (1997) Estimation of the sodium and chloride requirements for the young broiler chick. J Appl Poult Res 6:155–162. https://doi.org/10.1093/japr/6.2.155

Murakami A, Oviedo-Rondon E, Martins E, Pereira M, Scapinello C (2001) Sodium and chloride requirements of growing broiler chickens (twenty-one to forty-two days of age) fed corn-soybean diets. Poult Sci 80:289–294. https://doi.org/10.1093/ps/80.3.289

Mushtaq T, Mirza MA, Athar M, Hooge D, Ahmad T, Ahmad G, Mushtaq M, Noreen U (2007) Dietary sodium and chloride for twenty-nine-to forty-two-day-old broiler chickens at constant electrolyte balance under subtropical summer conditions. J Appl Poult Res 16:161–170. https://doi.org/10.1093/japr/16.2.161

Nagórna-Stasiak B, Lechowski HJ, Kowalczyk M (1999) Syteza vitaminy Cu indykow (synthesis of vitamin C in turkeys). Med Weter 55:195–198

Nimse SB, Pal D (2015) Free radicals, natural antioxidants, and their reaction mechanisms. R Soc Chem Adv 5:27986–28006. https://doi.org/10.1039/C4RA13315C

Noy Y, Frisch Y, Rand N, Sklan D (1994) Trace mineral requirements in turkeys. Worlds Poult Sci J 50:253–268

NRC (1994) Nutrient requirements of poultry. In: Nutrient requirements of domestic animals. National Academy Press, Washington, D.C.

Ognik K, Konieczka P, Mikulski D, Jankowski J (2020) The effect of different dietary ratios of lysine and arginine in diets with high or low methionine levels on oxidative and epigenetic DNA damage, the gene expression of tight junction proteins and selected metabolic parameters in Clostridium perfringens-challenged turkeys. Vet Res 51:50. https://doi.org/10.1186/s13567-020-00776-y

Opsahl W, Zeronian H, Ellison M, Lewis D, Rucker RB, Riggins RS (1982) Role of copper in collagen cross-linking and its influence on selected mechanical properties of Chick bone and tendon. J Nutr 112:708–716. https://doi.org/10.1093/jn/112.4.708

Palliyeguru MWCD, Rose SP, Mackenzie AM (2011) Effect of trypsin inhibitor activity in soya bean on growth performance, protein digestibility and incidence of sub-clinical necrotic

enteritis in broiler chicken flocks. Br Poult Sci 52:359–367. https://doi.org/10.1080/00071668.2011.577054
Perdomo J, Harms R, Arrington L (1966) Effect of dietary iodine upon egg production, fertility and hatchability. Proc Soc Exp Biol Med 122:758–760
Pesti GM, Rowland GN, Ryu KS (1991) Folate deficiency in chicks fed diets containing practical ingredients. Poult Sci 70:600–604. https://doi.org/10.3382/ps.0700600
Rama Rao SV, Raju MVLN, Panda AK, Poonam NS, Shyam Sunder G (2011) Effect of dietary α -tocopherol concentration on performance and some immune responses in broiler chickens fed on diets containing oils from different sources. Br Poult Sci 52:97–105. https://doi.org/10.1080/00071668.2010.548792
Rao RSV, Raju MVLN, Sharma RP, Nagalakshmi D, Reddy MR (2003) Lameness in chickens: alleviation by dietary manipulation. Poultry Int 42:56–61
Richards JD, Zhao J, Harrell RJ, Atwell CA, Dibner JJ (2010) Trace mineral nutrition in poultry and swine. Asian Australas J Anim Sci 23:1527–1534. https://doi.org/10.5713/ajas.2010.r.07
Robertson EI, Fiala GF, Scott ML, Norris LC, Heuser GF (1947) Response of anemic chicks to peteroylglutamic acid. Proc Soc Exp Biol Med 64:441–443. https://doi.org/10.3181/00379727-64-15822
Rodehutscord M, Wendt P, Strobel E (2003) Reducing the phosphorus concentration in diets for turkeys between 10 and 22 weeks of age. Br Poult Sci 44:591–597. https://doi.org/10.1080/00071660310001618316
Romanchik JE, Morel DW, Harrison EH (1995) Distributions of carotenoids and α-tocopherol among lipoproteins do not change when human plasma is incubated in vitro. J Nutr 125:2610–2617. https://doi.org/10.1093/jn/125.10.2610
Ross ML, Bryan DDSL, Abbott DA, Classen HL (2019) Effect of protein sources on performance characteristics of turkeys in the first three weeks of life. Anim Nutr 5:396–406. https://doi.org/10.1016/j.aninu.2019.09.002
Rucker RB, Rucker BR, Mitchell AE, Cui CT, Clegg M, Kosonen T, Uriu-Adams JY, Tchaparian EH, Fishman M, Keen CL (1999) Activation of Chick tendon lysyl oxidase in response to dietary copper. J Nutr 129:2143–2146. https://doi.org/10.1093/jn/129.12.2143
Ruiz N, Harms RH (1989) Riboflavin requirement of Turkey Poults fed a corn-soybean meal diet from 1 to 21 days of age. Poult Sci 68:715–718. https://doi.org/10.3382/ps.0680715
Ruiz N, Harms RH, Linda SB (1990) Niacin requirement of broiler chickens fed a corn-soybean meal diet from 1 to 21 days of age. Poult Sci 69:433–439. https://doi.org/10.3382/ps.0690433
Sahin K, Onderci M, Sahin N, Gursu MF, Kucuk O (2003) Dietary vitamin C and folic acid supplementation ameliorates the detrimental effects of heat stress in Japanese quail. J Nutr 133:1882–1886. https://doi.org/10.1093/jn/133.6.1882
Salyi G, Banhidi G, Szabo E, Sandor G, Ratz F (1993) Acute selenium poisoning in broilers. Magyar Allatoi-vosok Lapja 48:22–26
Sandercock D, Mitchell M (2004) The role of sodium ions in the pathogenesis of skeletal muscle damage in broiler chickens. Poult Sci 83:701–706. https://doi.org/10.1093/ps/83.4.701
Sell JL, Soto-Salanova MF, Palo P, Jeffrey M (1997) Influence of supplementing corn-soybean meal diets with vitamin E on performance and selected physiological traits of male turkeys. Poult Sci 76:1405–1417. https://doi.org/10.1093/ps/76.10.1405
Smith A, Rose S, Wells R, Pirgozliev V (2000) Effect of excess dietary sodium, potassium, calcium and phosphorus on excreta moisture of laying hens. Br Poult Sci 41:598–607. https://doi.org/10.1080/713654976
Spark M (2004) Untersuchungen zum Futterwert einer auf Molke produzierten Hefe (Kluyveromyces fragilis) als Eiweissfuttermittel für Absetzferkel. University of Veterinary Medicine Hannover, Hannover
Stein HH, Berger LL, Drackley JK, Fahey GC, Hernot DC, Parsons CM (2008) Nutritional properties and feeding values of soybeans and their co-products. In: Johnson LA, White PJ, Galloway R (eds) Soybeans, chemistry, production, processing, and utilization. AOCS Press, Urbana, pp 613–660
Sun Y, Kim SW (2017) Intestinal challenge with enterotoxigenic Escherichia coli in pigs, and nutritional intervention to prevent postweaning diarrhea. Anim Nutr 3:322–330

Swayne DE, Shlosberg A, Davis RB (1986) Salt poisoning in Turkey poults. Avian Dis 30:847–852. https://doi.org/10.2307/1590599

Tatara M, Krupski W, Jankowski M, Zduńczyk Z, Jankowski J, Studziński T (2011) Effects of dietary calcium content and vitamin D source on skeletal properties in growing turkeys. Br Poult Sci 52:718–729. https://doi.org/10.1080/00071668.2011.631984

Todorovic M, Mihailovic M, Hristov S (1999) Effects of excessive levels of sodium selenite on daily weight gain, mortality and plasma selenium concentration in chickens. Acta Vet (Yugoslavia) 49:313–319

Underwood EJ, Suttle NF (1999) The mineral nutrition of livestock. CABI Publications, Wallingford

Uni Z, Ferket R (2004) Methods for early nutrition and their potential. Worlds Poult Sci J 60:101–111. https://doi.org/10.1079/WPS20038

Venäläinen E, Valaja J, Jalava T (2006) Effects of dietary metabolisable energy, calcium and phosphorus on bone mineralisation, leg weakness and performance of broiler chickens. Br Poult Sci 47:301–310. https://doi.org/10.1080/00071660600741776

Vieira S, Penz A Jr, Pophal S, Godoy de Almeida J (2003) Sodium requirements for the first seven days in broiler chicks. J Appl Poult Res 12:362–370. https://doi.org/10.1093/japr/12.3.362

Villaverde C, Baucells MD, Manzanilla EG, Barroeta AC (2008) High levels of dietary unsaturated fat decrease α-tocopherol content of whole body, liver, and plasma of chickens without variations in intestinal apparent absorption. Poult Sci 87:497–505. https://doi.org/10.3382/ps.2007-00292

Ward NE, Jones J, Maurice DV (1985) Inefficacy of propionic acid for depleting laying hens and their progeny of vitamin B12. Nutr Rep Int 32:1325–1332

Watanabe F, Yabuta Y, Tanioka Y, Bito T (2013) Biologically active vitamin B12 compounds in foods for preventing deficiency among vegetarians and elderly subjects. J Agric Food Chem 61:6769–6775. https://doi.org/10.1021/jf401545z

Weaver VM, Welsh J (1993) 1,25-dihydroxycholecalciferol supplementation prevents hypocalcemia in magnesium-deficient chicks. J Nutr 123:764–771. https://doi.org/10.1093/jn/123.4.764

Whitacre ME, Combs GF Jr, Combs SB, Parker RS (1987) Influence of dietary vitamin E on nutritional pancreatic atrophy in selenium-deficient chicks. J Nutr 117:460–467. https://doi.org/10.1093/jn/117.3.460

Whitehead CC (1990) Biotin in animal nutrition. Animal nutrition and health. Vitamins and Fine Chemicals Division, Basel

Wilson H, Arrington L, Harms R (1968) High levels of dietary iodine for delaying sexual maturity of egg production type pullets. Poult Sci 47:1535–1539

World's Poultry Science Association (WPSA) (1985) Mineral requirements and recommendations for growing birds. Worlds Poult Sci J 41:252–258

Yen JT, Jensen AH, Baker DH (1977) Assessment of the availability of niacin in corn, soybeans and soybean meal. J Anim Sci 45:269–278. https://doi.org/10.2527/jas1977.452269x

Young RJ, Norris LC, Heuser GF (1955) The Chick's requirement for folic acid in the utilization of choline and its precursors betaine and methylaminoethanol. J Nutr 55:353–362. https://doi.org/10.1093/jn/55.3.353

Youssef IM, Westfahl C, Sünder A, Liebert F, Kamphues J (2008) Evaluation of dried distillers' grains with solubles (DDGS) as a protein source for broilers. Arch Anim Nutr 62:404–414. https://doi.org/10.1080/17450390802332985

Part IV

Overview on Diagnosis, Prevention and Control of Turkey Diseases

Diagnosis of Turkey Diseases

18

Awad A. Shehata and Hafez M. Hafez

Abstract

Turkeys are susceptible to various health disorders and infectious diseases, which can result in significant economic losses. Examples of the main problems in turkeys are respiratory manifestations, enteric disorders and diseases, locomotory disturbances, nervous manifestations, and a drop in egg production. The causes of these conditions can range from infectious and noninfectious causes such as housing structure, climate, stocking density, and hygiene. Each of these factors can impact the overall health of the flock, either positively or negatively, and should be carefully managed. In this chapter, we will discuss the causes of the main problem in turkey production, diagnostic approaches, and prevention and control strategies.

Keywords

Respiratory manifestations · Enteric disorders · Locomotory disturbances · Nervous manifestations · Drop in egg production

A. A. Shehata (✉)
TUM School of Natural Sciences, Bavarian NMR Center (BNMRZ), Structural Membrane Biochemistry, Technical University of Munich, Garching, Germany
e-mail: awad.shehata@tum.de

H. M. Hafez
Institute of Poultry Diseases, Faculty of Veterinary Medicine, Free University of Berlin, Berlin, Germany
e-mail: hafez.mohamed@fu-berlin.de

H. M. Hafez, A. A. Shehata (eds.), *Turkey Diseases and Disorders Volume 2*,
https://doi.org/10.1007/978-3-031-63322-5_18

18.1 Respiratory Diseases in Turkeys

Respiratory conditions greatly affect turkeys, which can lead to significant economic losses. These diseases can cause increased mortality rates, higher medication costs, lower egg production, reduced eggshell quality, and decreased hatchability. The causes of respiratory diseases can be attributed to several pathogens, alone and/or combined with other microorganisms or noninfectious factors such as climate and management issues. The most common avian respiratory tract infections are shown in Tables 18.1 and 18.2. For more details, please refer to Abdul-Aziz and Barnes (2018); Williams et al. (2018); Brugère-Picoux et al. (2020).

The severity of clinical signs, course of the disease, and mortality after viral infections are extremely variable and are influenced by the immune status of the birds, virulence, and the pathogenicity of the involved agents, as well as by many environmental factors, concurrent diseases, and the type of secondary infections. The diagnosis of viral diseases is not a straightforward business. The diagnosis consists of case history, management, and environmental investigation on the spot. In addition, clinical investigations and postmortem examination are important steps toward disease diagnosis. However, clinical signs and necropsies are mostly not the final steps of the diagnosis. The final diagnosis can only be reached by laboratory diagnosis.

Table 18.1 Possible causes of respiratory diseases in turkeys (Hafez 1990)

Noninfectious	Infectious
Managemental causes	**Viral causes**
Litter quality	Newcastle Disease
Stocking density	Avian Influenza
Ventilation rate	Avian pox (Visceral form)
Temperature	Avian Metapneumovirus
High ammonia level[a]	Infectious laryngotracheitis[b]
High dust concentration	**Bacterial causes**
Nutritional causes	*E. coli*
Vitamin A deficiency	*Mycoplasma*
	P. multocida
	Ornithobacterium rhinotracheitis
	Chlamydia
	Fungal causes
	A. fumitgatus
	Parasitic cause
	Cryptosporidium

[a] Infectious laryngotracheitis is less frequent in turkeys. However, it was reported in turkeys that displayed symptoms such as swollen sinuses, conjunctivitis, tracheitis, and expectoration of bloody mucus (Gulacti et al. 2007; Portz et al. 2008)

[b] The maximum ammonia levels in poultry houses have been set as 50 ppm by the Occupational Safety and Health Administration (OSHA). High levels of ammonia cause epithelial corneal erosions and loss of cilia of the respiratory epithelium

Table 18.2 The main viruses causing respiratory manifestations

Virus	Main clinical picture and lesions	Isolation	Identification
Avian influenza	• Respiratory manifestations • Hemorrhages on the shank • Septicemia	• Embryonated chicken eggs via allantoic sac	• Type identification (PCR/agar gel immunodiffusion test) • Subtyping (hemagglutination inhibition test/PCR) • Pathotyping (intravenous pathogenicity index/ sequencing of haemagglutinin cleavage site)
Newcastle	• Respiratory signs • Nervous signs (torticollis and opisthotonus) • Enteric disorders (diarrhea, proventricular, and ileocecal hemorrhages)	• Embryonated chicken eggs at 9–11 days old via allantoic sac route	• Hemagglutination inhibition • PCR and sequence analysis of cleavage site of F-gene • Pathotyping: Intracerebral pathogenicity index (ICPI) in day-old chicks, mean death time (MDT), or intravenous pathogenicity index (IVPI)
Other paramyxoviruses	• PMV 2,3, 6, and 7 are associated with mild respiratory manifestations and a drop in egg production	• Embryonated chicken eggs at 8–10 days old via allantoic sac route	• Hemagglutination inhibition • PMV-2 does not agglutinate the RBCs of chickens and turkeys • Not all PMV-6 agglutinate chicken red blood cells • PMV-3 produces intracytoplasmic and intranuclear inclusion bodies in glial cells
Infectious laryngotracheitis	• Less frequent in turkeys • Swollen sinuses, conjunctivitis, tracheitis, and expectoration of bloody mucus	• Embryonated chicken eggs 9- to 11-day-old via chorioallantoic membrane (CAM)	• Pock lesions on the chorioallantoic membrane • Intranuclear inclusion body • Immunofluorescent assay (IFA) • Agar gel immunodiffusion test • PCR (6 genotypes in chickens: I, II, III, IV, V, and VI)
Pox	• Cutaneous form on the snood • Wet form- oral cavity	• Embryonated chicken eggs 9–11 days old via chorioallantoic membrane (CAM)	• Pock lesions on the chorioallantoic membrane • Intracytoplasmic inclusion body • PCR/IFA • Serology: AGID/ELISA
Avian metapneumovirus	• Respiratory manifestation • Sinusitis	• Cell culture (vero cells/chicken embryo rough cell line)	• PCR • Serology: Serum neutralization test/IFA/ELISA

18.2 Enteric Disorders and Diseases

Enteric disorders are a significant group of disorders and diseases that affect poultry and continue to cause high economic losses worldwide. These disorders can result in increased mortality rates, decreased weight gain, increased medication costs, and negatively influenced feed conversion rates (Jindal et al. 2010). Various pathogens, including viruses, bacteria, and parasites, are responsible for enteric disorders. These pathogens can act alone (mono-causal), with other microorganisms (multi-causal), or in conjunction with noninfectious factors, such as feed and management-related issues (Table 18.3). However, determining whether the cause of enteric disorders in poultry is infectious or noninfectious is challenging under field conditions.

18.2.1 Noninfectious Causes

The digestive system of a bird is highly dependent on its nutrition. Any malfunction of the digestive system can lead to poor digestion, reduced gut movement, and inadequate nutrient absorption. In the same way, low-quality feed and poor

Table 18.3 Possible causes of enteric diseases in turkeys

Noninfectious	Infectious
Managemental causes	**Viral causes**
Litter quality	Newcastle Disease
Stocking density	Coronavirus enteritis
Ventilation rate	Hemorrhagic enteritis
Temperature	Rotavirus
Distribution of feeders and drinkers	Astrovirus
	Avian influenza
Nutritional causes	Reovirus
Vitamin A deficiency	**Bacterial causes**
	E. coli
	P. multocida
	Salmonella
	Clostridia
	Fungal causes
	Candida albicans
	Parasitic cause
	Coccidia
	Histomonas
	Hexamita
	Ascaridia
	Heterakis gallinarum
	Cestodes
	Cryptosporidium

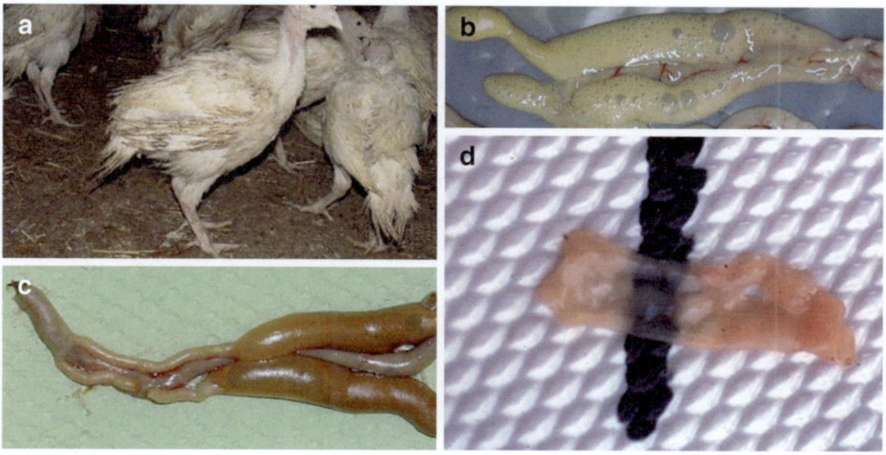

Fig. 18.1 Enteric disorders in turkeys. (**a**) Diarrhea and decreased weight gain and emaciation, (**b**, **c**) enteritis, (**d**) thinning of the intestinal wall © Hafez

nutrition can increase the risk of enteric disorders or even directly cause them. Nutritional factors such as feed palatability, feed ingredient quality, feed formulation, and pellet quality play a crucial role in maintaining gut health. Incorrect feeding techniques such as sudden changes or restrictions in feed intake, and fiber percentages in the feed can also lead to changes in the intestinal flora and/or enzymatic activity, resulting in digestive disorders and diarrhea (Fig. 18.1). Excessive levels of mycotoxins and biogenic amines in the feed can also cause enteric disorders. Dust in feed can impact its palatability and reduce feed intake and poorly stored feed can contain fungal spores and rancid fat, negatively affecting the vitamin content in the feed and reducing feed intake. Inadequate feeder space and false distribution can cause competition among birds, leading to variations in feed intake within the flock. Undesirable substances such as mycotoxins in feed can adversely affect the immune system, increase the risk of infectious diseases, and reduce the effectiveness of vaccines.

Proper rearing management is crucial for poultry production's well-being, productivity, and profitability while ensuring animal welfare. Rearing management encompasses various factors that impact the birds' health, such as house structure, climatic conditions (including ventilation, temperature, and litter condition), stocking density, feed and water supply, hygienic conditions, and the knowledge and expertise of the stockman. These factors are interdependent and can either promote or hinder the flock's health condition. To achieve optimal performance outcomes, turkey flock managers must integrate effective environmental, husbandry, nutrition, and disease control programs. Rearing management should prioritize meeting the birds' needs, promoting production, and preventing disease. Any disruptions can cause stress, leading to reduced bird resistance, increased susceptibility to infections, and weakened immune responses to vaccines.

18.2.2 Infectious Causes

Infectious agents like viruses, bacteria, fungi, and parasites can cause intestinal disorders. These agents can enter poultry farms through different routes, both vertically and/or horizontally. Vertically transmitted infections like *Salmonella* and *E. coli* can be a problem, along with improper hatchery management. However, these agents can also be transmitted horizontally via direct contact between infected and noninfected birds or indirectly via contaminated feed, water, equipment, environment, or dust through ingestion or inhalation. Table 18.4 summarizes the diagnostic methods of enteric viral diseases in turkeys.

Table 18.4 The main viruses causing enteric disorders

Virus	Main clinical picture and lesions	Isolation	Identification
Astrovirus	• Associated with diarrhea and high mortality	• Embryonated chicken eggs via the yolk sac route • Although chicken astroviruses can be isolated on cell culture, such as chicken embryo liver cells, turkey astroviruses cannot be isolated on cell culture	• Electron microscopy • IFA/SNT and ELISA • In situ hybridization • Typing: Two types of turkey astroviruses are known, TastV-1 and TastV-2
Coronavirus Enteritis	• Associated with enteric disorders in turkeys and can be detected in the intestinal contents and bursa Fabricius tissues	• Embryonated turkey eggs via the amniotic route	• Electron microscopy • IFA/SNT and ELISA • Immunohistochemistry • Hemagglutination inhibition test (TCoV agglutinates guinea pig and rabbit erythrocytes) • TCoV is classified as group 3 Coronaviruses with avian infectious bronchitis virus
Hemorrhagic Enteritis	• Bloody dropping and sudden death • Swollen intestines, filled with bloody contents • Splenomegaly and hepatomegaly • Petechial hemorrhages can also be seen in several tissues	• Chicken embryo liver (CEL) cells from SPF chickens • The main cytopathic effects on CEL cells are rounded cell degeneration after the first and fourth passages	• Histologically: intranuclear inclusion bodies in the macrophages and lymphocytes of the spleen • AGID, PCR, and ELISA
Reovirus	• A progressive condition affecting mainly tom turkeys at 12–16 weeks old, causing enteritis, immunosuppression, viral arthritis/tenosynovitis, interstitial myocarditis	• Embryonated chicken eggs via the yolk sac route or on CAM • Chicken embryo liver cells, chicken kidney cells • The Yolk sac route is preferred for the initial isolation	• Electron microscopy • PCR • Immunohistochemistry • ELISA
Rotavirus	• Associated with enteric disorders • Replicates primarily in jejunum and ileum	• Primary culture of chicken and turkey kidney and chicken embryonic liver cells • MA104 cell line • Trypsin activation is essential for replication	• Electron microscopy • IFA/SNT and ELISA

18.3 Locomotory Disturbances

Leg weakness or lameness is a significant factor leading to poor welfare among poultry and causing considerable economic losses in meat-type poultry around the globe. Lameness in birds can often be linked to pain, which can severely limit their ability to move around and access food and water. These can negatively impact their health and production and reduce the quality of animal products. In addition, poor welfare can negatively influence all aspects of a bird's well-being (Kierończyk et al. 2017; Szafraniec et al. 2020). Locomotory disturbances in poultry are a complex issue caused by various noninfectious and infectious factors (Table 18.5. For more details, please see Chap. 1. Figures 18.2 and 18.3 show some locomotory disorders in turkeys.

Table 18.5 Possible causes of locomotory disturbances in turkeys

Noninfectious	Infectious
Managemental causes	**Viral causes**
Litter quality	Reovirus
Stocking density	Marek's disease
Light program	Osteopetrosis
Nutritional causes	**Bacterial causes**
Vitamin B2 deficiency	*E. coli*
Perosis	*Mycoplasma synoviae*
Rickets/osteomalacia	*P. multocida*
	Enterococci
	Staphylococcus
	Streptococcus
	Fungal causes
	Aspergillus fumigatus
	Parasitic cause
	Cryptosporidium

Fig. 18.2 Arthritis in turkeys © Hafez

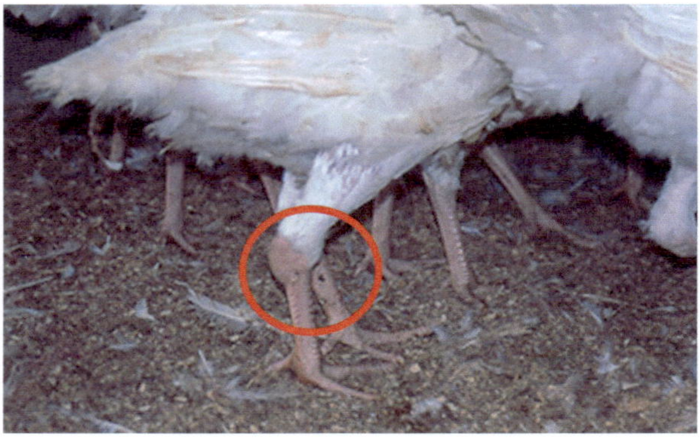

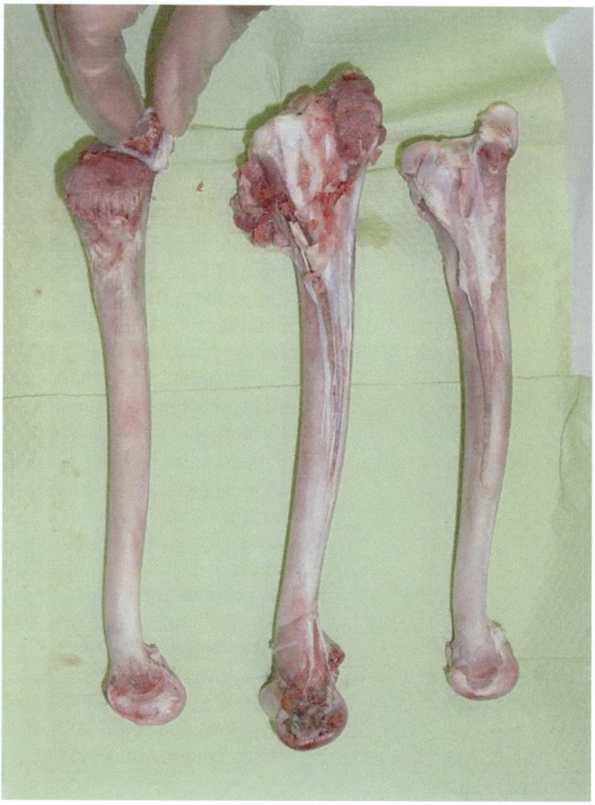

Fig. 18.3 Tibial torsion in turkeys © Hafez

18.4 Nervous Manifestations in Turkeys

Several potential causes of nervous manifestations in turkeys have been reported (Table 18.6). Differential diagnosis is based on the age of the birds, susceptibility, accompanied symptoms, and laboratory diagnosis.

Newcastle disease and avian influenza viruses also cause nervous manifestations in poultry. Neurological signs are predominant in waterfowl. The presumptive and laboratory diagnosis is described in Table 18.6.

Table 18.6 Possible causes of nervous manifestations in turkeys

Etiology	Diagnosis
Viral causes	
Newcastle disease	• NDV causes respiratory, nervous manifestations and enteric disorders • The presumptive and laboratory diagnosis is described in Table 18.2
Avian influenza	• Neurological signs are predominant in waterfowl • The presumptive and laboratory diagnosis is described in Table 18.2

(continued)

Table 18.6 (continued)

Etiology	Diagnosis
Marek's disease	• Clinically MD has four forms: • Cutaneous form (nodular enlargement of the feather follicle; the fully productive form occurs in the feather follicles) • Neural form (transient paralysis syndrome is associated with a brain infection. Partial or complete paralysis for 1–2 days. Thickening of peripheral nerves) • Visceral lymphomas (including diffuse enlargement of the bursa of Fabricius, lymphomas in myocardium and lung, during the early phase, it causes bursal atrophy due to multiplication in B-lymphocytes) • Ocular form (iridociliary lymphomas, loss of iris pigmentation, and irregular eye pupil) • Histopathology: The predominant is lymphocytes, pleomorphic cell population
Avian encephalomyelitis	• Clinical signs in newly hatched chicks (vertical transmission) • Clinical signs at the second week of age (horizontal transmission) • Clinically: Neurological signs in baby chicks and cataracts in adults • No postmortem lesions • Histopathologically: Gliosis, lymphocytic cuffing, and central chromatolysis of neurons
Israel turkey meningoencephalitis	• Drowsiness, incoordination, torticollis, tremors, progressive paresis, and paralysis • Diagnosis is based on IFA, hemagglutination inhibition test, neutralization test, and PCR
Bacterial causes	
Fowl cholera (chronic form)	• Torticollis or twisting of the neck • Diagnosis: table
Riemerella anatipestifer	• A disease of young ages (Table 18.8)
Botulism	• No postmortem lesions • Diagnosis is based on mice bioassay
Staphylococcus	• Table 18.8
Spirochaetosis	• Paralysis • Presence of ticks • Enlargement and mottled spleen
Salmonella arizonae	• Paralysis, torticollis, and opisthotonus
Nutritional causes	
Thiamin deficiency	• Stargazing posture
Vitamin A deficiency	• Associated with abnormal gait in growing turkeys
Vitamin B2 deficiency	• Curled-toe paralysis (the toes bent inward and downward) • On necropsy: Thickening of sciatic nerves

Table 18.6 (continued)

Etiology	Diagnosis
Vitamin E deficiency	• Characterized by neurological signs including muscle weakness, ataxia, backward retraction of the head and neck, rapid contraction, and relaxation of the legs • Nutritional encephalomalacia (lesions in brains)
Parasitic cause	
Baylisascaris sp. (Cerebrospinal nematodiasis)	• Paralysis • Cerebrospinal nematodiasis has also been reported in several mammalian and several avian species, including chickens, turkeys, ostriches, emus, pigeons, doves, quails, pheasants, partridges, blue jays, robins, and psittacines (Diab et al. 2012). Raccoon is the natural host
Toxicological causes	
Organophosphorus toxicity	• Characterized by paralysis, profuse salivation • Degenerative changes in peripheral nerves and spinal cord • The diagnosis is based on the determination of acetylcholinesterase in the brain and plasma
Lead poisoning	• Demyelination in peripheral nerves and the presence of INIB in tubular cells of the kidney

Table 18.7 Disease complex and/or lesion and some possible involved causes

Omphalitis	• *E. coli*, *Pseudomonas aeruginosa*, *Enterococcus*, *Proteus*
Salpingitis	• *E. coli*, *Riemerella anatipestifer*, *Gallibacterium anatis* (less frequent)
Spondylitis (Vertebral osteoarthritis)	• *Staphylococcus aureus*, *E. coli*, *E. cecorum*
Orchitis	• *E. coli*, *Chlamydia*, *Staphylococcus* (less frequent), *Salmonella* (less frequent), Pasteurella (less frequent)
Neonatal septicemia	• *E. coli*, *Pseudomonas aeruginosa*, *Enterococcus*, *Salmonella*
Viruses causing a drop in egg drop	• Avian influenza, Newcastle disease, Paramyxovirus 3, avian metapneumovirus
Viruses of public health concerns	• Avian influenza, Newcastle disease
Viruses causing high mortality	• Avian influenza, Newcastle disease, avian metapneumovirus, avian encephalomyelitis

18.5 Other Health Problems in Turkeys

The main disease complex and/or lesion and some possible involved causes are shown in Table 18.7.

18.6 Diagnosis of Bacteria in Turkeys

Establishing a strategy to control and prevent diseases begins with diagnosis, which can be a complex process for disease complexes. This involves gathering case history, conducting on-site management and environmental investigations, as well as clinical investigations and postmortem examinations. Clinical signs and lesions are rarely sufficient to make a definite diagnosis. As a result, laboratory diagnosis becomes necessary. The final diagnosis of infectious diseases can only be made through laboratory diagnosis, which involves isolating and identifying the causative agent and possibly detecting DNA or RNA using PCR, followed by typing. Serological examinations to detect antibodies can also be conducted. It's important to note that an accurate diagnosis is essential for effective disease control and prevention. In general, however, many factors such as governmental regulations, the goal of examinations, cost-benefit analysis, equipment facilities, availability of reagents, and experiences of the staff are influenced and, in some instances, limit the choice of the laboratory methods (Hafez and Hess 1999). Nowadays, ELISA is the preferred method for detecting antibodies due to its quick and dependable results.

The laboratory diagnosis of bacterial diseases in turkeys is shown in Figs. 18.4 and 18.5 and Tables 18.8, 18.9, 18.10, and 18.11. Gram-negative bacteria, such as *E. coli*, *Salmonella*, *Bordetella*, *Pasteurella*, and *Pseudomonas aeruginosa*, cause turkeys' problems (Fig. 18.4 and Table 18.8). Several factors should be considered for the isolation of bacteria, including the selection of media, incubation conditions, and incubation time.

MacConkey agar, which contains bile salts and crystal violet to prevent the growth of most gram-positive bacteria, is a widely recognized medium for isolating gram-negative *Enterobacteriaceae* and other gram-negative bacteria (Jung and Hoilat 2022). To isolate *Salmonella* spp. and *Campylobacter* spp., specific media are utilized. For instance, modified semisolid Rappaport Vassiliadis agar (MSRV) and modified charcoal-cefoperazone-deoxycholate agar (mCCDA), respectively. The incubation condition is also a determinant factor for bacterial isolation. Both *Brachyspira* and *Clostridium* spp. require strict anaerobic conditions for growth (Mappley et al. 2014; Xu et al. 2021). However, *Avibacterium paragallinarum* is cultivated under strict microaerobic conditions (Blackall and Soriano-Vargas 2020). Certain types of bacteria, such as *Mycoplasma* spp. *and Mycobacterium avium*, require longer incubation time.

Various biochemical reactions are used in phenotypic identification tests, which can be carried out using commercial systems like analytical profile index (API) systems. These tests can even help differentiate between different subspecies. For instance, in *Pasteurella multocida*, the classification of the three subspecies, *subsp. multocida, septica, or gallicida*, is based on their ability to ferment sorbitol and dulcitol (Hunt Gerardo et al. 2001).

Biochemical methods for testing bacterial species can be both time-consuming and resource-intensive. Additionally, differences in characteristics among members of the same species or the inability of some bacteria to ferment certain carbohydrates

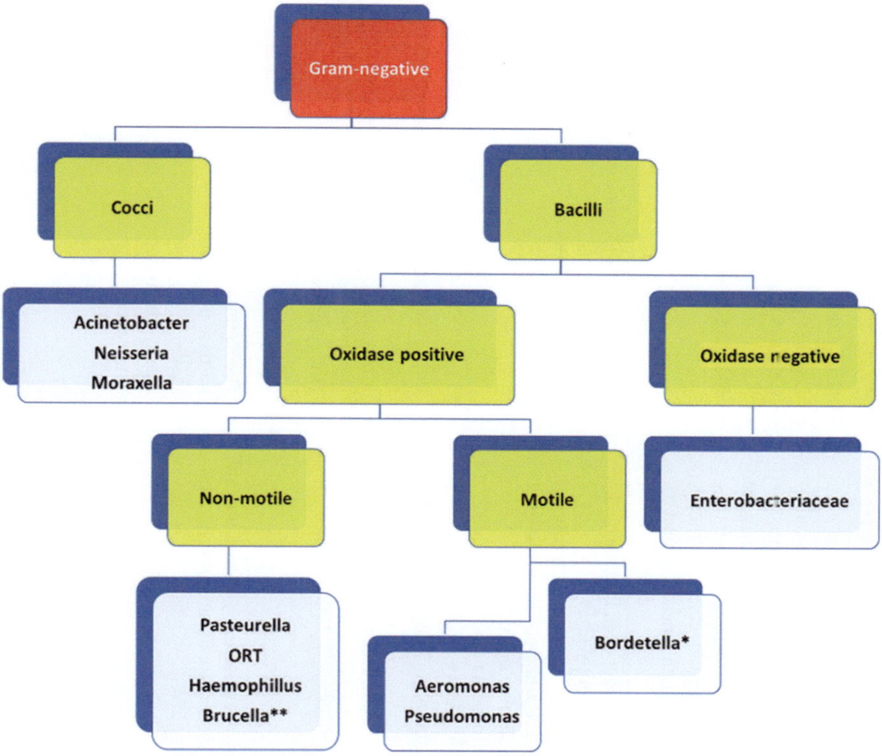

Fig. 18.4 Laboratory diagnosis of gram-negative bacteria in turkeys. The oxidase test is the key test for gram-negative bacteria. * = Bordetella pertussis and B. para-pertussis are non-motile. ** = *Brucella* cannot produce acids from glucose

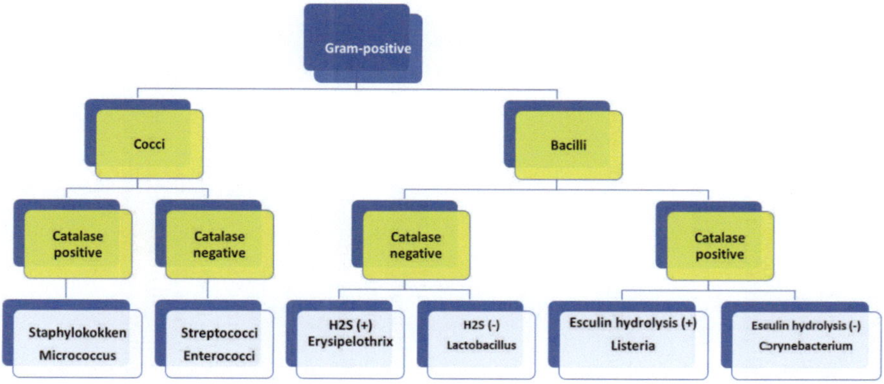

Fig. 18.5 Laboratory diagnosis of Gram-positive in turkeys. The oxidase test is the key test for Gram-negative bacteria. * = Bordetella pertussis and B. para-pertussis are non-motile. ** = Brucella cannot produce acids from glucose

Table 18.8 Presumptive diagnosis, isolation, and identification of **gram-negative bacteria in turkeys**

Bacteria	Main clinical picture and lesions	Isolation	Identification
Bordetella	• Respiratory disease in young turkeys, Transmitted vertically • High morbidity rates characterize uncomplicated outbreaks • Softening and distortion of tracheas	• MacConkey agar (pinpoint translucent) • Blood agar (three forms of colonies: (i) small translucent and pearl-like, (ii) raised brown-tinged center, (iii) rough colonies represent the nonpathogenic bacteria) • Strict aerobe (in liquid media, it must be aerated)	• Oxidase positive • Catalase positive • Agglutinate Guinea pig • Microagglutination test
Campylobacter	• Infection is usually subclinical	• Selective-Blood agar (i.e., Preston media, Campy-cefex, Campy-CVA) • Selective-Charcoal-containing media (Charcoal-Cefoperazone-Deoxycholate, Charcoal-Selective media) • Chromogenic media (Campylobacter-chromogenic plating medium)	• Hydrolysis of hippurate and indoxyl (only *C. jejuni*) • Molecular detection
Chlamydia psittaci	• Hepatomegaly, Splenomegaly • Fibrinous exudate on the air sacs and peritoneum • Fibrinous pericarditis	• Chicken embryo (6–7 days old via yolk sac route) • Cell culture (Vero, HeLa, BGM, L-929)	• Impression smear (Giemsa stain) • ELISA • Molecular identification and typing (7 genotypes: A, B, C, D, E, and E-B), Type D is common in turkeys • Serology (ELISA, direct complement fixation test, IFA)
E. coli	• Several disease conditions: colisepticemia coligranulomatosis (Hjärre's disease)., enteritis, omphalitis, salpingitis, swollen head syndrome, cellulitis, orchitis, osteomyelitis / synovitis, and panophthalmitis	• MacConkey agar (lactose-positive, hot pink colonies), some colonies do not form pink colonies) • Blood agar • Klingeler's iron agar • Triple sugar iron agar	• Biochemical tests (lactose positive, oxidase negative) • PCR • Serotyping based on O-antigen

ORT	• Pleuropneumonia	• Does not grow on MacConkey agar • Sheep blood agar 5% (microaerobic) • 90% of ORT strains are resistant to gentamicin, and polymyxin (can be used as selective media)	• Highly pleomorphic gram-negative • Biochemical (catalase negative, oxidase positive, no color on triple sugar agar, and indole negative) • AGID for serotyping (to date, 18 serotypes, designated A to R) • PCR
Pasteurella	• Pleuropneumonia (thick layer of fibrinous exudate on the surface) • Acute form (hemorrhagic septicemia) • Chronic (nervous signs, arthritis, synovitis of sternal bursa, conjunctivitis, sinusitis, swelling of the wattle)	• Dextrose starch agar containing 5% • Chicken serum or blood agar • Does not grow on MacConkey	• Biochemical • PCR • MALDI-TOF • Capsular types (5 serotypes; A, B, D, E, and F)
Riemerella anatipestifer	• Young birds are most susceptible • Respiratory signs, lameness, and central nervous symptoms (falling back, rolling onto their sides, unable to stand)	• Brain-Heart-Glucose-Infusion (BHI), Mueller-Hinton broth II with 2% lysed horse blood, chocolate agar, and trypticase soy agar • Does not grow on MacConkey	• Not easily identified based on biochemical tests • Produces oxidase, catalase, and phosphatase • IFA on impression smears • Serotyping can be done using AGID (21 serotypes) • Molecular biological methods
Salmonella	• Fibrinous pericarditis, hepatic serositis, splenomegaly, and omphalitis) • *S. arizonae* is more frequent in turkeys and causes meningoencephalitis and panophthalmitis. poults • Eggshell penetration (paratyphoid *Salmonella*) • Waterfowl resist *S. Pullorum*	• Xylose lysine deoxycholate, xylose lysine tergitol-4 agar (black colonies) • MacConkey agar	• Biochemical tests • Serotyping based on Flagellar and somatic antigens (*S. typhimurium* is Group B, and *S. pullorum* and Enteritidis are Group D) • All salmonellae are motile except *S. pullorum Gallinarum* • A rapid whole-blood test is not a satisfactory serological test for turkeys
Pseudomonas aeruginosa	• Omphalitis • Keratitis and penetration of cornea	• Characteristic grapelike odor • Green diffusible pigment on nutrient agar	• Oxidase positive • Motile

Table 18.9 Presumptive diagnosis, isolation, and identification of **gram-positive bacteria**

Bacteria	Main clinical picture and lesions	Isolation	Identification
Erysipelothrix rhusiopathiae	• Frequent at 18–20 weeks old • Associated also with de-toeing • Erythema and edema of infected skin • Septicemia • Arthritis • Vegetative valvular endocarditis • The snood and margin of dewlap are swollen and necrotic	• Blood agar (alpha hemolysis) • Tripple sugar agar (black discoloration) • Isolation can be enhanced using enriched media, that is, brain heart infusion with 5% serum) • Embryonated chicken eggs via the yolk sac route • Mouse inoculation	• Biochemical (catalase negative, H2S positive) • IFA • PCR • Serotyping using double agar gel precipitation test (26 serotypes)
Listeria monocytogenes	• Necrotizing myocarditis • Septica = emia and encephalitis	• Blood agar • Blood agar containing 0.05% potassium tellurite • Lithium chloride-phenylethanol-moxalactam • Polymyxin-acriflavine-lithium-chloride-ceftazidime-esculin-mannitol egg yolk broth	• Histopathology (extensive necrotizing myocarditis) • Biochemical tests and MALDI-TOF • PCR and multilocus sequence typing (MLST), phage typing
Mycobacterium avium	• Emaciation • Nodules in bone marrow, spleen, liver, and thymus	• Löwenstein-Jensen Medium • Middlebrook 7H10 • Proskauer-Beck liquid. • Incubation for a minimum of 4 weeks	• Acid-fast stained smear and histopathologic section, calcification, and fibrosis are uncommon • Negative for niacin, Tween80 hydrolysis, nitrate, urease, and tellurite • Molecular identification • ELISA • Tuberculin test

Staphylococcus aureus	• Osteomyelitis, arthritis • Femoral head necrosis • Involves mainly the proximal tibiotarsus and proximal femur and is less common in proximal tarsometatarsus, distal femur, distal tibiotarsus, proximal humerus, and vertebrae	• Bovine blood agar (β-hemolysis) • Selective media (Staphylococcus medium 110, Mannitol-salt agar, tellurite glycine agar)	• Catalase positive • Coagulase positive • Latex agglutination • PCR
Streptococci	• Acute septicemic and chronic form • Septicemia • *Streptococcus gallolyticus* subsp. gallolyticus is characterized by an **enlarged and mottled spleen,** primarily observed in pigeons and turkey poults • *S. gallolyticus subsp. asteurianus* causes meningitis, sepsis, and endocarditis in turkeys, goslings, and ducks	• Blood agar • α-hemolysis, β-hemolysis	• Catalase is positive and ferments a variety of sugars • PCR
Enterococci	• Locomotory disturbances such as myositis, arthritis, osteomyelitis, tenosynovitis, spondylitis, femoral head necrosis, and tendinitis. *E. faecalis* and *E. faecium* can cause omphalitis, endocarditis, and amyloidosis • *E. hirae* and *E. durans* were isolated from young birds suffering from encephalomalacia	• Blood agar: variable hemolysis (α-) or no (γ-hemolysis) • Selective media: colistin and nalidixic acid media. Enterococci may produce	• MALDI-TOF, multiplex PCR, or PCR and sequencing the 16S RNA gene • Biochemical tests (mannitol, sorbitol, arabinose, sucrose, or raffinose)

Table 18.10 Presumptive diagnosis, isolation, and identification of anaerobic bacteria turkeys

Bacteria	Main clinical picture and lesions	Isolation	Identification
Clostridium perfringens (necrotic enteritis)	• Lesions are mainly in jejunum and ileum, Turkish-towel-like • Cholangiohepatitis	• Blood agar (inner beta and outer alpha hemolysis) • Egg yolk agar (lecithinase positive) • Neomycin-polymyxin-Agar	• PCR • ELISA for detection of toxins (6 Toxin types: A-G)
Clostridium botulinum (Botulism)	• Nervous manifestations such as leg paralysis and drop neck	• Chopped meat-glucose-starch medium • Egg yolk agar (lipase positive)	• Toxin detection through ELISA • Mice bioassay (intraperitoneal) for demonstration of botulinum neurotoxin (8 toxins: A-H)
Clostridium septicum	• Dark reddish-purple to green, weepy areas of the skin • Affected areas usually include the abdomen, breast, wings, or legs • Extensive blood-tinged oedema, with or without gas (crepitus), subcutis • Infection may extend into underlying musculature, which may be discolored and contain oedema and gas	• Pre-reduced tryptose phosphate medium or Phenylethanol agar (swarms the entire plate) • Egg yolk agar (lecithinase and lipase negative)	• PCR • MALDI-TOF • IFA

Table 18.11 Presumptive diagnosis, isolation, and identification of *Mycoplasma* in turkeys

Bacteria	Main clinical picture and lesions	Isolation	Identification
Mycoplasma gallisepticum	• Massive accumulation of mucus in the infraorbital sinuses • Drop in egg production	• Frey's Medium • PPLO broth • Mycoplasma synoviae requires media supplemented with nicotinamide adenine dinucleotide (NAD) and cysteine • *Mycoplasma meleagridis* does not ferment glucose (no change in mycoplasma broth containing glucose and phenol red)	• PCR • Serum plate agglutination (SPA) test (highly efficient for detection of IgM in recent infection) • *M. Iowae* does not elicit an immune response • Hemagglutination inhibition test (more specific than SPA)
Mycoplasma iowae	• Late embryonic mortality • Skeletal abnormalities, including chondrodystrophy		
Mycoplasma synoviae	• Eggshell apex abnormalities • Breast blister • Swelling of foot pads • Infectious synovitis and subclinical respiratory manifestations		
Mycoplasma meleagridis	• Late Embryonic mortality • Skeletal abnormalities, including chondrodystrophy • Caseated fibrinolytic airsacculitis at the time of hatch • Airsacculitis/spondylitis		

can lead to unreliable or inconsistent testing results, as seen in the cases of *Gallibacterium anatis* and *Riemerella anatipestifer* (Christensen and Bisgaard 2010).

It is worth noting that even commercially available databases for bacterial identification may not always have up-to-date classification data, which can hinder accurate and dependable identification, as demonstrated with *Ornithobacterium rhinotracheale* (van Empel and Hafez 1999).

Biochemical methods for testing bacterial species can be both time-consuming and resource-intensive. Additionally, differences in characteristics among members of the same species and/or the inability of some bacteria to ferment certain carbohydrates can lead to unreliable or inconsistent results, as seen in the cases of *Gallibacterium anatis* and *Riemerella anatipestifer*. It is worth noting that even commercially available databases for bacterial identification may not always have up-to-date classification data, which can hinder accurate and dependable identification, as demonstrated with *Ornithobacterium rhinotracheale*, reviewed by (Liebhart et al. 2023).

References

Abdul-Aziz T, Barnes HJ (2018) Gross pathology of avian diseases. The American Association of Avian Pathologists, Omnipress, Madison, WI. 53704

Blackall PJ, Soriano-Vargas E (2020) Infectious Coryza and related bacterial infections. In: Swayne DE (ed) Diseases of poultry, 1st edn. Wiley, pp 859–873

Brugère-Picoux J, Shivaprasad HL, Venne D, Bouzouaïa M (2020) Manual of poultry diseases, version anglaise. Association francaise pour l'avancement des Sciences (AFAS), Paris, France

Christensen H, Bisgaard M (2010) Phylogenetic relationships of Riemerella anatipestifer serovars and related taxa and an evaluation of specific PCR tests reported for R. Anatipestifer. J Appl Microbiol 108(5):1612–1619. https://doi.org/10.1111/j.1365-2672.2009.04558.x

Diab SS, Uzal FA, Giannitti F, Shivaprasad HL (2012) Cerebrospinal nematodiasis outbreak in an urban outdoor aviary of cockatiels (*Nymphicus hollandicus*) in southern California. J Vet Diagn Invest 24(5):994–999. https://doi.org/10.1177/1040638712455797

van Empel PC, Hafez HM (1999) Ornithobacterium rhinotracheale: a review. Avian Pathol 28(3):217–227. https://doi.org/10.1080/03079459994704

Gulacti I, Eroksuz Y, Bulut H, Ceribasi AO (2007) Outbreak of clinical infectious laryngotracheitis in Turkey. Vet Rec 160(16):554–555. https://doi.org/10.1136/vr.160.16.554

Hafez HM (1990) Serological surveillance for antibodies against different avian infectious agents on Turkey flocks naturally infected with Turkey rhinotracheitis. J Veterinary Med Ser B 37(1–10):369–376. https://doi.org/10.1111/j.1439-0450.1990.tb01071.x

Hafez H, Hess M (1999) Modern techniques in diagnosis of poultry diseases. Archiv für Gefluegelkunde 63:237–245

Hunt Gerardo S, Citron DM, Claros MC, Fernandez HT, Goldstein EJ (2001) Pasteurella multocida subsp. multocida and P. multocida subsp. septica differentiation by PCR fingerprinting and alpha-glucosidase activity. J Clin Microbiol 39(7):2558–2564. https://doi.org/10.1128/JCM.39.7.2558-2564.2001

Jindal N, Patnayak DP, Chander Y, Ziegler AF, Goyal SM (2010) Detection and molecular characterization of enteric viruses from poult enteritis syndrome in turkeys. Poult Sci 89(2):217–226. https://doi.org/10.3382/ps.2009-00424

Jung B, Hoilat GJ (2022) MacConkey medium. In: StatPearls; StatPearls Publishing: Treasure Island, FL

Kierończyk B, Rawski M, Józefiak D, Świątkiewicz S (2017) Infectious and non-infectious factors associated with leg disorders in poultry—a review. Ann Anim Sci 17(3):645–669. https://doi.org/10.1515/aoas-2016-0098

Liebhart D, Bilic I, Grafl B, Hess C, Hess M (2023) Diagnosing infectious diseases in poultry requires a holistic approach: a review. Poultry 2(2):252–280. https://doi.org/10.3390/poultry2020020

Mappley LJ, La Ragione RM, Woodward MJ (2014) Brachyspira and its role in avian intestinal spirochaetosis. Vet Microbiol 168(2–4):245–260. https://doi.org/10.1016/j.vetmic.2013.11.019

Portz C, Beltrão N, Furian TQ, Júnior AB, Macagnan M, Griebeler J, Lima Rosa CAV, Colodel EM, Driemeier D, Back A, Barth Schatzmayr OM, Canal CW (2008) Natural infection of turkeys by infectious laryngotracheitis virus. Vet Microbiol 131(1–2):57–64. https://doi.org/10.1016/j.vetmic.2008.02.029

Szafraniec GM, Szeleszczuk P, Dolka B (2020) A review of current knowledge on Staphylococcus agnetis in poultry. Animals 10(8):1421. https://doi.org/10.3390/ani10081421

Williams SM, Dufour-Zavala L, Jackwood MW, Lee MD, Lupiani B, Reed WM, Speckman E, Woolcock PR (2018) Laboratory manual for the isolation, identification and characterization of avian pathogens, 6th edn. American Association of Avian Pathologists, Jacksonville, FL

Xu W, Wang H, Liu L, Miao Z, Huo Y, Zhong Z (2021) Prevalence and characterization of Clostridium perfringens isolated from different chicken farms in China. Anaerobe 72:102467. https://doi.org/10.1016/j.anaerobe.2021.102467

Vaccination and Treatment of Turkeys' Diseases

19

Awad A. Shehata and Hafez M. Hafez

19.1 Introduction

Vaccination is one of the most effective ways to protect farmers, employers, and poultry against infectious diseases. For several years, poultry vaccines have played a crucial role in promoting the health and welfare of fowls while also boosting production. Vaccination is a cost-effective and essential tool to protect poultry from infectious diseases, minimize losses, and limit the need for antibiotics treatment in poultry farms.

Vaccines work by stimulating the immune system. However, several factors must be considered before using the vaccines, including government regulations, the epidemiological situation in your area, the purpose of vaccination, vaccine availability, and cost-benefit analysis. With huge advancements in vaccine production technology over the years, several types of vaccines were developed, such as recombinant, subunit, reverse genetic, and nucleic acid vaccines. The cost of vaccines can be significantly reduced while ensuring better efficacy. These technologies also allow for quick intervention to combat the constant mutation of microorganisms. Additionally, efficient vaccines against bacterial infections can help reduce the use of antibiotics and prevent the development of resistant bacteria. Elements of vaccination are shown in Fig. 19.1.

A. A. Shehata (✉) · H. M. Hafez
TUM School of Natural Sciences, Bavarian NMR Center (BNMRZ), Structural Membrane Biochemistry, Technical University of Munich, Garching, Germany

Institute of Poultry Diseases, Faculty of Veterinary Medicine, Free University of Berlin, Berlin, Germany
e-mail: awad.shehata@tum.de; hafez.mohamed@fu-berlin.de

H. M. Hafez, A. A. Shehata (eds.), *Turkey Diseases and Disorders Volume 2*,
https://doi.org/10.1007/978-3-031-63322-5_19

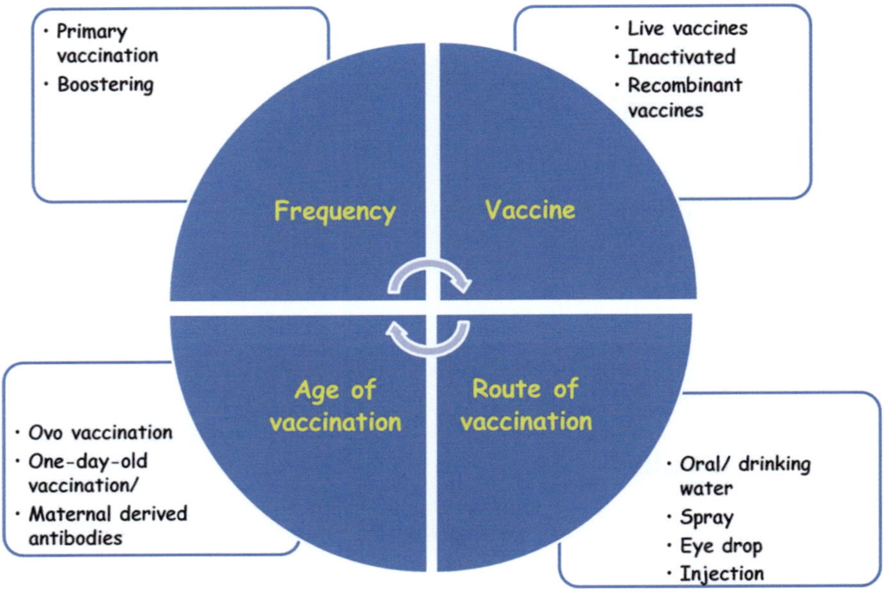

Fig. 19.1 Elements of vaccination

19.2 Types of Vaccines

Several types of vaccines as inactivated, live attenuated, or recombinant are known.

19.2.1 Inactivated Vaccines

Inactivated or killed vaccines contain viruses whose pathogenicity has been inactivated via physical and chemical means, but their protein coat structure remains immunogenic. Ultraviolet radiation and heat inactivate the viruses physically, while formalin is used chemically (Furuya et al. 2010). Killed vaccines provide long-term immunity to flocks against infectious diseases such as Newcastle disease and avian influenza, as well as bacteria such as *Salmonella* and fowl cholera. A schematic diagram of inactivated vaccine development is shown in Fig. 19.2.

Several success factors for inactivated vaccines have been described.:

(i) **Antigenic matching and antigenic load**. Antigenic matching between the vaccine and circulated field virus/bacteria is critical in achieving optimal vaccine efficacy. Therefore, selecting a seed master with a good antigenic match to the field virus is crucial; the closer the antigenic match between vaccines and the circulated field microorganisms, the better efficacy of the vaccine (Wood et al. 1985). Accordingly, seed updating is a prerequisite for obtaining efficient vaccines, particularly influenza vaccines. A rapid and high-level

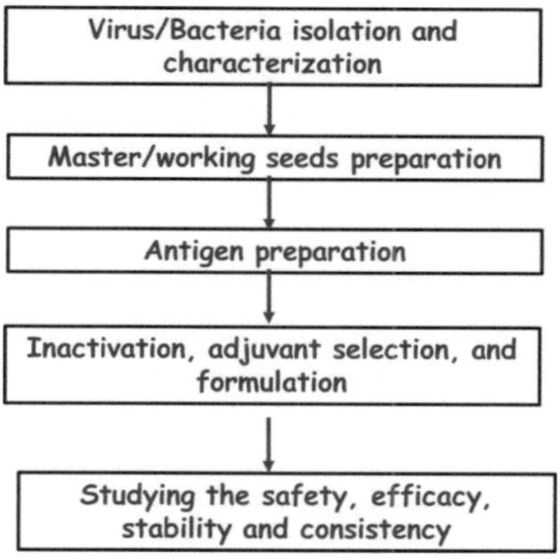

Fig. 19.2 Schematic diagram of inactivated vaccine development

immune response against avian influenza could be achieved using higher amounts of antigen (Sasaki et al. 2009).

(ii) **Sterility**. In manufacturing inactivated vaccines, the harvested allantoic fluid is inactivated with formaldehyde. The time required must be sufficient to ensure freedom from a live virus and microorganisms. The active ingredient is usually emulsified with mineral or vegetable oil. The exact formulations are generally commercial secrets. Inactivated antigens must be tested for completion of inactivation in SPF eggs and/or birds. Also, formalin residues must be determined in the final product.

(iii) **Adjuvant**. A potent adjuvant to enhance the immunogenicity of the inactivated vaccines is important (Aucouturier et al. 2001). A perfect oil adjuvant enhances the immune response. The Montanide ISA71 adjuvant has been shown to stimulate both humoral and cell-mediated immune responses in different animal species, including chickens (Jang et al. 2010, 2011).

(iv) **Stability.** Stability is defined as the ability of a vaccine to retain its chemical, physical, microbiological, and biological properties within specified limits throughout its shelf-life. Depending on the nature of the antigen and the manufacturing process, stability-indicating parameters should be selected on a case-by-case basis.

(v) **Efficacy.** The efficacy of each component of multivalent or combined vaccines must be demonstrated. It must be shown that there is no interference between the components, that is, one component does not cause a significant decrease in another component's immunological response or viability.

19.2.2 Live Vaccines

Live vaccines contain attenuated or avirulent viruses that maintain their immunogenic antigenicity for stimulating the body's immune response. Generally, live vaccines stimulate both cellular and humoral immune responses. Examples of live vaccines are HB1, clone 30, and Lasota against the Newcastle disease virus

(Wambura and Kataga 2011). Several live bacterial vaccines exist, for example, *Salmonella*, *E. coli*, and *Mycoplasma.*

19.2.3 Recombinant Vaccines

Genetically engineered vectors can be used as vaccines. One advantage of viral vector vaccines is that they can trigger both antibody- and cell-mediated immune responses without requiring an adjuvant. This makes them safe, as arthropods cannot transmit them and are not excreted in body fluids. Additionally, they can produce antigens in the appropriate configuration and deliver multiple antigens simultaneously.

19.2.4 Autogenous Vaccines

Autogenous vaccines are defined as *"inactivated immunological veterinary medicinal products which are manufactured from pathogens and antigens obtained from an animal or animals in an epidemiological unit and used for the treatment of that animal or those animals in the same epidemiological unit or for the treatment of an animal or animals in a unit having a confirmed epidemiological link"* (EU 2019/6). These vaccines are recommended in case of the following conditions: (1) Commercial vaccines are unavailable. (2) Commercially licensed vaccines do not have the required antigenic variants for the circulated pathogen and cannot provide cross-protection. (3) Commercial vaccines fail to provide the expected protection and have been proven ineffective.

19.3 Commonly Used Routes of Administration of Vaccines

19.3.1 Oral/Drinking Water

Drinking water is the most commonly used method for administering live vaccines to birds. While this method is convenient, there are some drawbacks. Improper and unequal distribution of vaccines can occur, as well as inconsistency and variability in water quality. Impurities or residues in the water can cause the vaccine to become inactive, and if water is withdrawn before vaccination, the birds may suffer from water starvation. Several precautions should be considered: (1) Stopping any water treatment for 48 h before vaccination; (2) using clean systems with cold, chlorine-free water only; (3) using the amount of water to be consumed within 2 h; (4) encouraging the birds to drink by keeping them thirsty before vaccination; (5) using stabilizers such as skim milk, 3–4 g/L; (6) lowering the water lines to ensure uniform water consumption; and (7) opening the freshwater source only when all the water has been consumed.

19.3.2 Spray Method

Administering live vaccines via spray is a practical method that reduces the need for handling, provides effective mucosal immunity, and is cost-effective. However, there is a chance of respiratory reactions occurring due to small particle size, and the deposition of particles is influenced by factors such as relative humidity, temperature, and hygiene. To ensure proper vaccine application, the following precautions are recommended: (1) cleaning the equipment thoroughly and free from disinfectants; (2) calibrating the droplet size to 150–200 microns and test the sprayer with water before use; (3) keeping the sprayer 50–100 cm over the birds; (4) spraying during cooler hours with the ventilation turned off and lights on, and all openings in the house should be closed; and (5) After spraying, wait 15–20 min before resuming ventilation.

19.3.3 Eye Drop Method

This method is very efficient. However, it is highly labor intensive. It can be used for some vaccines against NDV. Precautions of the eye drop vaccination method include: (1) make sure to use bottles and droppers that have been approved for use, (2) calibrate the dropper and ensure the volume is diluted properly, (3) use colored diluent for monitoring vaccinations, and (4) be sure to vaccinate at a moderate speed to ensure high-quality work.

19.3.4 Injection

The intramuscular or subcutaneous is also a very efficient vaccination method; however, it requires high labor. The injection can be used for inactivated vaccines against ND, fowl cholera, NDV, etc. The following precautions are also recommended for the injection method: (1) follow the guidelines of the vaccine manufacturer; (2) check the storage condition, vaccine type, and lot number; (3) warm vaccines for 1–2 h at 35–37 °C to make injection easier and less stressful for the birds; (4) calibrate injectors; (5) use full ventilation to avoid stressing the birds; (6) deposit the vaccine after the needle has penetrated to its full length; (7) change needles every few hundred birds to prevent contamination; (8) ensure the needle is directed into the muscle and avoid blood vessels and internal organs; and (9) avoid self-injection (Fig. 19.3).

19.3.5 Thigh-Stick Method or Wing-Web

This method is efficient and can provide 95–100% protection. It is commonly used for avian pox and fowl cholera. The thigh-stick method is used frequently in

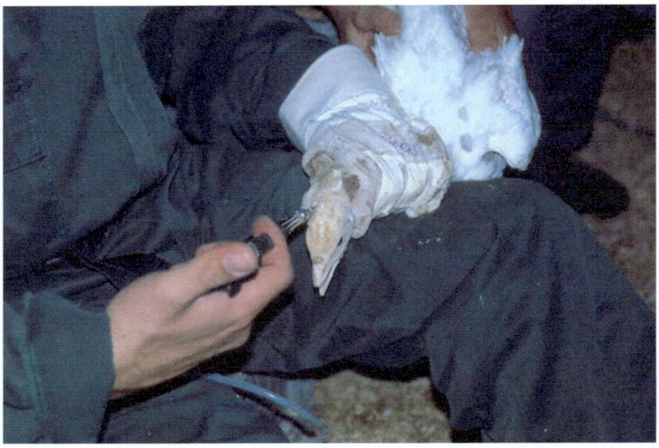

Fig. 19.3 Vaccination of turkeys © Hafez

turkeys. This method is highly labor-intensive as it needs individual handling. Birds should be examined for "vaccinal takes" in case of pox vaccination 7–10 days post-vaccination.

19.3.6 In Ovo Vaccination

The in ovo vaccination method is utilized in hatcheries for administering live and live cell-mediated vaccines to provide the hatched birds early protection, particularly against Marek's disease. The amniotic sac route is commonly used for the in ovo vaccination. This method stimulates both innate and adaptive immune responses. However, this method comes with challenges, such as using costly equipment, requiring training for vaccinators, and a risk of embryo death due to potential fungal or bacterial contamination through the egg holes.

19.4 Suggested Vaccination Program for Turkeys

Vaccination programs for individual farms will vary depending on the area's local conditions, disease status, and individual preferences. Examples of different vaccination programs are shown in Tables 19.1 and 19.2.

Table 19.1 Suggested vaccination program against TRT and ND in meat turkeys

Age/week	Vaccine	Type of vaccine	Application method
01	1. TRT	Live	DW/spray
03	2. TRT	Live	DW
04	1. ND	Live	DW
08	3. TRT	Live	DW
09	2. ND	Live	DW
14	4. TRT[a]	Live	DW
16	3. ND[a]	Live	DW

[a]Only in cocks. *TRT* Turkey rhinotracheitis, *ND* Newcastle, *DW* Drinking water (Hafez and Jodas 1997)

Table 19.2 Suggested vaccination program for turkey breeder (Hafez and Jodas 1997)

Vaccine	Age of vaccination	Type	Application route
Pasteurella. multicide (Cholera)	7th, 12th, and 26th weeks	Inactivated	S/C or I/m
Erysipelas	12th and 16th	Inactivated	S/C or I/m
Newcastle (ND)	5th, seventh, 11th, 16th day	**Live**	Drinking water
	26th week	Inactivated	S/C or I/m
Paramyxovirus-3	26th of age	Inactivated	S/C or I/m
Turkey Rhinotracheitis (TRT)	1st, third, 8th,14th, 16th weeks	**Live**	Drinking water
Turkey Rhinotracheitis (TRT)	26th week	Inactivated	S/C or I/m
Avian encephalomyelitis (AE)	23rd week	**Live**	Drinking water

19.5 Preventive Measures to Avoid Vaccination Failure

When designing a vaccination program for turkeys, it's crucial to take into account the following to prevent vaccination failure.

19.5.1 Vaccinal Seed and Vaccine Formulation

The vaccinal seed should exhibit a high homology with the circulated pathogen. Records of all batches of vaccines and their standard tests of vaccine potency should be maintained. Moreover, the antigen titer should be optimal so that the vaccine may provide the proper immunity level. Before vaccine usage, it's important to check the expiration date mentioned on it. It's important to avoid using expired vaccines.

19.5.2 Proper Storage of Vaccines

Storage conditions directly impact the effectiveness of vaccines (Siddique et al. 2016). To maintain their stability and viability for extended periods, vaccines need to be kept at proper cold temperatures because they lose their potency over time. It is critical to store vaccines at temperatures below 4 °C to ensure proper storage and cold chain temperature (Evans and Pope 1995). Developing countries face challenges maintaining cold chain systems during transportation due to several factors, such as loss of electric power, substandard refrigeration systems, overchilling, etc. Additionally, extra chilling of oil-based vaccines leads to the crystallization of adjuvant materials such as aluminum hydroxide salts, resulting in decreased vaccine potency.

When vaccines are exposed to direct sunlight, the antigens in the vial can be destroyed, which may cause the vaccine to become ineffective. Therefore, it is important to avoid exposing vaccines to sunlight during the formulation process for oral or parental vaccines. When administering oral vaccines, opening the cap of the vaccine vial inside water is recommended. Mixing vaccines in drinking water should be done in a room or a shaded area. To prevent sunlight from affecting the vaccine during transportation, it is best to use black or colored bags and cartons.

On the other hand, freeze-dried vaccines need to be kept at low temperatures, preferably in a refrigerator at 4 °C, and during transportation, cooling or ice blocks should be used to maintain low temperatures. Freezing and thawing must be avoided. These vaccines should be removed from the refrigerator/freezer only when needed and used immediately. Live vaccines for poultry flocks should be used within 2 h of reconstitution. Once reconstituted, their potency decreases rapidly. Reconstituted vaccines should be used as soon as possible, and unused vaccines can be stored in the refrigerator for 6 h. After 6 h, they should be discarded.

Thermostable vaccines can withstand temperatures between 2 and 8 °C and are more useful when maintaining a cold chain is difficult and cost-effective. These vaccines have some resistance to both cold and hot environments. Thermostable vaccines may solve the challenges associated with cold chain and storage temperatures (Alders and Spradbrow 2001). These vaccines can preserve their effectiveness and immunizing properties for up to a year at temperatures between 2 and 8 °C and up to 3 months in dried form at temperatures up to 28 °C (Ideris et al. 1987). Thermostable vaccines may be administered through various routes, including intraocular, intranasal, injection, and oral (through drinking water or feed) methods (Tu et al. 1998; Wambura et al. 2000).

19.5.3 Use of Stabilizers

Stabilizers are substances that are added to a vaccine to increase the shelf life of the vaccine. The skim milk at the rate of 3–4 g/L can also be added as a suitable alternative stabilizer (Nwanta et al. 2005).

19.5.4 Use of Immune Stimulants

Various substances have been utilized for immune stimulation in poultry, including vitamin E, selenium, and levamisole (Koller 1982; Bashir 1994). Selenium supplementation has been found to enhance the humoral immune response in chicks (Droke and Loerch 1989; Arshad et al. 2005) and increase natural resistance to antigenic stimuli (Colnago et al. 1984; Madron and Vrzgulova 1988). Additionally, using selenium in feed increases humoral antibody titers (Schrauzer 2000).

19.5.5 Booster Dose

To achieve successful immunization, some vaccines need a booster dose (Bosha and Nongo 2012). This booster dose should be given at least 10–20 days after the initial dose. The initial dose is necessary to prepare the body for the vaccine, while the booster is essential for maximum protection against the used antigen. If a booster dose is not given, the antibody titers may be low, leading to vaccine failure. Studies have shown that priming with a live attenuated vaccine, followed by a booster of a killed vaccine and a second booster with a live vaccine, provides the best protection against Newcastle disease (Afzal et al. 2015). However, regular subsequent inoculations are also necessary.

19.5.6 Avoiding Stress and Use of Immunostimulants

The birds must be healthy to achieve an effective vaccine response in poultry. Any stress should be avoided before administering the vaccine. The environment and sheds should have normal temperatures before vaccination, and the birds should be in good physical condition. Stress can negatively affect the immune response, so it is not recommended to vaccinate birds during stressful conditions. To minimize stress, they should add vitamins and minerals to their drinking water before, during, and after vaccination.

To ensure the health of poultry, it is important to regularly analyze the commercial feed provided to them and check for toxins. During humid conditions, fungi can grow on feed ingredients, and their metabolites can enter the poultry's body, leading to immune suppression, reduced growth, hypersensitivity, and decreased feed intake. Regular checks can help prevent these issues.

Taking vitamin and mineral supplements can help improve the body's immune response by affecting immune cells or changing metabolic and endocrine functions. This leads to faster production of antibodies and achieving a protective level of antibodies in less time. Vitamin E and selenium have been shown to modulate the immune response and prevent vaccine failure. Research has found that vitamin E may enhance the immune response to antigens in cockerels, but too much of it may depress specific immune responses. Giving excess vitamins, amino acids, minerals, and their combinations can enhance disease resistance by stimulating humoral and cellular immunity and phagocytosis. Birds need optimal vitamin nutrition to ensure optimal immune response and disease resistance. Adding higher levels of vitamins A, C, E, and selenium can improve the immune response of birds to vaccination and reduce the chances of vaccination failure in broiler poultry flocks.

19.5.7 Deworming Before Vaccination

It is recommended to deworm adult birds at least 15 days before vaccination. Sick birds should receive proper treatment and only be vaccinated after fully recovering. Vaccinations should only be given to healthy flocks.

19.5.8 Monitoring of Subclinical Infections

Regarding poultry, diseases like coccidiosis can lead to subclinical infections. The birds may seem healthy, but the infection can last long. It is important to regularly monitor the birds' appearance, the color and consistency of their droppings, any unusual sounds they make, and signs of respiratory distress. Only after the staff in charge has determined that the birds are in good health should they be vaccinated.

19.5.9 Consideration of Maternal Antibodies

When breeding or raising poultry, the adult flocks are vaccinated against viruses common in the area. As a result, newly hatched chicks have maternal antibodies from their mothers in their blood. For the best results, it is recommended to administer the ND vaccine to birds at least 7 days old (Nwanta et al. 2005).

19.5.10 Continuous Surveillance

It is important to regularly monitor the spread of diseases and collect data on their patterns. Blood/serum and fecal samples can be collected and sent to a laboratory for disease diagnosis. Tracheal and cloacal swabs can also be taken and sent to a laboratory to isolate pathogens. Checking antibody titters against applied vaccines on a routine basis can help maintain an optimum level of protection against the disease.

19.6 Treatment of Bacterial Diseases in Turkeys

Knowing which bacteria are involved before using antimicrobials can contribute to developing antibiotic resistance in bacteria is important. However, in cases where preventative measures are unsuccessful in preventing the spread of infection,

Table 19.3 Treatment of bacterial infections using antimicrobials

Avian pathogen	Potential effective antibiotics
Bordetella avium	• Tetracyclines can be used; however, it is difficult to be treated via drinking water
Clostridium perfringens	• Bacitracin, penicillin, lincomycin, erythromycin • Polyether ionophores suppress *C. perfringens*
E. coli	• Treatment should be based on a susceptibility test • Streptomycin, neomycin, tetracyclines, oxytetracycline, chlortetracycline, and sulfonamides
Erysipelas	• Penicillin is the antibiotic of choice • Inherently resistant to sulfonamides
Mycoplasma	• Tylosin, tetracyclines, erythromycin • Tiamulin is effective but neurotoxic when combined with sulfonamides and or ionophores • Inherently resistant (intrinsic) to penicillin and sulfonamides
Pasteurella multocida	• Tetracyclines, sulfonamides, sulfaquinoxaline, sulfadimethoxine, and erythromycin • Inherently resistant (intrinsic) to lincomycin
Salmonella	• Lincomycin, lincomycin/lincospectin, neomycin, and tetracyclines • Treatment of salmonellosis is prohibited in the EU • Resistant to erythromycin
Staphylococcus and *Streptococcus* infections	• Benzylpenicillin, penicillin, lincomycin, and erythromycin

treatment in several cases is necessary to control bacterial infectious diseases, maintain the health and welfare of the flock, and minimize economic losses. Ensuring an accurate diagnosis, proper product selection, appropriate dosage, adequate duration, and ongoing effectiveness monitoring is crucial. Concurrent sensitivity testing should also be conducted, as the resistance tests of microorganisms can vary by region and over time. In the event of treatment failure, corrective action should be taken.

A potential cause of treatment failure is the pH of water. Mixing acidic (low pH) products with basic (high pH) products in water may produce precipitants that will plug the drinking water.

Examples of acidic antibiotics include tetracyclines, oxytetracycline, lincomycin, chlortetracycline, and amprolium (Table 19.3). However, tylosin, sulfonamides, and bacitracin are basic products. Neomycin and penicillin are neutral. It's important to consider the hardness and pH level of drinking water used for oral medication as it may cause incompatibilities (Brugère-Picoux et al., 2020).

References

Afzal F, Saliha U, Fawad N, Ahmed S, Rehman HU, Munawar J, Naheed G, Siddique B (2015) Isolation, characterization of Newcastle disease virus and comparative efficacy of different vaccine regimes in broiler birds. J Anim Plant Sci 25:971–976

Alders RG, Spradbrow PB (2001) SADC planning workshop on Newcastle disease control in village chickens. In: Proceedings of an international workshop, Maputo, Mozambique, 6–9 March

Arshad M, Siddique M, Ashraf M, Khan HA (2005) Effect of selenium supplementation on antibody titres against infectious bursal disease vaccine in broiler chicks. Pak Vet J 25:203–204

Aucouturier J, Deville S, Perret C, Vallée I, Boireau P (2001) Assessment of efficacy and safety of various adjuvant formulations with a total soluble extract of Trichinella spiralis. Parasite 8(2 Suppl):S126–S132. https://doi.org/10.1051/parasite/200108s2126

Bashir FK (1994) Effect of vitamin E deficiency and excess on immune system of broiler chicken. M. Sc Thesis. Department of Veterinary Pathology, College of Veterinary Sciences

Bosha JA, Nongo NN (2012) Common breaches in poultry vaccine handling and administration in Makurdi metropolis: a recurrent phenomenon in the tropics. J Vet Sci 9:11–16

Colnago GL, Jensen LS, Long PL (1984) Effect of selenium and vitamin E on the development of immunity to coccidiosis in chickens. Poult Sci 63(6):1136–1143. https://doi.org/10.3382/ps.0631136

Droke EA, Loerch SC (1989) Effects of parenteral selenium and vitamin E on performance, health and humoral immune response of steers new to the feedlot environment. J Anim Sci 67(5):1350–1359. https://doi.org/10.2527/jas1989.6751350x

Evans M, Pope M (1995) Vaccine handling and storage in general practice. Health Trends 27:124–126

Furuya Y, Regner M, Lobigs M, Koskinen A, Mullbacher A, Alsharifi M (2010) Effect of inactivation method on the cross-protective immunity induced by whole "killed" influenza a viruses and commercial vaccine preparations. J Gen Virol 91(6):1450–1460. https://doi.org/10.1099/vir.0.018168-0

Hafez HM, Jodas S (1997) Putenkrankheiten (Turkey Dieses: Enke Verlag, Stuttgart-Germany). In: Woernle H, Jodas, S (eds) (2001) Poultry diseases, 2nd edn. Ulmer Verlag

Ideris A, Ibrahim AL, Spradbrow PB, Hung HC (1987) Development of food pellet Newcastle disease vaccine. In: Copland JW (ed) Newcastle disease in poultry. A new food pellet vaccine. ACIAR, Canberra, pp 20–23

Jang SI, Lillehoj HS, Lee SH, Lee KW, Park MS, Bauchan GR, Lillehoj EP, Bertrand F, Dupuis L, Deville S (2010) Immunoenhancing effects of Montanide ISA oil-based adjuvants on recombinant coccidia antigen vaccination against Eimeria acervulina infection. Vet Parasitol 172(3–4):221–228. https://doi.org/10.1016/j.vetpar.2010.04.042

Jang SI, Lillehoj HS, Lee SH, Lee KW, Lillehoj EP, Bertrand F, Dupuis L, Deville S (2011) Montanide™ ISA 71 VG adjuvant enhances antibody and cell-mediated immune responses to profilin subunit antigen vaccination and promotes protection against Eimeria acervulina and Eimeria tenella. Exp Parasitol 127(1):178–183. https://doi.org/10.1016/j.exppara.2010.07.021

Koller L (1982) Chemically induced immune-modulation. J Am Vet Med Assoc 181:1102–1106

Madron P, Vrzgulova N (1988) Vitamin E and selenium increase the natural resistance of farm animals. Veterinarskvi 38:369–371

Nwanta J, Umoh J, Abdu P, Ajogi I, Ajogi I, Egege S (2005) Comparison of the cost of unvaccinated and oral vaccinated local chickens with a Malaysian thermostable Newcastle disease vaccine (Ndv$_4$hr) in Kaduna state, Nigeria. Bull Anim Health Prod Afr 53(3):203–210. https://doi.org/10.4314/bahpa.v53i3.32711

Sasaki T, Isoda N, Soda K, Sakamoto R, Saijo K, Hagiwara J, Kokumai N, Ohgitani T, Imamura T, Sawata A, Lin Z, Sakoda Y, Kida H (2009) Evaluation of the potency, optimal antigen level and lasting immunity of inactivated avian influenza vaccine prepared from H5N1 virus. Jpn J Vet Res 56(4):189–198

Schrauzer GN (2000) Selenomethionine: a review of its nutritional significance, metabolism and toxicity. J Nutr 130(7):1653–1656. https://doi.org/10.1093/jn/130.7.1653

Siddique F, Iqbal A, Manzoor M (2016) Thermostable vaccine: a new horizon in poultry industry: a review. Adv Anim Vet Res 3:42–46

Tu TD, Phuc KV, Dinh NT, Quoc DN, Spradbrow PB (1998) Vietnamese trials with a thermostable Newcastle disease vaccine (strain I2) in experimental and village chickens. Prev Vet Med 34(2–3):205–214. https://doi.org/10.1016/s0167-5877(97)00065-2

Wambura PN, Kataga S (2011) Putative protective antibody response following oral vaccination of multi-age free ranging helmeted Guinea fowls (Numida meleagris) with Newcastle disease virus strain I-2 coated on oiled rice. Trop Anim Health Prod 43(1):99–102. https://doi.org/10.1007/s11250-010-9659-2

Wambura PN, Kapaga AM, Hyera JM (2000) Experimental trials with a thermostable Newcastle disease virus (strain I2) in commercial and village chickens in Tanzania. Prev Vet Med 43(2):75–83. https://doi.org/10.1016/s0167-5877(99)00089-6

Wood JM, Kawaoka Y, Newberry LA, Bordwell E, Webster RG (1985) Standardization of inactivated H5N2 influenza vaccine and efficacy against lethal A/Chicken/Pennsylvania/1370/83 infection. Avian Dis 29(3):867–872

Key Factors for Successful Organic Turkey Production

20

Shereen Basiouni, Xochitl Hernandez-Velasco, Guillermo Tellez-Isaias, and Wolfgang Eisenreich

Abstract

Changing trends in food consumption can distinctly influence consumers' perspectives on farming form, type, and construction and production of feed, processing, and animal welfare. Organic farming has grown 70% in Europe over the last 10 years. Germany has the largest market for organic foods in the EU. Consumers may assume that foods labeled "natural" are organic, hormone, and antibiotic-free; however, "organic" should not be considered a synonym for natural products. This chapter will shed light on the key factors for organic poultry production.

Keywords

Organic farms · Biosecurity · Chronic stress · Alternatives to antimicrobials

S. Basiouni (✉)
Institute of Molecular Physiology, Johannes-Gutenberg University, Mainz, Germany
e-mail: sbasioun@uni-mainz.de

X. Hernandez-Velasco
Departamento de Medicina y Zootecnia de Aves, Facultad de Medicina Veterinaria y Zootecnia (FMVZ), UNAM, Cd. de Mexico, Mexico
e-mail: xochitlh@fmvz.unam.mx

G. Tellez-Isaias
Department of Poultry Science, University of Arkansas Agricultural Experiment Station, Fayetteville, USA
e-mail: gtellez@uark.edu

W. Eisenreich (✉)
Structural Biochemistry of Membranes, Bavarian NMR Center, Technical University of Munich (TUM), Garching, Germany
e-mail: wolfgang.eisenreich@mytum.de

H. M. Hafez, A. A. Shehata (eds.), *Turkey Diseases and Disorders Volume 2*,
https://doi.org/10.1007/978-3-031-63322-5_20

20.1 Introduction

A “natural” food is defined as follows: “a product containing no artificial ingredient or added color and is only minimally processed” (Gernat et al. 2021). The main requirement to produce “organic” is the production of food animals, without the use of antimicrobials. Once birds are given any antimicrobials, they become automatically ineligible to be sold as organic, which is challenging for some producers. Additionally, the ban on antimicrobials as growth promotors in conventional farms force the poultry industry to change to meet the ongoing changes in the market and market preferences and “keep up” with consumer demands (Gernat et al. 2021; Hafez and Shehata 2021).

The rearing and production of antibiotics-free turkeys is not much different from conventional production but requires more attention to detail. Taking proactive measures to prevent diseases and establishing ethical, profitable, and appropriate programs and protocols are required to stop the use of antibiotics. No single program can guarantee positive outcomes. Antibiotics may be necessary due to unfavorable conditions or disease outbreaks, therefore educating producers on animal care practices and implementing good environmental programs to stop the use of antibiotics.

To successfully raise turkeys without antibiotics, reducing or eliminating exposure to disease-causing agents and stressful conditions is urgent. The process of hatching, processing, and delivery of poults is particularly challenging for young turkeys, as they are susceptible to growth, disease, environment, management, and nutritional factors. Proper management can help to reduce the use of antibiotics. The main key factors in reducing antibiotics in poultry farms are the implementation of biosecurity, water sanitation, creating the right microenvironment, restoring healthy gut, and avoiding chronic stressors. In poultry production without the use of antimicrobials, animal welfare becomes a challenge when birds get sick and require antimicrobial treatment. Since oxidative stress and chronic inflammation are mediated by chronic inflammation, supplementing poultry with antioxidants and anti-inflammatory might mitigate the chronic stress in animals.

20.2 Quality of Baby Chicks (Poults)

The quality of day-old chicks (poults) plays a crucial role in achieving optimum production outcomes. To ensure good quality of poults, starting with a healthy breeder flock is essential. The egg quality, as well as the health and nutritional well-being of the breeders, significantly impact the quality of the day-old poults. Usually, young breeders produce poults that are more susceptible to brooding conditions and are more likely to die within the first week of life (Brugère-Picoux et al. 2020). Turkey breeders should be vaccinated against several diseases to protect the poults based on maternal antibodies. For example, vaccination of the breeder against viral arthritis and avian encephalomyelitis. IgG antibodies usually descend in the egg yolk, while the IgM and IgA descend in the egg albumin. The egg size and quality are also determinants of poult quality. The day-old chick will generally be 2/3 of the original egg weight, which is, in turn, positively associated with slaughter weight.

The residual yolk (8–10% of the total poult weight) is proportional to the egg's size. However, bigger eggs tend to have poorer shell quality, negatively influencing hatchability and poult quality. Also, bigger eggs are more likely to overheat during incubation, producing poor poult quality.

The major strategy to control vertically transmitted diseases such as *Salmonella* should now be directed to clean the production chain from the top. It is important to collect all eggs at least three times per day and immediately disinfect them on the farm to prevent rapid penetration of microorganisms. Two common methods are used to disinfect turkey-hatching eggs under field conditions: fumigation and dipping in a detergent or disinfectant solution.

Fumigation involves using formaldehyde gas for at least 20 min with a concentration of 35 mL formalin mixed with 17.5 g potassium permanganate and 20 mL water per m^3 space. The temperature during fumigation should be maintained at a minimum of 20–24 °C and relative humidity at 70%. The eggs should be placed in trays that allow the fumigant to contact as much of the shell surface as possible. After fumigation, hygienic measures should be taken to prevent recontamination. However, formaldehyde gas can be unpleasant and pose health hazards to the operator, so some owners prefer to use wet treatments. Different sanitizing solutions, such as chlorine, glutaraldehyde, and quaternary ammonium compounds, are available.

Egg dipping in detergents or disinfectants is highly effective in greatly reducing or eliminating the bacteria from the shell when performed correctly. However, there is little or no effect on those bacteria that have already penetrated the shell. Attention also must be directed to the temperature of the detergent, which must be higher than the egg temperature. Dipping turkey-hatching eggs in disinfectant and/or antibiotics using differential temperature dipping (TDD) or pressure differential dipping (PDD) to control egg-transmitted *Salmonella* and other bacterial pathogens has been widely investigated and is of great value.

Hatchery hygiene, including the right temperature, relative humidity, and ventilation, also determines chick quality. Hatcheries must be designed to permit only a one-way traffic flow from the egg entry room through egg trays, incubation, hatching, and holding rooms to the van loading area. The ventilation system must prevent the recirculation of contaminated air. Trays used for hatchery should be thoroughly cleaned and disinfected before eggs are placed on them. Fumigation and disinfection programs should not be used to replace cleanliness but to support it. All eggs should be sanitized on arrival at the hatchery (presetting treatment) using fumigation.

Additionally, fumigation can be carried out after setting. This provides final disinfection following handling, transport, and various environmental contaminations during the storage of hatching eggs. Further fumigations are mostly carried out immediately after the transfer of hatching eggs from the setter to the hatch.

Some turkey hatcheries in Germany wash the hatching eggs after delivery with disinfectants and water, then dry them with hot air, then dipping in 1000 ppm Gentamicin and/or 500 ppm Enrofloxacin under reduced pressure of 500 mbar for

5 min. The partial vacuum is then released, and the eggs are allowed to soak the antibiotics at atmospheric pressure for 10 min. After removal from the dip, eggs are allowed to drain and dry before setting in the incubator.

Precautions should be followed since dipping solutions can become excessively contaminated with resistant microorganisms such as Pseudomonads and organic material. Thorough and continuous bacteriological monitoring of the dip solution is also required. The concentration of the antibiotics must be examined regularly and renewed routinely. By using enrofloxacin, the pH value of the dipping solution can be corrected during storage. The use of egg dipping in antimicrobials should be critically evaluated because of the irregular uptake of dip solution, uneven distribution of the active substance in the egg compartments, and lack of standardization in the dipping technique.

20.2.1 Keeping Optimal Microenvironment

During the period between the end of the 19th day of incubation and 4–5 days after hatching, the embryo that was once poikilothermic transforms into homeothermic chicks (Brugère-Picoux et al. 2020). When raising turkey poults, they need to be kept in an environment with additional heat for about 4–6 weeks. The stocking density and live weight per m^2 are shown in Table 20.1. Implementing a high management level is crucial to encourage the poults to eat and drink. Creating an environment that encourages healthy water and feed intake is crucial. Suppose day-old poults are exposed to extreme temperature changes (either too cold or too hot). In that case, it can hinder early consumption and ultimately delay the development of their intestinal tract and immune system. Studies have shown that maintaining a poult body temperature between 39.6 and 40.0 °C (103–104 °F) results in the best outcomes for consistent starts. Brooding young turkey poults is the most important phase of raising turkeys.

High levels of ammonia in poultry houses can negatively affect the health and safety of birds. It can cause respiratory diseases and even blindness, impacting bird production and increasing costs. Additionally, ammonia concentrations above 20 ppm can contribute to the spread of viruses, such as Newcastle and TRT. Poultry houses also emit greenhouse gases such as carbon dioxide, nitrous oxide, methane, and hydrogen sulfide. The maximum ammonia levels in turkey houses have been set as 50 ppm by the Occupational Safety and Health Administration (OSHA). High

Table 20.1 Stocking density and live weight per m^2

Age/week	Hens/m^2	Tom/m^2
0–6	11	11
7–12	5.5	4–4.5
13–16	5	3–3.5
>17	4.5	2.5–3
Maximum kg/m^2 at slaughter age	56	64

levels of ammonia cause epithelial corneal erosions and loss of cilia of the respiratory epithelium. It's important to keep carbon dioxide (CO_2) and carbon monoxide (CO) levels in turkey barns, as they can greatly affect bird health and performance (Mohammad Al-Kerwi et al. 2022). Monitoring and keeping CO levels below 25 ppm and CO_2 below 2500 ppm is crucial to avoid these negative effects.

20.2.2 Biosecurity

The first preparation should ensure the bird's rearing space and all-in-all-out and cleaned between flocks for the best possible results. Biosecurity is the first line of defense in securing a clean environment and the health of the young flock. A program that limits and controls exposure to beetles, flies, rodents, wild birds, and human traffic is essential. These programs and plans should be reviewed regularly and revised when needed.

Preventing and controlling infectious diseases requires careful planning and effective management practices. These measures aim to prevent the introduction and spread of infectious diseases by implementing surveillance programs and preventive vaccinations. Environmental management, including biosecurity measures, cleaning, and disinfection, is crucial in limiting exposure to infectious agents. In cases where there is a suspected or confirmed outbreak with significant public health concern and economic impacts, such as with highly pathogenic avian influenza (HPAI) and Newcastle diseases (ND) in the EU, eradication policies involving the culling of infected and contact flocks are enforced by legislation. The same applies to breeder flocks with *Salmonella enteritidis* or *Salmonella typhimurium* infections. In some cases, such as *Mycoplasma* in breeder flocks, the industry is responsible for preventing the spread of vertically transmitted infections.

Biosecurity is a set of management practices to prevent and/or reduce the potential for introducing and spreading infectious agents onto and between farms.

Farm owners, managers, and workers should be involved in designing and implementing the biosecurity plan. Controlling the movement of people, animals, equipment, and vehicles in and out of the farm and within the farm area is essential. Cleaning, disinfection, and vector control must be integrated into a comprehensive disease control program. They should include houses, equipment, and surroundings. The procedure should be tailored to meet the particular needs. Knowledge about microorganisms, physical and chemical agent sensitivity, and transmission mode are essential.

Ensuring that no other animal species besides the intended livestock are present at the farm is important. Employees should also refrain from keeping or working with other domestic animals at home to prevent potential hazards, such as *Salmonella*. For safety reasons, dogs, cats, and any other type of poultry should not be allowed in the barns. It is recommended to keep the turkey houses locked and prevent any unwanted visitors from entering the poultry buildings. Only necessary vehicles like feed trucks should be permitted on-site and sprayed with disinfectant on their wheels and underside. To prevent transmission and cross-contamination, it is crucial to conduct regular staff examinations and identify any carriers. These precautions will help maintain a safe and healthy environment for all on the farm.

When creating a program for cleaning and disinfecting a barn, it's important to include a schedule, the type and concentration of disinfectant, the desired level of cleanliness, and regular microbiological monitoring. It's crucial to disinfect both water and feed lines to ensure a thorough cleaning. Overnight sanitation with 2000 ppm of sodium hypochlorite is effective for water lines, while a special feed flush with 2% propionic acid can disinfect feed lines. Rodents, especially rats and mice, can contaminate poultry houses, so it's essential to implement a well-planned and routine rodent control program inside and outside the barn. Utilizing bait stations with single-dose anticoagulants is the most effective method of killing rodents. Traps placed at the entrance and in feed rooms are also helpful, and monitoring rodent activity and bait consumption is necessary for a successful program (Gazdzinski 2004).

Darkling beetles are also important carriers of some pathogens and have become a major pest in turkey barns, especially in brooding facilities. An insecticide should be applied when the birds are moved out of the barn to control them. After cleaning and disinfection, the house should be left empty for 2–4 weeks before a new flock is placed. Litter for new flocks should be of good quality and stored in barns with good control over rodents, wild birds, and other animals. Litter contamination during storage can be a major source of several microbes in the turkey industry. Precautions should be taken during storage, especially with straw, which can contaminate the field. Therefore, it is recommended to quarantine straw for at least 3 months instead of using fresh straw. It is important to note that litter moisture is directly related to several health disorders in turkeys.

20.2.3 Water Sanitation Is One of the Cornerstones of Success

The importance of on-farm water sanitation cannot be overstated. Even farms that rely on public water systems, minerals, and differing pH levels can lead to high levels of harmful bacteria. Additionally, using certain products such as probiotics, citric acid, organic acid, vaccines, and milk replacers, as well as naturally occurring substances like iron, manganese, and sulfur, can all contribute to developing a biofilm in the water system. This biofilm can provide a protective environment for bacteria and molds, rendering some water sanitation programs ineffective. Therefore, starting with clean water lines and maintaining a consistent water sanitation program is critical to prevent undesired growth and exposure. While there may be times when products are necessary that cannot be treated with water, such as on-farm vaccinations, minimizing exposure is essential to reducing waterline contamination.

The first step toward establishing a good water sanitation program is to clean the water lines between flocks. This is best done using a commercial peroxide sanitizer when the birds are moved out and again 24–48 h before placing the new flock. This process removes unwanted biofilm that harbors unwanted bacteria. Ensure all the cleaner is removed from the lines and replaced with the daily water sanitizer before poult placement. This will keep the source from repopulating. If this process has never been done before, multiple applications may be needed to clean the lines properly the first time.

To ensure proper sanitation, it is important to use an approved daily water sanitizer and regularly monitor it. A consistent monitoring program should be in place to maintain proper sanitation. Chlorine is a common sanitation method, but its effectiveness depends on the pH level, which should be between 5.5 and 6.5. If the pH level exceeds this range, an acid should be introduced to lower it to the desired level. This is important because a pH level of 7 or higher will produce hypochlorite ions, which are ineffective at sanitizing water. Acidification to a pH of 6.5 or lower will produce hypochlorous acid, which is 80 times more effective as a sanitizer. It is important to inject the acid and chlorine separately into the water lines and never mix them in the same container to avoid producing chlorine gas. To measure the effectiveness of water sanitation, the oxidative reductive potential (ORP) levels of the treated water should be $\geq$750 mV in all waterlines if an oxidizing sanitizer like chlorine is used. ORP levels and pH measurements can be analyzed on the spot with an electronic device that can analyze the water sample at the farm.

Another effective water sanitizer option is chlorine dioxide, which is pricier but less corrosive than chlorine. To measure its levels, test strips are required as ORP does not apply to this material. Collecting water samples and sending them to a lab to fully comprehend water quality concerns related to all water sources is advisable. Certain water problems may require professional assistance beyond the scope of a water sanitizer.

20.2.4 Litter Management

Maintaining dry and cake-free litter conditions during the growing cycle can be challenging without regular mechanical tilling or adding fresh bedding. It's crucial to manage drinkers to keep the litter dry, especially during brooding. Whether using bell or nipple line drinkers, correctly adjusting system heights and pressure is essential. If the system height is too low or the flow rate is too high, birds may wastewater, resulting in wet litter on the floor.

20.2.5 Prevention of Chronic Stress on Poultry

Particular attention should be taken to chronic stress that reduces animal performance and increases the animal's susceptibility to infections. Some factors have been reported: dysbiosis, leaky gut, heat stress, mycotoxins, endotoxins, and oxidized diet, Fig. 20.1 (Shehata et al. 2022a).

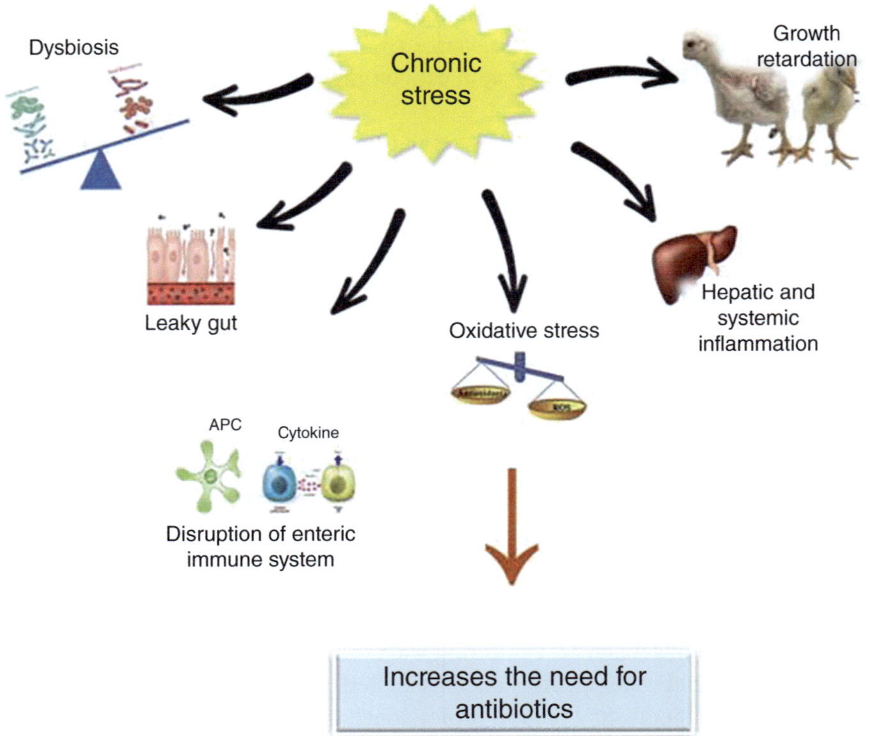

Fig. 20.1 Chronic stress in poultry and the link with antimicrobial resistance acquisition. Upon antibiotic application, the resistant gene reservoir will be developed due to the eradication of susceptible strains from the gastrointestinal tract (Shehata et al. 2022a)

20.2.5.1 Dysbiosis and Leaky Gut Syndrome

Interactions between the host (poultry species) and intrinsic or extrinsic factors that influence gut health are shown in Fig. 20.2. The gut microbiota is considered a hidden metabolic "organ," due to its significant impact on the host's metabolism, physiology, nutrition, and immune function. The relationship between the intestinal microbiota and their respective host animals is dynamic and mutually beneficial (synbiosis). This complex interaction is crucial for maintaining good health, and any imbalance in the gut microbiota can lead to several metabolic disorders. In poultry production, intestinal health and proper functioning are essential for optimal nutrient absorption and growth, directly affecting animal performance. The gut microbiota is composed mainly of bacteria, fungi, and protozoa, and during dysbiosis, microbial metabolites can cause oxidative stress and inflammation (Shehata et al. 2022b).

Also, during dysbiosis (a disturbance in the gut microbiota due to an imbalance between beneficial and opportunistic microbiota), lipopolysaccharide (LPS) production by gram-negative bacteria increases and induces local and systematic inflammation by stimulating the intestinal epithelial cells and macrophages and

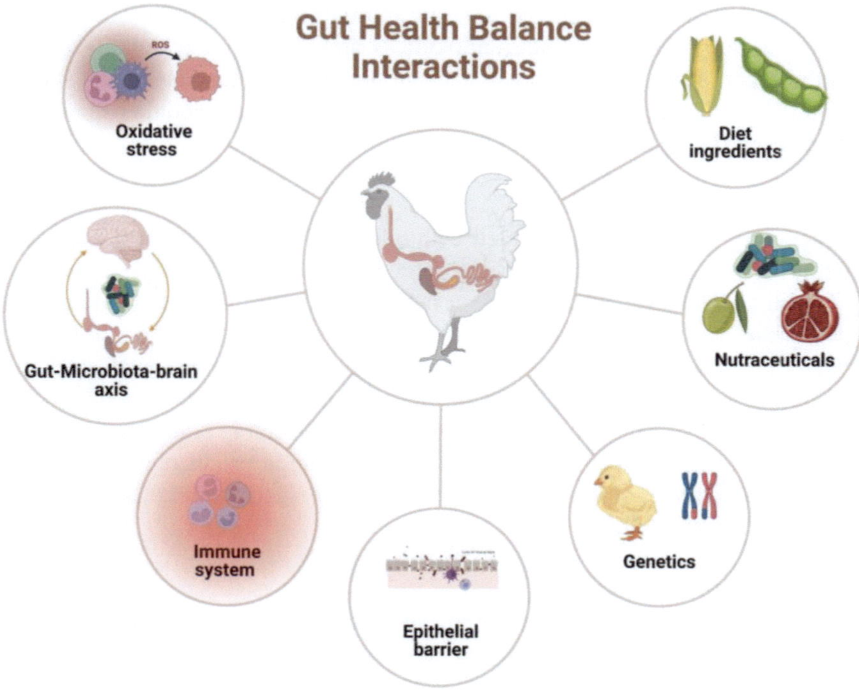

Fig. 20.2 Interactions between the host (poultry species) and intrinsic or extrinsic factors that influence gut health (created with BioRender.com) (Tellez-Isaias et al. 2023)

causing disruption of the tight junction or even leaky gut syndrome. The impaired intestinal barrier function, commonly known as "leaky gut," is a condition in which the small intestine lining becomes damaged, leading to infiltration of luminal contents such as bacteria and their associated components, including toxins, to pass between epithelial cells. These conditions subsequently lead to cell damage and/or inflammation of the intestine, characterized by increased levels of bacteria-derived endotoxins in the blood. This inflammatory process consumes significant nutrients and negatively affects metabolic responses, particularly immune-metabolic and endocrine responses (Basiouni et al. 2022). As a result, animal performance is severely reduced.

20.2.5.2 Mycotoxins and Endotoxins

Fungi and molds can grow and produce mycotoxins on food, grains, and animal diets as suitable substrates. These mycotoxins are strain-specific and produced by various fungi, particularly molds, making them a global issue due to massive economic losses. Negative health impacts of mycotoxins include reduced feed intake, weight gain, feed efficiency, growth performance, immunity, hatchability, increased mortality, organ damage, carcinogenicity, teratogenicity, and decreased egg output. Mycotoxins also induce lipid peroxidation and alter cellular redox signaling, antioxidant status, and membrane integrity through ROS generation. In addition, mycotoxins suppress intracellular antioxidants and reduce glutathione levels. On the other hand, ROS production can activate pro-inflammatory mediators such as TNF-α, IL-1α, IL-1β, and IL-6 and NO expression, which can modulate the inflammatory response (Tellez-Isaias et al. 2023).

20.2.5.3 Oxidized Diet

Many poultry diets are supplemented with oils rich in polyunsaturated fatty acids (PUFA) because they are a great energy source and can improve the taste and quality of pellets. However, PUFAs have a higher oxidation rate than saturated fats, meaning they can go bad faster. The unstable double bonds in PUFAs allow more oxygen to react at those points, which factors like light, transition metals, and temperature during feed production and storage can cause. This reaction produces reactive substances that can be harmful and cause an undesirable odor. The process of lipid peroxidation can also produce harmful degradation products like peroxides, aldehydes, and polar compounds, which can cause oxidative stress (Saracila et al. 2021; Tellez-Isaias et al. 2023).

20.2.5.4 Heat Stress

Heat stress is one of the most challenging stressors associated with poultry production in tropical and subtropical regions and central and eastern Europe. Poultry farms are susceptible to high ambient temperatures due to their feathers, lack of sweat glands, and high production. Heat stress causes several adverse effects on the intestinal mucus layer, tight junction, enteric immune system, and antioxidant system. Heat stress is a key contributor to systemic oxidative stress by causing a redox imbalance between the pro- and antioxidants in favor of prooxidants. Oxidative stress disrupts the intestinal barrier, alters the cellular processes, and subsequently increases intestinal permeability, facilitating bacteria's translocation from the gut. Other pathological problems associated with heat stress include (i) hepatic inflammation due to activation of nuclear factor-kappa B (NF-κ B) by bacteria and LPS; (ii) systemic inflammation due to the increase of pro-inflammatory mediators such as interleukin (IL-1β), IL-6, and tumor necrosis factor α (TNFα); (iii) decreased feed intake due to hypothalamic inflammation and inhibition; and (iv) growth retardation due to muscle proteolysis stimulation, muscle protein synthesis inhibition, and decreased feed intake (Saracila et al. 2021).

20.2.6 Nutraceuticals Supplementation to Reduce Antibiotics

Complementary (along with) and alternative (in place of) medicine (CAM) is the term for medical products and practices that are not part of standard medical care. In poultry, essential oils, plant extracts, probiotics, prebiotics, and fermentation products have been claimed to stimulate digestive enzymes, improve gut histology, modulate the immune system, and have antibacterial, anticoccidial, antifungal, or antiviral activity (Shehata et al. 2022b). Several alternative feed additives to antimicrobial growth promoters (AGP) to improve intestinal health have been described. Interestingly, several phytogenic substances having antioxidant and/or anti-inflammatory effects in poultry have been described (Shehata et al. 2022a). Polyphenols can suppress oxidative stress through two mechanisms: (i) reactive oxygen species (ROS) scavenger by several mechanisms, such as electron donation, hydrogen donation, and transition metal chelators, and (ii) upregulation of ROS removing enzyme synthesis, which in turn stimulates the expression of antioxidant enzymes. Polyphenols exhibiting antioxidant effects in poultry are curcumin, epigallocatechin-3-gallate, quercetin, resveratrol, cinnamon oil, turmeric rhizome, and pro-anthocyanidins. Several phytogenic substances have been used in poultry and exhibited an anti-inflammatory effect. Examples of phytogenic substances that are used in poultry are *Salix* spp. (*Salicaceae*), *Boswellia* spp. (*Burseraceae*), Cinnamaldehyde, *Zingiber officinale* (*Zingiberaceae*), capsaicin, curcumin, piperamides, and thymol. These phytogenic substances act by one or more of the following mechanisms: (i) inhibition of the release of inflammatory mediators such as IL-1, IL-6, and TNF alpha; (ii) inhibition of leukotrienes; (iii) inhibition of cyclooxygenase (COX) and lipoxygenase (LOX); and (iv) inhibition of arachidonic acid.

20.3 Conclusions

In conclusion, infectious diseases in turkeys are mostly associated with severe economic losses; early recognition and monitoring programs are essential in managing infectious diseases. However, a universal solution for preventing and controlling infectious diseases does not exist. Generally, the measures mentioned above alone is of little value unless improvements in all aspects of management and biosecurity accompany them. In addition, effective education programs must be implemented to increase public awareness of the necessary measures to protect against food-borne infections like *Salmonella* and or *Campylobacter* in food products from turkeys. Finally, research must continue to find additional control and preventive measures.

The interest in digestive physiology and the role of microorganisms has generated data whereby human and animal well-being can be enhanced and the risk of disease reduced. New molecular techniques allow an accurate assessment of the flora composition, resulting in improved strategies for elucidating mechanisms, given the recent international legislation and domestic consumer pressures to withdraw growth-promoting antibiotics and limit antibiotics available for treating bacterial infections. Probiotics, postbiotics, and phytogenic substances can offer a good alternative option.

References

Basiouni S, Tellez-Isaias G, Latorre DG, Graham DB, Petrone-Garcia MW, El-Sweedi H, Yalçın S, Wahab A, Visscher C, May-Simera LM, Huber C, Eisenreich W, Shehata AA (2022) Anti-inflammatory and antioxidative phytogenic substances against secret killers in poultry: current status and prospects, vol 10. Vet Res

Brugère-Picoux J, Shivaprasad HL, Venne D, Bouzouaïa M (2020) Manual of poultry diseases, version anglaise. Association francaise pour l'avancement des Sciences (AFAS), Paris, France

Gazdzinski P (2004) 5th International symposium on turkey diseases: Berlin, 16–19 June 2004. DVG-Service-GmbH, Giessen

Gernat AA, Santos FBO, Grimes JL (2021) Alternative antimicrobials in the Turkey industry—challenges and perspectives. Ger J Vet Res, vol 1, pp 37–47

Hafez HM, Shehata AA (2021) Turkey production and health: current challenges. Ger J Vet Res 1(1):3–14. https://doi.org/10.51585/gjvr.2021.0002

Mohammad Al-Kerwi MS, Mardenli O, Mahdi Jasim MR, Al-Majeed MA (2022) Effects of harmful gases emitted from poultry houses on productive and health performance. IOP Conf Ser: Earth Environ Sci 1060(1):012082. https://doi.org/10.1088/1755-1315/1060/1/012082

Saracila M, Panaite TD, Papuc CP, Criste RD (2021) Heat stress in broiler chickens and the effect of dietary polyphenols, with special reference to willow (Salix spp.) Bark supplements—a review. Antioxidants (Basel) 10(5):686. https://doi.org/10.3390/antiox10050686

Shehata AA, Attia Y, Khafaga AF, Farooq MZ, El-Seedi HR, Eisenreich W, Tellez-Isaias G (2022a) Restoring healthy gut microbiome in poultry using alternative feed additives with particular attention to phytogenic substances: challenges and prospects. Ger J Vet Res 2(3):32–42. https://doi.org/10.51585/gjvr.2022.3.0047

Shehata AA, Yalçın S, Latorre JD, Basiouni S, Attia YA, Abd El-Wahab A, Visscher C, El-Seedi HR, Huber C, Hafez HM, Eisenreich W, Tellez-Isaias G (2022b) Probiotics, prebiotics, and phytogenic substances for optimizing gut health in poultry. Microorganisms 10(2):395. https://doi.org/10.3390/microorganisms10020395

Tellez-Isaias G, Eisenreich W, Petrone-Garcia VM, Hernandez-Velasco X, Castellanos-Huerta C-H, Tellez G Jr, Latorre JD, Bottje WG, Senas-Cuesta R, Coles ME, Hargis BM, El-Ashram S, Graham BD, Shehata AA (2023) Effects of chronic stress and intestinal inflammation on commercial poultry health and performance: a review. Ger J Vet Res 3(1):38–57. https://doi.org/10.51585/gjvr.2023.1.0051/

MIX
Papier aus verantwortungsvollen Quellen
Paper from responsible sources
FSC® C105338

If you have any concerns about our products,
you can contact us on
ProductSafety@springernature.com

In case Publisher is established outside the EU,
the EU authorized representative is:
Springer Nature Customer Service Center GmbH
Europaplatz 3, 69115 Heidelberg, Germany

Printed by Libri Plureos GmbH
in Hamburg, Germany